Satellite Terrestrial Networks: Technologies, Security and Applications

Satellite Terrestrial Networks: Technologies, Security and Applications

Guest Editor

Jiaxin Zhang

MDPI

Basel • Beijing • Wuhan • Barcelona • Belgrade • Novi Sad • Cluj • Manchester

Guest Editor
Jiaxin Zhang
School of Information and
Communication Engineering
Beijing University of Posts
and Telecommunications
Beijing
China

Editorial Office
MDPI AG
Grosspeteranlage 5
4052 Basel, Switzerland

This is a reprint of the Special Issue, published open access by the journal *Electronics* (ISSN 2079-9292), freely accessible at: https://www.mdpi.com/journal/electronics/special_issues/UCQM4W4VK5.

For citation purposes, cite each article independently as indicated on the article page online and as indicated below:

Lastname, A.A.; Lastname, B.B. Article Title. *Journal Name* **Year**, *Volume Number*, Page Range.

ISBN 978-3-7258-5525-4 (Hbk)
ISBN 978-3-7258-5526-1 (PDF)
https://doi.org/10.3390/books978-3-7258-5526-1

Contents

About the Editor

Jiaxin Zhang

Jiaxin Zhang is with the School of Information and Communication Engineering at Beijing University of Posts and Telecommunications. His research interests include satellite–terrestrial integrated networks, wireless resource management, and next-generation mobile communication systems. He has led and participated in several national and international projects on broadband satellite communications, cross-domain networking, and intelligent network control. His work has appeared in leading journals and conferences in the fields of communications and networking. He has contributed to the development of innovative methods for beam hopping, interference mitigation, and secure transmission, and has collaborated with both academic and industry partners to advance practical applications of satellite communication technologies. His current focus is on the design of resilient and intelligent satellite–terrestrial infrastructures to support future 6G systems and global connectivity.

Preface

The Reprint on Satellite Terrestrial Networks presents a timely collection of research devoted to the integration of space and ground communication infrastructures. Its scope covers network architectures, intelligent resource management, spectrum sharing, security mechanisms, routing strategies, and application-oriented solutions. The aim is to provide readers with a consolidated view of the technological foundations and emerging directions that are shaping this field. Our motivation for compiling this Reprint is the growing importance of global, secure, and resilient connectivity in the era of low-Earth-orbit constellations and beyond. The contributions demonstrate both theoretical advances and practical approaches to challenges such as interference coordination, latency reduction, and robust service delivery. This Reprint is addressed to researchers, engineers, and practitioners from academia and industry who are engaged in the design, evaluation, and deployment of next-generation communication systems. We hope that it will inspire further work and collaboration toward the realization of ubiquitous satellite–terrestrial integration.

Jiaxin Zhang
Guest Editor

electronics

MDPI

ditorial

ditorial: Satellite Terrestrial Networks: Technologies, Security and Applications

iaxin Zhang

School of Information and Communication Engineering, Beijing University of Posts and Telecommunications, Beijing 100876, China; jxz@bupt.edu.cn

The ongoing evolution toward sixth-generation (6G) communications necessitates the deep integration of terrestrial and non-terrestrial infrastructures into the seamless Space–Air–Ground–Sea Integrated Network. Low Earth orbit (LEO) mega-constellations, combined with advanced terrestrial networks, offer unprecedented opportunities for global coverage, resilient connectivity, and low-latency services. However, this paradigm also introduces formidable challenges in terms of spectrum efficiency, dynamic resource management, mobility support, security, and end-to-end quality of service. This Special Issue of *Electronics*, "Satellite Terrestrial Networks: Technologies, Security and Applications", brings together twelve research contributions that collectively advance the state of the art in these domains.

Recent research has highlighted that STNs will play a decisive role in the evolution from 5G toward 6G networks, where the integration of non-terrestrial components is no longer optional but rather a fundamental design principle [1]. A number of survey works have emphasized that efficient spectrum sharing and interference management remain among the most critical technical challenges [2]. Furthermore, the rapid growth of low Earth orbit (LEO) mega-constellations has accelerated the need for scalable routing and resource allocation schemes capable of coping with highly dynamic topologies [3,4].

In addition to communication aspects, computing and intelligence are increasingly being integrated into STNs. The paradigm of edge computing, initially studied in terrestrial mobile networks, has been extended to LEO satellite systems, enabling low-latency task execution and real-time data processing [5]. Meanwhile, machine learning methods, including reinforcement learning and graph neural networks, have demonstrated their potential in optimizing resource allocation, routing, and beam management in dynamic environments [6,7]. Signal processing research has also advanced toward robust synchronization and anti-jamming techniques, which are crucial for ensuring reliability in contested environments [2].

Security and resilience are equally pressing issues in STNs. Beyond conventional cryptographic measures, recent studies have proposed physical-layer security mechanisms and frequency-hopping strategies to defend against jamming and eavesdropping [8]. At the system level, the rise of cloud-native platforms and network function virtualization provides new opportunities for the flexible deployment and orchestration of communication and computing functions across satellite and terrestrial segments [9].

Against this background, the twelve studies included in this Special Issue provide concrete solutions and experimental validations that align with these global research trends. Collectively, they offer both theoretical insights and practical frameworks that bring satellite–terrestrial networks closer to deployment in real-world scenarios.

Motivated by the growing role of artificial intelligence in communication networks, Sun et al. (Contributor 1) introduce SCNOC-Agentic, an LLM-driven framework for satellite network operation and control. The system integrates intent refinement, multi-agent workflows, long-term memory, and graph-based retrieval, achieving significant improvements in network task planning, fault analysis, and resource optimization. Zhang et al. (Contributor 2) address the resource management problem in beam-hopping-based satellite systems, proposing a progressive decomposition framework for joint beam scheduling and power/frequency allocation. Their approach, supported by geographic isolation mechanisms, ensures efficient coexistence between GEO and LEO systems while enhancing throughput. Song et al. (Contributor 3) tackled the challenge of Doppler frequency-offset estimation in LEO systems. By leveraging beam pointing information instead of GNSS support, the authors developed an accurate and lightweight estimation method suitable for highly dynamic non-terrestrial networks.

Machine learning-based methods are also investigated for interference detection in GNSS. Baldini and Bonavitacola (Contributor 4) evaluate the impact of signal bit-depth on the accuracy of wireless interference classification, demonstrating the trade-offs between hardware sampling constraints and ML model performance. Edge computing and task offloading are studied by Li et al. (Contributor 5), who propose a joint optimization framework for task scheduling and communication resource allocation in LEO satellite edge networks. Their results show improvements in energy efficiency, task completion rates, and overall system cost. Pan et al. (Contributor 6) focus on burst spread spectrum signals and introduced a novel signal structure that eliminates synchronization overheads while improving acquisition reliability, stealth, and resource efficiency.

The coexistence problem between terrestrial IMT base stations and satellite services is examined by Jia et al. (Contributor 7). Through a power control and intervention algorithm combined with deep reinforcement learning, they demonstrate how harmful interference can be mitigated while preserving terrestrial throughput. Wang et al. (Contributor 8) contribute to routing in satellite self-organizing networks, proposing SQL-CBRP, a cluster-based routing protocol evaluated on the OMNeT++ platform. Their results show reduced delay and packet loss compared to classical methods under high loads. Han et al. (Contributor 9) further advance routing optimization by introducing a deep reinforcement learning-based multipath routing algorithm for LEO mega-constellations. Their multipath discovery and graph-neural-network-based traffic scheduling achieve considerable improvements in throughput, reliability, and delay over shortest-path schemes.

In the context of beam management, Liu and Pan (Contributor 10) investigate optimal beamwidth settings for quasi earth-fixed cells in LEO systems, showing that the dynamic adjustment of beamwidth maximizes uplink coverage probability and outperforms 3GPP baseline schemes. Anti-jamming mechanisms are studied by Yu et al. (Contributor 11), who propose a capacity enhancement method for frequency-hopping systems. By adapting hopping rates, the method improves capacity without compromising jamming resistance. Finally, Shi et al. (Contributor 12) introduce ComEdge, a cloud-native platform for integrated computing and communication in satellite–terrestrial networks. By employing microservices, containers, and service mesh technologies, the platform provides a flexible environment for real-world deployment and demonstrates the feasibility of AI-driven network resource management.

The diversity of these studies highlights the breadth of ongoing research in STNs. Several themes are evident across the contributions: the need for intelligent automation through machine learning and AI, the challenge of efficient coexistence between heterogeneous systems, the focus on robustness against interference and jamming, and the shift toward practical, deployment-ready platforms. Nonetheless, significant challenges remain, including the validation of the proposed solutions under real-world conditions, the balance between AI-driven efficiency and explainability, the energy constraints of spaceborne platforms, and the adaptation of algorithms to time-varying and uncertain environments. Addressing these challenges will be crucial for building scalable, secure, and efficient satellite–terrestrial networks that are capable of supporting next-generation applications.

Conflicts of Interest: The author declares no conflicts of interest.

List of Contributions:

1. Sun, W.; Sun, C.; Zhang, Y.; Yin, Z.; Kang, Z. SCNOC-Agentic: A Network Operation and Control Agentic for Satellite Communication Systems. *Electronics* **2025**, *14*, 3320. https://doi.org/10.3390/electronics14163320.
2. Zhang, J.; Li, W.; Li, Y.; Wang, H.; Li, S. A Framework for Joint Beam Scheduling and Resource Allocation in Beam-Hopping-Based Satellite Systems. *Electronics* **2025**, *14*, 2887. https://doi.org/10.3390/electronics14142887.
3. Song, Y.; Xu, J.; Sun, C.; Li, X.; An, S. A Doppler Frequency-Offset Estimation Method Based on the Beam Pointing of LEO Satellites. *Electronics* **2025**, *14*, 2539. https://doi.org/10.3390/electronics14132539.
4. Baldini, G.; Bonavitacola, F. A Machine Learning Evaluation of the Impact of Bit-Depth for the Detection and Classification of Wireless Interferences in Global Navigation Satellite Systems. *Electronics* **2025**, *14*, 1147. https://doi.org/10.3390/electronics14061147.
5. Li, J.; Chai, R.; Gui, K.; Liang, C. Joint Task Offloading and Resource Scheduling in Low Earth Orbit Satellite Edge Computing Networks. *Electronics* **2025**, *14*, 1016. https://doi.org/10.3390/electronics14051016.
6. Pan, S.; Yin, L.; Tan, Y.; Wang, Y. Structure Design and Reliable Acquisition of Burst Spread Spectrum Signals Without Physical Layer Synchronization Overhead. *Electronics* **2024**, *13*, 4586. https://doi.org/10.3390/electronics13234586.
7. Jia, M.; Meng, S.; Wang, H.; Tang, Z.; Jin, Z. A Power Control and Intervention Algorithm for Co-Existing IMT Base Stations and Satellite Services. *Electronics* **2024**, *13*, 4108. https://doi.org/10.3390/electronics13204108.
8. Wang, G.; Zhang, J.; Zhang, Y.; Liu, C.; Chang, Z. Performance Evaluation of Routing Algorithm in Satellite Self-Organizing Network on OMNeT++ Platform. *Electronics* **2024**, *13*, 3963. https://doi.org/10.3390/electronics13193963.
9. Han, C.; Xiong, W.; Yu, R. Deep Reinforcement Learning-Based Multipath Routing for LEO Megaconstellation Networks. *Electronics* **2024**, *13*, 3054. https://doi.org/10.3390/electronics13153054.
10. Liu, C.-T.; Pan, J.-Y. Optimal Beamwidth for Maximizing Uplink Coverage Probability in Quasi Earth-Fixed LEO Satellite Communication System. *Electronics* **2024**, *13*, 1349. https://doi.org/10.3390/electronics13071349.
11. Yu, Z.; Hao, Z.; Yao, W.; Jia, M. A Capacity Enhancement Method for Frequency-Hopping Anti-Jamming Communication Systems. *Electronics* **2023**, *12*, 4457. https://doi.org/10.3390/electronics12214457.
12. Shi, H.; Zhang, X.; Wu, P.; Chen, J.; Zhang, Y. ComEdge: Cloud-Native Platform for Integrated Computing and Communication in Satellite–Terrestrial Network. *Electronics* **2023**, *12*, 4252. https://doi.org/10.3390/electronics12204252.

References

1. Kodheli, O.; Lagunas, E.; Maturo, N.; Sharma, S.K.; Shankar, B.; Montoya, J.F.M.; Duncan, J.C.M.; Spano, D.; Chatzinotas, S.; Kisseleff, S.; et al. Satellite Communications in the New Space Era: A Survey and Future Challenges. *IEEE Commun. Surv. Tutor.* **2021**, *23*, 70–109. [CrossRef]
2. Chen, L.; Lagunas, E.; Chatzinotas, S.; Ottersten, B. The Next Generation of Beam Hopping Satellite Systems: Dynamic Beam Illumination with Selective Precoding. *IEEE Trans. Wirel. Commun.* **2023**, *22*, 2666–2682. [CrossRef]
3. Zhang, J.; Wang, Y.; Li, S.; Zhao, L.; Chen, R. LEO Mega Constellations: Review of Development, Impact, Surveillance, and Governance. *Space Sci. Technol.* **2022**, *2022*, 9865174. [CrossRef]

4. Cao, X.; Li, Y.; Xiong, X.; Wang, J. Dynamic Routings in Satellite Networks: An Overview. *Sensors* **2022**, *22*, 4552. [CrossRef] [PubMed]
5. Taleb, T.; Samdanis, K.; Mada, B.; Flinck, H.; Dutta, S.; Sabella, D. On Multi-Access Edge Computing: A Survey of the Emerging 5G Network Edge Cloud Architecture and Orchestration. *IEEE Commun. Surv. Tutor.* **2017**, *19*, 1657–1681. [CrossRef]
6. Chen, Y.; Cao, H.; Wang, L.; Chen, D.; Liu, Z.; Zhou, Y.; Shi, J. Deep Reinforcement Learning-Based Routing Method for Low Earth Orbit Mega-Constellation Satellite Networks with Service Function Constraints. *Sensors* **2025**, *25*, 1232. [CrossRef] [PubMed]
7. Kedrowitsch, A.; Black, J.; Yao, D. Resilient Routing for Low Earth Orbit Mega-Constellation Networks. In Proceedings of the NDSS Workshop on Space Security (SpaceSec), San Diego, CA, USA, 1 March 2024. Available online: https://www.ndss-symposium.org/ (accessed on 26 September 2025).
8. Kumar, R.; Arnon, S. Review of Physical Layer Security in Integrated Satellite–Terrestrial Networks. *Electronics* **2024**, *13*, 4414. [CrossRef]
9. Jiang, W. Software Defined Satellite Networks: A Survey. *Digit. Commun. Netw.* **2023**, *9*, 1243–1264. [CrossRef]

electronics
MDPI

Article

ComEdge: Cloud-Native Platform for Integrated Computing and Communication in Satellite–Terrestrial Network

Haoyang Shi, Xing Zhang *, Peixuan Wu, Jingkai Chen and Yufei Zhang

School of Information and Communication Engineering, Beijing University of Posts and Telecommunications, Beijing 100876, China; shihaoyang@bupt.edu.cn (H.S.); peixuan_wu@bupt.edu.cn (P.W.); chenjingkai@bupt.edu.cn (J.C.); zyf1999@bupt.edu.cn (Y.Z.)
* Correspondence: hszhang@bupt.edu.cn

Abstract: Leveraging technological advancements such as containers, microservices, and service mesh, cloud-native edge computing (CNEC) has become extensively discussed and applied in both academia and industry. The integration of mobile edge computing and communication is crucial for the future communication architecture in order to fully utilize distributed and fragmented communication resources and computing power. The potential for cloud-native integration can help merge mobile edge computing and communication, enhancing network flexibility and resource utilization. This paper investigates the implementation plan for extending cloud-native capabilities to integrated computing and communication (INCCOM) in the satellite–terrestrial network. We construct an experimental verification platform called ComEdge in a real-world setting. Subsequently, we analyze the architecture, functional characteristics, and deployment of the platform in a real-world environment. Furthermore, we explore the solution of deep reinforcement learning in the deployment of cloud-native core network and conduct a preliminary verification of the platform's potential to enable artificial intelligence in a real production environment, which will provide guidance to both academic and industry sectors. Finally, we conduct an analysis on the challenges and opportunities encountered by the cloud-native INCCOM network system.

Keywords: satellite–terrestrial network; cloud-native edge computing; integrated computing and communication; deep reinforcement learning; core network

1. Introduction

In the past few years, the ground communication network has further expanded and developed with the increase in communication volume worldwide. However, the ground communication network cannot offer equal coverage to remote areas or the edge network due to insufficient resource planning, which leads to unassured communication service quality in these areas. At the same time, satellite communication systems leverage the benefits of high altitude and multicast/broadcast capabilities while complementing traditional ground networks to enable novel mobile communication networks [1]. Increasingly, organizations are initiating projects on the satellite–terrestrial hybrid network (STHN). Notably, OneWeb, O3b, SpaceX, and Telesat are among the companies proposing satellite-based Internet solutions [2]. As shown in Figure 1, owing to its intrinsic benefits encompassing wide scope, efficient processing capabilities, and versatile nature, STHN finds application across numerous practical domains, intelligent transportation systems [3], military tasks, disaster relief [4], etc.

On the one hand, cloud-native computing inherits from the development of cloud computing and has the characteristics of Infrastructure as a Service (IaaS), Platform as a Service (PaaS), and Software as a Service (SaaS) [5]. Additionally, cloud-native computing achieves high reliability, elasticity, manageability, and observability of applications through means such as containerization, microservices architecture, declarative API, automated deployment, dynamic configuration, and auto-scaling, better adapting to the rapid changes in and complexity of cloud environments.

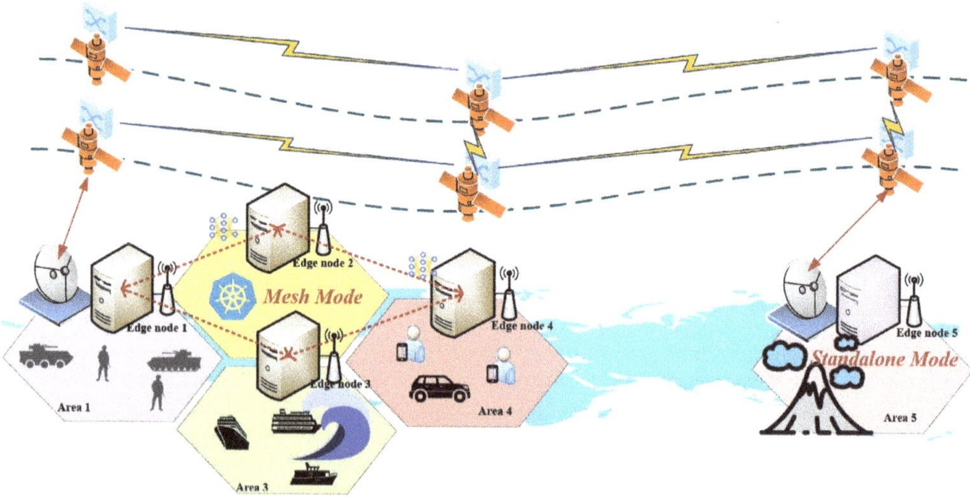

Figure 1. ComEdge platform scenario.

On the other hand, future universal computation requires an open and integrated network architecture that deeply integrates computing power, communication, and industry intelligence in a decentralized network. This new network architecture will transform traditional cloud AI into edge AI and then support network for AI (NET4AI) [6], enabling a variety of AI applications such as multi-task transfer learning on edge devices [7] and the integration of edge cloud-native and intelligent wireless communication. Cloud-native and edge intelligence are the core parts of NET4AI. Artificial intelligence can achieve better results in edges that are closer to data and have greater elasticity compared to single cloud computing.

In this context, the trend of cloud-native computing in the telecommunications industry is on the rise. Compared to almost any other service, telecommunication services have higher requirements for resilience, security, and performance [8]. The Third Generation Partnership Project (3GPP) initially introduced the Service-Based Architecture (SBA) paradigm [9] for 5G Core (5GC). This paradigm involves the decomposition of monolithic Network Functions (NFs) into multiple microservices, offering finely grained functionalities. These microservices are designed for deployment as Virtual Machines (VMs) or lightweight containers (e.g., Docker) in a cloud-native fashion, making efficient use of cloud computing resources. Consequently, NFs can be dynamically scaled both horizontally (out/in) and vertically (up/down) to accommodate fluctuations in signaling traffic from a multitude of devices. The emergence of open-source core network platforms such as Open5GS [10] and Free5GC [11] further assists research in realizing cloud-native core network.

While the emerging mobile core network can enhance deployment flexibility and scalability through cloud-native environments [12,13], several challenges need to be addressed in practical deployments at multiple edge nodes close to users. First, due to limited computing and data storage resources at edge nodes, the cloud-native network functions (CNFs) should be distributed across multiple edge nodes and run in response to terminal service requests following communication protocol procedures. It is foreseeable that edge-distributed core networks will consume more communication resources than remote data center-based core networks. Furthermore, running CNFs on edge nodes incurs costs for container deployment and operation. Therefore, a cloud-native edge core network deployment scheme should be designed to reduce operational costs and control traffic overhead.

The paper is structured as follows: The challenge of cloud-native INCCOM is introduced in Section 2. The ComEdge experimental platform is detailed in Section 3. Following this, Section 4 discusses the deployment challenges and solutions for the cloud-native mobile core network in ComEdge. Section 5 provides an analysis of the challenges and opportunities encountered by the INCCOM network system. Lastly, Section 6 offers the paper's conclusion.

2. The Challenge of Cloud-Native INCCOM

As mobile edge computing (MEC) and AI continue to evolve, computing power has become ubiquitous. Networks must now offer intelligent services that enable efficient collaboration between cloud, edge, and terminal computing power [14]. The upcoming 6G network is aimed at achieving seamless global coverage, spanning across land, sea, air, and space. It will also enable the establishment of a hierarchical network which will facilitate centralized and distributed collaboration. The network will support distributed edge autonomy and management, alongside unified and simpler protocols. The integration of computing and communication systems will enable seamless collaboration among cloud, edge, network, and computing power. Communication and computing elements will understand and collaborate with each other, thereby achieving real-time and precise discovery of available computing power, flexible and dynamic computing, as well as connection service scheduling. These advancements will offer ubiquitous services, optimizing the allocation of computing resources, and enhancing the utilization efficiency of communication and computing resources. At the same time, cloud-native architecture is becoming a prevailing trend in modern software architecture design patterns. The objective is to enhance the elasticity, reliability, and scalability of applications. Cloud-native architecture highlights container-based design, microservices, continuous delivery, and automated management in order to improve application performance and ease of management. There is an urgent need to design and implement a smart cloud-native network platform that integrates centralized and distributed architectures and blends communication and computing. The main challenges are as follows.

The primary challenge is how to effectively control and utilize fragmented resources on demand, encompassing efficient scheduling of distributed computing resources, complete utilization of fragmented data silos, appropriate utilization of diverse communication methods, and effective consolidation of decentralized and heterogeneous models. Task scheduling, node management, and structure merging are essential aspects to consider when scheduling fragmented resources. Distributed frameworks such as Hadoop [15] and Spark [16] have attained maturity and are extensively used in fields like data processing and machine learning, providing the ability to manage and schedule computing power in distributed environments. The problems of data silos and model consolidation can be addressed by utilizing machine learning frameworks such as federated learning, transfer learning, and lifelong learning.

The second challenge for INCCOM involves ensuring network determinacy and security. The primary distinction between determinate and non-determinate networks is the assurance of transmission time. Non-determinate networks may encounter issues like lost, delayed, or disordered data packets which can affect real-time applications. In contrast, determinate networks utilize techniques that predict the time of data packet transmission and guarantee the reliability and stability of real-time applications. In cluster networks experiencing real-time topology changes, data security, communication security, and model security are crucial considerations. Blockchain technology and federated learning have potential in enhancing network and data security, making them a vital aspect of network security.

It is crucial to note that the deployment method of the core network plays a significant role in the performance of satellite-ground integrated communication systems in both the present and future stages. However, the orchestration of cloud-native network functions in the core network remains an unresolved issue, with very few works dedicated specifically to addressing this problem. In the context of the satellite-ground fusion network, ref. [17–19] discuss several novel core network deployment and management architectures. Ref. [17] proposes the vision of deploying certain critical network functions in the core network on satellites to reduce latency. Ref. [18] further elaborates on this vision by distributing all or some of the network functions of the 5G core network across multiple satellites in different orbits and leveraging the "service set" mechanism in 5G to alleviate the significant signaling pressure between satellites and the ground. Ref. [19] addresses the issue of signaling storms introduced by satellite mobility and presents the in-orbit stateless core network case. It is important to note that the focus above is on the overall architectural research of the mobile core network in the satellite-ground fusion network scenario, with a lack of research on actual deployment algorithms for distributed core networks. In a typical network environment, the authors of refs. [20,21] implemented a cloud-native Service Function Chaining (SFC) framework that provides traffic steering mechanisms for establishing end-to-end network services. In [22], the authors propose a horizontal scaling algorithm that leverages Control Theory to dynamically adjust the number of instances of the Access and Mobility Management Function (AMF) based on traffic load. To conserve resources required for User Plane Function (UPF) instances, the authors of ref. [23] propose a UPF instance scaling mechanism based on deep reinforcement learning. Although these efforts reduce the traffic overhead in the control plane of the core network deployed in the form of virtual machines, containers, or slices, there is a lack of consideration for the relationship between fluctuating service requests, deployment costs of the core network at multiple edge nodes, and inter-node control signaling overhead.

Clearly, to simultaneously minimize deployment costs and control signaling overhead while adapting to the dynamic nature of user service requests, an intelligent CNFs deployment architecture is required. This architecture provides deployment strategies for CNFs by learning the spatiotemporal relationships between user service requests, deployment costs, and control signaling overhead.

3. ComEdge Experimental Platform

With the rapid advancement of architectural technology, the cloud-native concept has gained widespread acceptance. Various technologies, including service meshes, microservices, containers, and Kubernetes, have become the standard for cloud edge architecture in the 5G era. Many industry, academic, and research organizations are engaged in the study of cloud-native edge computing, with examples such as OpenYurt [24], launched by Alibaba, EdgeFoundry [25] from the Linux Foundation, and KubeEdge [26] by Huawei. Following an evaluation of different architectural options, we chose KubeEdge to establish our platform. KubeEdge seamlessly connects cloud and edge applications and resources, creating a unified computing architecture that brings about better data processing efficiency and user experience. This section presents a cloud-native INCCOM platform developed using KubeEdge.

3.1. Platform Scenario

Figure 1 illustrates the cloud-native edge intelligence scenario where the functions of each node can be flexibly deployed on various real-world devices, such as drones, vehicles, ships, satellites, and cloud hosts in different scenarios. The physical communication methods used within the cluster are also varied, including but not limited to fiber optic, wifi, microwave, as well as different types of satellites. Moreover, despite the various underlying communication methods used, the cloud-native edgemesh [12] technology shields the complex underlying topology, enabling the platform to be agile in the allocation and scheduling of business, tasks, and resources. The majority of nodes in the system are equipped with containerized 5G base stations, and the master node deploys a containerized 5G core network, granting access to numerous 5G terminals through the platform's southbound interface. The platform's quick deployment on nodes with varied architectures and forms, added to the flexibility and maneuverability of actual physical nodes, alongside the integration of satellite links, greatly expands the various application scenarios of the platform.

In the actual construction of the platform, we use five physical nodes with different architectures, computing power, and communication capabilities as the main cluster parts of ComEdge. The actual situation of each physical node is shown in Table 1.

Table 1. The actual situation of each physical node.

Node	CPU	Memory	Communication Method
node1	X86/64 bit, 16 core	126 GB	wifi; ethernet
node2&3	X86/64 bit, 8 core	8 GB	wifi
node4	ARM/64 bit, 4 core	8 GB	ethernet
node5	ARM/64 bit, 4 core	4 GB	wifi;ethernet

Meanwhile, the various nodes in the platform can operate in two working modes as needed based on the actual situation, and some devices of the ComEdge platform are shown in Figure 2.

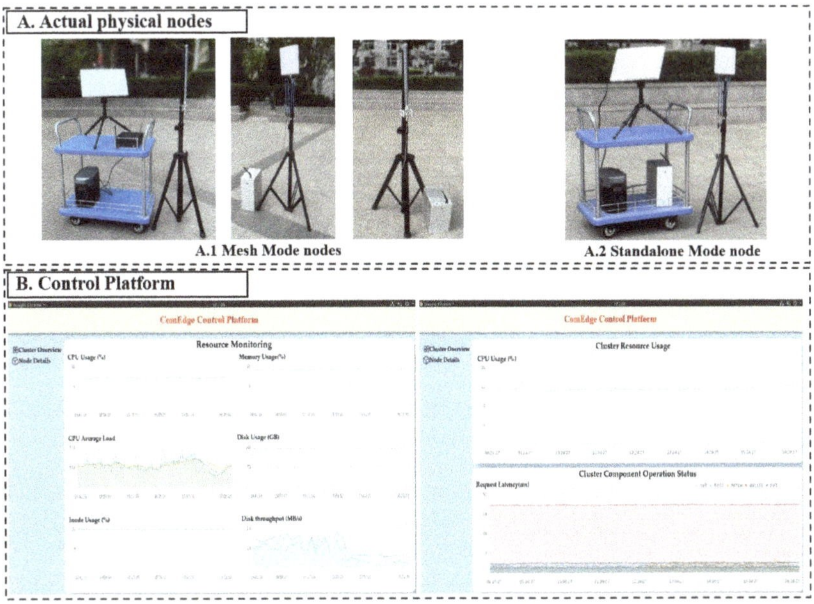

Figure 2. Actual deployment of ComEdge platform.

Mesh Mode: A network is formed between Nodes 1, 2, 3, and 4 through a wifi hotspot, with Node 1 creating the hotspot, and Nodes 2, 3, and 4 wirelessly connecting, while Node 5 joins through a wired connection, thus forming a fully functional high-availability cluster. Our platform deploys multiple master nodes to establish a high-availability cluster that adopts a master-slave architecture, with the master node in physical node 1 acting as the primary node, while the other master nodes act as secondary nodes. The primary node bears the responsibility of processing all requests and forwarding them to the secondary nodes, which receive and perform the corresponding operations. In case of failure or ineffectiveness of the primary node, the secondary nodes can automatically take over its role to ensure the high availability and reliability of the cluster. The multiple master nodes collaborate with each other and manage all components and resources in the cluster.

Standalone Mode: The standalone mode refers to the case when only one physical node is present in the system or when a task is designated to a single physical node. It proves useful in scenarios where tasks are simplistic or necessitate specific requirements for edge devices.

3.2. Platform Architecture

The platform architecture, depicted in Figure 3, consists of three main parts. Nodes integrated with the 5G communication system connect to various terminal devices via the southbound interface, depending on the scenario, such as a plethora of IoT sensors, vehicles in the vehicular network, various communication devices across military and civilian applications, including mobile phones and smart watches. Additionally, nodes integrated with satellite antennas establish a communication link with high-throughput Earth-orbiting satellites via the northbound interface, establishing a connection with the data network. The platform can be deployed on general architectures, including X86 and ARM operating systems. The system resources are virtualized, and various businesses and data can be carried out in a containerized way.

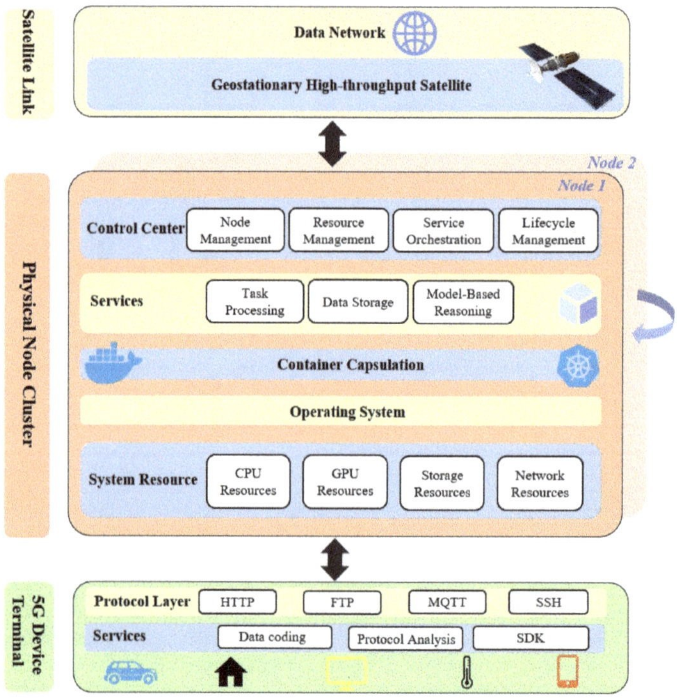

Figure 3. Architecture of ComEdge platform.

The control panel of the platform cluster includes the following important features:

- Node management: Node management is an extremely important part of cluster operation and maintenance. It mainly includes operations such as adding, deleting, maintaining, monitoring, and auto-scaling nodes.
- Resource management: Resource management achieves flexible configuration and scheduling of compute and storage resources through the use of container resource limitations, requests, and pod scheduling strategies to achieve optimal resource utilization efficiency.
- Service orchestration: Service orchestration is achieved by managing and coordinating a group of related containers as an application. The main goal of service orchestration

is to coordinate aspects such as container storage, networking, deployment, scaling, and upgrading to create highly available and scalable services, and to provide rich service discovery and routing mechanisms.

- Lifecycle management. Lifecycle management functionality allows users the management and monitoring of the entire lifecycle of containers, including container creation, operation, monitoring, updating, and destruction. Through comprehensive container management, lifecycle management ensures stable operation and high availability of containers, improving application efficiency and reliability.

The services of each node in the platform cluster include the following aspects. (1) Task processing: the tasks to be processed can be resource-intensive tasks distributed over a certain period of time, or they can be short, lightweight, and stateless tasks. (2) Data storage: it refers to how to store data in containers and manage the data in the containers. Considering the characteristics of containers, data persistence is usually not needed, and only short-term storage is required. (3) Model derivation: Edge intelligence is achieved by deploying federated learning, transfer learning [27], reinforcement learning, and other algorithms on various edge devices.

4. Deployment of Cloud-Native Core Network in ComEdge

In addressing the deployment of a cloud-native edge core network on the ComEdge platform, we formulated the following reasonable research hypotheses. Due to the ability to fully leverage the dynamic temporal relationships between user service requests, core network deployment and operational costs, DQN-CNFDA can utilize fewer resources while effectively reducing system overhead. Other approaches that consider only a single factor exhibit relatively poorer performance in terms of platform deployment, operational costs, and resource utilization.

To validate our research hypotheses, in this section, we first introduce the key processes and deployment examples of the core network. Afterward, we formulate the deployment optimization problem for the cloud-native edge core network, aiming to minimize the deployment and control traffic overhead of the core network.

4.1. Example Analysis

To further elucidate the actual processes of core network deployment costs and inter-node control signaling overhead in multiple edge nodes, it is necessary to provide a comprehensive analysis with illustrative examples of the key control plane processes that affect the placement cost of cloud-native edge core networks, such as user registration and session establishment. Furthermore, to gain a more intuitive understanding of the distributed core network deployment approaches, it is necessary to provide examples and compare different static edge core network deployment schemes.

Based on the initial registration and session establishment, these two fundamental processes clearly demonstrate the control signaling interactions among various network functions in the core network when fulfilling user service requests. Figure 4 illustrates the procedures for the initial registration and session establishment in the context of 5G network deployment. During the UE's initial registration, the Authentication and Mobility Management Function (AMF) authenticates the UE and communicates QoS/billing profiles to the Session Management Function (SMF). Following this, the SMF selects a User Plane Function (UPF) to serve as the anchor gateway for session establishment. To initiate uplink data transmission, the UE first establishes a radio connection with the base station, which subsequently sends a service request to the AMF. The AMF, in turn, replicates session states to the base station for effective QoS enforcement. For the delivery of downlink traffic, it is imperative for the anchor gateway to notify the AMF of data arrival. Subsequently, the AMF notifies the base station to initiate paging for the UE. If this paging process succeeds, the UE repeats the aforementioned procedure to establish the session anew. Based on the above process, it is possible to quantitatively determine the control signaling overhead between CNFs.

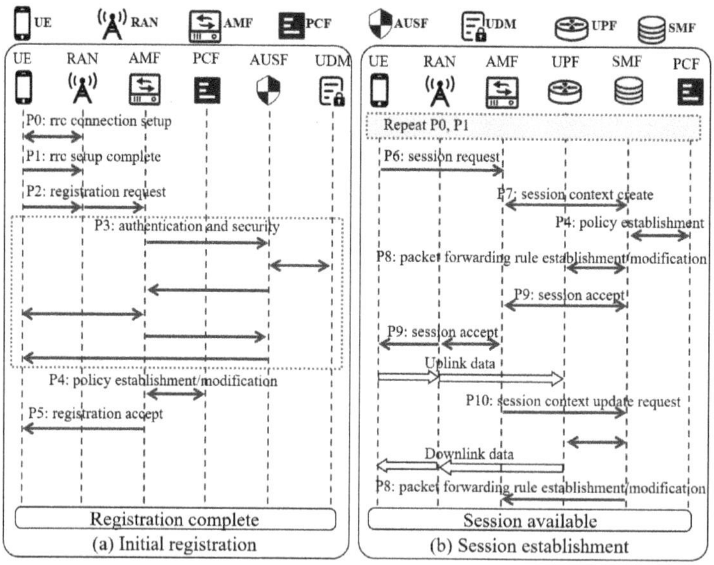

Figure 4. Partial 5G Signaling Process.

With the rise of edge computing and related infrastructure development, deploying the core network at the edge has become a major trend. The edge core network is positioned closer to users, which leads to a reduction in data transmission delay and the upload of sensitive data. This proximity helps conserve network bandwidth, enhance data transmission efficiency, and alleviate network congestion while simultaneously safeguarding privacy and data security. In additon, utilizing edge nodes as backup nodes reduces the risk of single point failure and enhances system reliability. Lastly, in comparison to large data centers, the edge core network significantly minimizes energy consumption and operational costs. Then, a brief description is provided for three different examples of edge core network deployment.

In Figure 5a, a relatively complete core network is deployed only on Edge Node 2. The direct consequence of this is that users outside Area 2 need to incur higher edge control overhead for initial registration and session establishment processes. Additionally, the resource utilization of the entire edge system is severely imbalanced. This is a scenario that we should try to avoid in practical deployments. In Figure 5b, a relatively complete core network is distributedly deployed on Edge Nodes 1, 2, and 3. In this case, the resource utilization of the edge system is more balanced, and it also improves, to some extent, the back-haul control overhead for users in Areas 1, 2, and 3. However, this deployment approach still has drawbacks when the number of users increases, especially when there is a significant increase in users in Areas 3 and 4. This can lead to more frequent interactions between AMF, SMF, UPF, and thus consume a substantial amount of inter-node communication resources. Compared to the first two scenarios, the third deployment scenario is more reasonable. In the above processes, network functions that interact frequently with the terminal, such as AMF and UPF, are deployed on almost every edge node, while network functions related to authentication, such as UDM and AUSF, are concentrated on one node. This ensures load balancing between nodes and effectively controls the control signaling overhead of the system.

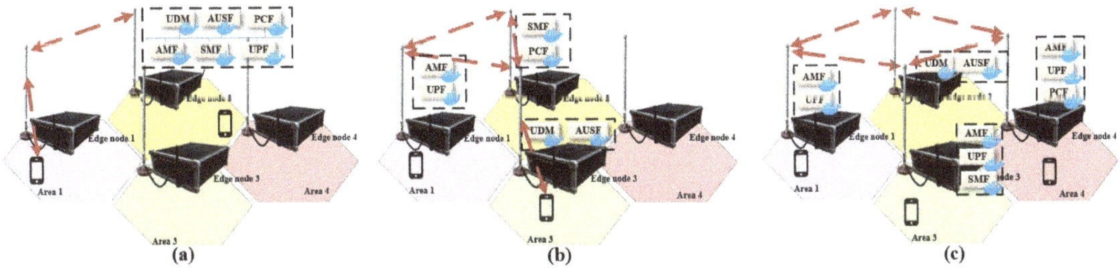

Figure 5. Three Examples of edge core network deployment. (**a**) A single core network is deployed on a single edge node. (**b**) A single core network is deployed on multiple edge nodes. (**c**) Multiple core networks are deployed on multiple edge nodes.

It should be noted that the number of users and the volume of requests in the wireless coverage areas of various edge nodes are dynamically changing. The above three examples of edge core network deployments are all static solutions and cannot adapt to dynamic real-world situations. In the following subsection, we introduce a DQN-based CNF deployment algorithm.

4.2. DQN-Based CNF Deployment Algorithm (DQN-CNFDA)

We developed a CNF deployment framework based on DQN to address the intricate relationships among service requests, provisioning costs, and control traffic overhead. This framework is designed to reduce costs by learning spatiotemporal patterns within service requests. Once the deployment of each CNF is completed, achieving user registration services is possible through CNFs interacting within or between nodes following the communication protocol process. To adaptively place CNFs in response to fluctuations in service requests, our approach, DQN-CNFDA, learns the optimal policy by gathering rewards through trial-and-error interactions with the environment [28].

To minimize the operation costs of the cloud-native edge core network, we begin by quantifying the communication between nodes and calculating the associated cost within the core network. We simplify the model by solely considering the communication cost between nodes, disregarding the cost within nodes. The meanings of each notation are explained in Table 2.

Table 2. The meanings of each notation.

Notation	Description
E	Set of edge nodes
e	Edge nodes index
N_E	Number of edge nodes
u_e^t	Number of available resource units in edge node e at time t
S	Set of services
s	Service index
F_s	Set of CNFs in service s
f	CNF index in set of CNFs F_s
p	CNF type index
$x_{p,f,e}^t$	CNF f with type p is provisioned on edge node e at time t or not
$n_{s,f,e}^t$	Number of service s request for CNF f to edge e at time t

Formula (1) indicates the occurrence of communication between CNF $f \in F_s$ and CNF $f+1 \in F_s$ across the edge nodes within the time interval t. $M^t(f,s)$ equals zero if CNF f and $f+1$ are provisioned the same edge node; otherwise, $M^t(f,s)$ equals one.

$$M^t(f,s) = \sum_{p}\sum_{p'}\sum_{f\in F_s}\sum_{f'\in F_s}\sum_{e\in E}\sum_{e'\in E} min(1, |e-e'| \cdot x_{p,f,e}^t \cdot x_{p',f',e'}^t). \quad (1)$$

D_{CM} in (2) represents the cost of individual communication. Meanwhile, C^t_{CM} represents the overall cost of communication between nodes necessary to accomplish the service core network.

$$C^t_{CM} = \sum_{s \in S} \sum_{f \in F_s} M^t(f,s) \sum_{e \in E} n^t_{s,f,e} \cdot D_{CM}. \tag{2}$$

D_{OP} in (3) represents the deployment cost of an individual CNF, and C^t_{OP} represents the deployment cost of the entire edge core network across all the nodes.

$$C^t_{OP} = \sum_{f \in F_s} \sum_{e \in E} x^t_{p,f,e} \cdot D_{OP}. \tag{3}$$

The objective of optimization can be described as

$$\min \quad C^t_{OP} + C^t_{CM}. \tag{4}$$

Resources on edge nodes are limited. CNFs deployed should not exceed the number of available resource units on edge nodes.

$$\sum_{p} \sum_{f \in F_s} x^t_{p,f,e} \leq u^t_e, \quad \forall t, e \in E. \tag{5}$$

We use the available resource units u^t_e of each edge node e within time t as the state space s_t. A represents the action space and an action a_t at time t is one of the combinations of edge clouds, where the number of combinations of edge nodes is 2^{N_E}.

$$s_t = \{u^t_1, u^t_2, u^t_3, u^t_1, \ldots, u^t_{N_E}\}. \tag{6}$$

Reward r_t at time t is defined as

$$r_t = -C^t_{OP} - C^t_{CM}. \tag{7}$$

Algorithm 1 introduces the CNFs deployment algorithm based on DQN. The weight θ is initialized in Line 1. In Lines 4–10, the model explores placement schemes for CNFs using various action probabilities of δ. In Line 11, the reward r_t and the new state s_{t+1} are obtained. Established experience replay D in Line 12 weakens the correlation between data, leading to an increased stability of the neural network. Lines 13–20 perform random minibatch of transitions from D in order to compute gradients. The algorithm applies gradient descent by updating the gradients based on the difference between the target value (corresponding to the old parameters) and the Q value (corresponding to the new parameters) to the current parameters. The old parameters are updated with the new parameters every C steps.

4.3. Evaluation Results

The ComEdge platform's main experimental environment consists of five edge nodes as shown in Table 1. The available resource units for edge nodes are set to the minimum memory value occupied by all CNFs. The main observations in the experiments focus on the two core network processes introduced in Section 4.1: registration and session establishment. Based on Alibaba Cloud's unit costs for CNF's operation and control traffic [29], D_{OP} and D_{CM} are, respectively, set to 0.875 RMB per hour and 0.233 RMB per hour. In addition, we implemented DQN-CNFDA using PyTorch and conducted replays on the ComEdge platform with real data from different cells [30].

The experiments compared the differences in cost and resource utilization among the following edge core network deployment schemes: (1) DQN-CNFDA described in Section 4; (2) the minimum deployment cost scheme (MDCS) mentioned in [31]; (3) engineering-experience-based CNF deployment scheme (ENE-CNFDS); (4) random CNF deployment scheme (R-CNFDS). Among these, Scheme 2 represents a category of current edge core network deployments (as described in [31]), which primarily focus on minimizing the deployment cost of network functions while not paying sufficient attention to control traffic overhead. Scheme 3 represents the approach described in [17–19], where either the entire or a portion of core network functions are statically deployed at satellite edge nodes. This approach solely addresses the significant control traffic overhead issue caused by frequent user-triggered registration process handovers in the ground core network, while neglecting the critical factor of deployment costs in the edge core network. The experimental results of the four aforementioned schemes in the ComEdge platform are presented in Figure 6.

Algorithm 1 DQN-based CNFs Deployment Algorithm (DQN-CNFDA)

Initialize action-value function Q with random weights θ
Initialize target action-value function $\hat{Q}$ with weights $\theta' = \theta$
Initialize replay memory D to capacity N
for episode = 1, M **do**
  Initialize state s
  for $t = 1, T$ **do**
    Generates the probability $\delta \in [0, 1]$
    if $\delta < \varepsilon$ **then**
      Randomly selects action $a_t \in A$
    else
      $a_t \leftarrow arg \max_a Q(s_t, a_t; \theta)$
    end if
    Set r_t and s_{t+1}
    Store transition (s_t, a_t, r_t, s_{t+1}) in buffer D
    Select random minibatch of transitions (s_t, a_t, r_t, s_{t+1}) from D
    if episode ends at time $i + 1$ **then**
      Set $y_i = r_i$
    else
      $y_i = r_i + \gamma \max_{a_{i+1}} \hat{Q}(s_{i+1}, a_{i+1}; \theta')$
    end if
    Perform a gradient descent step on $(y_i - Q(s_i, a_i; \theta))^2$ with respect to the network parameters θ
    Every C steps reset $\hat{Q} = Q$
  end for
end for

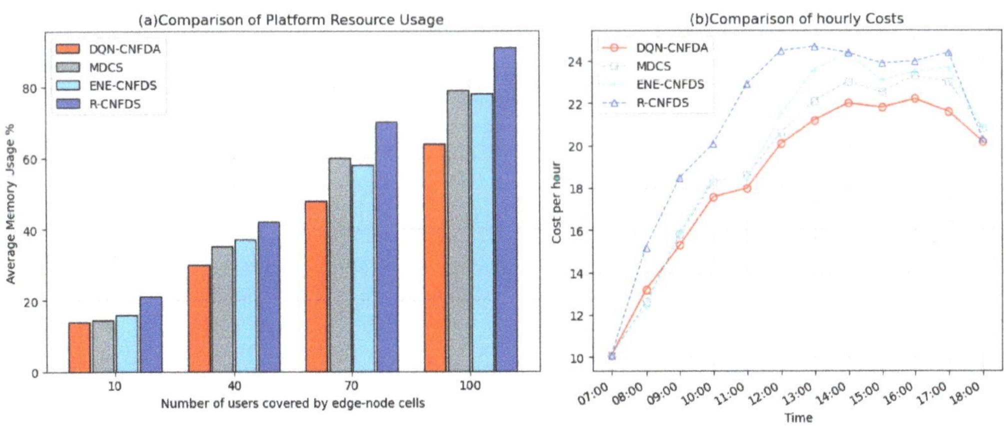

Figure 6. The results of the experiments conducted on the ComEdge platform.

Figure 6a illustrates a comparison of platform resource usage for different schemes as the number of users in the wireless coverage area of edge nodes increases. We use the more intuitive platform average memory usage. Consistent with subjective inference, when the number of users within the coverage of edge nodes increases, the number of running containers in the entire platform also increases accordingly, leading to an increase in platform memory usage. R-CNFDS neglects the control traffic overhead between nodes, resulting in inefficient resource utilization. This situation becomes even more severe as the number of users increases. It is worth noting that when the number of users is small, there is little difference in resource utilization between MDCS, ENE-CNFDS, and DQN-CNFDA. With fewer user requests within the node coverage, the required number of containers to be deployed is limited, resulting in relatively good performance for ENE-CNFDS, even though it does not consider deployment costs as effectively. Additionally, the lower volume of control signaling between nodes does not significantly impact MDCS in terms of control traffic. However, as the number of users increases (e.g., when a single node covers 70 to 100 users), DQN-CNFDA outperforms MDCS and ENE-CNFDS in effectively managing the relationship between deployment costs, control traffic, and dynamic user requests. It efficiently scales network functions on each node, leading to more pronounced advantages in system resource utilization.

Figure 6b illustrates a comparison of the costs associated with distributed deployment and operation of the core network on edge nodes using various schemes. All the required CNFs are launched simultaneously at 7:00, resulting in equal costs for all four schemes at the begin time. The operational costs of the edge core network increase gradually as the number of edge users and time progresses. The number of edge users gradually increases over time, leading to a corresponding increase in the operating cost of the edge core network. From 7:00 to 18:00, DQN-CNFDA successfully minimized the costs. The advantage of deep reinforcement learning in deploying CNFs on the ComEdge platform was fully demonstrated. It is worth noting that DQN-CNFDA can predict the placement of network functions based on user's service request behavior, while R-CNFDS does not involve any learning from user behavior, leading to a significant gap between the two. Furthermore, MDCS and ENE-CNFDS each consider minimizing deployment costs and controlling the inter-node signaling quantity, which reduces the gap with DQN-CNFDA. However, due to the lack of utilization of the relationship between dynamic user requests, deployment costs, and control traffic, MDCS and ENE-CNFDS exhibited significant cost fluctuations (from 13:00 to 17:00). Therefore, using MDCS and ENE-CNFDS results in higher deployment and operational costs. Specifically, the DQN-based scheme achieved a maximum cost reduction of up to 6.9% compared to the engineering-experience-based CNFs deployment scheme.

Based on the experimental results in the aforementioned platform, the proposed DQN-CNFDA better grasps the relationship between dynamic and time-varying user service requests, edge-core network deployment, and operational costs compared to engineering experience and other algorithms. Consequently, it reduces system expenses and enhances resource utilization efficiency.

5. Opportunities and Challenges

5.1. Intelligent Cluster

In the future integrated network, the terminals, base stations, core networks, etc., will evolve towards intelligence and become intelligent network elements. Collaboration among multiple intelligent entities can be aimed at a single task or multiple tasks. Its connotation includes task decomposition and combination, goal analysis and modeling, model training and inference, parameter iteration and sharing, etc. At the same time, for tasks specific to the physical world, resource scheduling and performance optimization are carried out for integrated computation and fusion to ensure the reliability of task execution.

Traditional mobile communication networks also utilize collaboration mechanisms, but these mechanisms are focused on connected collaboration with limited scope, minimal data volume, and collaboration restricted to same-level network elements. However, in the integrated computation network facing AI, multi-intelligent agent collaboration involves a large scale of collaboration, interactive data volume, and collaboration cost, including the possibility of multi-level horizontal and vertical collaboration. As a result, existing collaboration mechanisms cannot be reused. The challenges include selecting collaboration objects, mining and extracting collaboration information, and determining collaboration modes and mechanisms while controlling collaboration costs.

5.2. Telecommunications Industry

In the future, society will experience an intelligent era of universal connectivity. The next-generation mobile communication network will bridge the gap between humanity and the digital world. The evolution of the core network is crucial in the process of integrated networking. The core network acts as a convergence point for network businesses and applications, driving future development. Additionally, the core network is the center of the entire network topology, connecting various terminals and access networks, with a domino effect on the entire network. To adapt to the integration of communication and computation, the following areas need improvement in the process of core network evolution:

- Firstly, the increasing number of network functions leads to more service interactions between network functions, weaker security mechanisms for service interactions, and complex operational steps for traffic switching configurations when switching between new and old versions of services.
- Secondly, the observability of network function services is poor, the visualization of topology logic is weak, the storage of service operation logs is not standardized, and performance indicators are not standardized, which results in a large amount of manual operations in maintaining network function services and locating and troubleshooting issues.
- Finally, operators already have a pool of virtualized resources in the network function virtualization (NFV) stage, which are mostly built and distributed in the form of virtual machine resources. Therefore, it is necessary to consider reusing and migrating virtualized resources into a cloud-native telecommunications infrastructure, running the 5G network in the form of containerized resources, and further improving the flexibility and resource utilization of the network.

6. Conclusions

Due to the growing advancements in cloud-native edge computing and the necessity of effective collaboration amongst cloud, edge, and terminal computing power, the INCCOM network set the path for the next-generation network evolution. The main work of this paper is as follows. We begin by discussing cloud-native INCCOM in the context of a satellite–terrestrial hybrid network. Subsequently, we introduce the ComEdge experimental verification platform that we created and tested in a real-world setting. We then proceed to analyze the platform's architecture, functional characteristics, and the deployment status in actual production environments. Additionally, we explore the solution of deep reinforcement learning in the deployment of cloud-native core network and conduct a preliminary verification of the platform's potential to enable artificial intelligence in a real production environment. Lastly, we analyze the challenges and opportunities facing the integrated network from the perspective of intelligent clusters as well as the telecommunications industry.

Author Contributions: Conceptualization, H.S.; Methodology, H.S.; Software, P.W., J.C. and Y.Z.; Resources, X.Z.; Data curation, H.S., P.W. and J.C.; Writing—original draft, H.S.; Supervision, X.Z. All authors have read and agreed to the published version of the manuscript.

Funding: This work is supported by the National Key Research and Development Program of China 2021YFB2900504.

Data Availability Statement: All relevant data are within the paper.

Conflicts of Interest: The authors declare no conflict of interest.

References

1. Wang, P.; Zhang, J.; Zhang, X.; Yan, Z.; Evans, B.G.; Wang, W. Convergence of satellite and terrestrial networks: A comprehensive survey. *IEEE Access* **2019**, *8*, 5550–5588. [CrossRef]
2. Karavolos, M.; Nomikos, N.; Vouyioukas, D.; Mathiopoulos, P.T. HST-NNC: A novel hybrid satellite-terrestrial communication with NOMA and network coding systems. *IEEE Open J. Commun. Soc.* **2021**, *2*, 887–898. [CrossRef]
3. Xiong, G.; Zhu, F.; Dong, X.; Fan, H.; Hu, B.; Kong, Q.; Kang, W.; Teng, T. A kind of novel ITS based on space-air-ground big-data. *IEEE Intell. Transp. Syst. Mag.* **2016**, *8*, 10–22. [CrossRef]
4. Hubenko, V.P.; Raines, R.A.; Mills, R.F.; Baldwin, R.O.; Mullins, B.E.; Grimaila, M.R. Improving the global information grid's performance through satellite communications layer enhancements. *IEEE Commun. Mag.* **2006**, *44*, 66–72. [CrossRef]
5. Li, F. Cloud-native database systems at Alibaba: Opportunities and challenges. *Proc. Vldb Endow.* **2019**, *12*, 2263–2272. [CrossRef]
6. Yang, T.; Ning, J.; Lan, D.; Zhang, J.; Yang, Y.; Wang, X.; Taherkordi, A. Kubeedge wireless for integrated communication and computing services everywhere. *IEEE Wirel. Commun.* **2022**, *29*, 140–145. [CrossRef]
7. Chen, Q.; Zheng, Z.; Hu, C.; Wang, D.; Liu, F. On-edge multi-task transfer learning: Model and practice with data-driven task allocation. *IEEE Trans. Parallel Distrib. Syst.* **2019**, *31*, 1357–1371. [CrossRef]
8. TU Group. Cloud Native Thinking for Telecommunications. 2020. Available online: https://github.com/cncf/telecom-user-group/blob/master/whitepaper/cloud_native_thinking_for_telecommunications.md (accessed on 27 March 2020).
9. 3GPP. System Architecture for the 5G System. 3GPP TS 23.501 V15. 3.0. 2018. Available online : https://www.etsi.org/deliver/etsi_ts/123500_123599/123501/15.03.00_60/ts_123501v150300p.pdf (accessed on 20 September 2018) .
10. Open5GS: Open Source Project of 5GC and EPC (Release 16). Available online: https://open5gs.org/ (accessed on 11 June 2021).
11. Free5GC: An Open-Source Project for 5th Generation (5G) Mobile Core Networks. Available online: https://free5gc.org/ (accessed on 3 May 2021).
12. Imadali, S.; Bousselmi, A. Cloud native 5g virtual network functions: Design principles and use cases. In Proceedings of the 2018 IEEE 8th International Symposium on Cloud and Service Computing (SC2), Paris, France, 18–21 November 2018; pp. 91–96.
13. Luong, D.H.; Thieu, H.T.; Outtagarts, A.; Ghamri-Doudane, Y. Predictive autoscaling orchestration for cloud-native telecom microservices. In Proceedings of the 2018 IEEE 5G World Forum (5GWF), Silicon Valley, CA, USA, 9–11 July 2018; pp. 153–158.
14. Tang, X.; Cao, C.; Wang, Y.; Zhang, S.; Liu, Y.; Li, M.; He, T. Computing power network: The architecture of convergence of computing and networking towards 6G requirement. *China Commun.* **2021**, *18*, 175–185. [CrossRef]
15. Apache Hadoop Project Develops Open-Source Software for Reliable, Scalable, Distributed Computing. Available online: https://hadoop.apache.org/ (accessed on 16 July 2022).
16. Spark: Unified Engine for Large-Scale Data Analytics. Available online: https://spark.apache.org/ (accessed on 20 August 2022).
17. Cui, H.; Zhang, J.; Geng, Y.; Xiao, Z.; Sun, T.; Zhang, N.; Liu, J.; Wu, Q.; Cao, X. Space-air-ground integrated network (SAGIN) for 6G: Requirements, architecture and challenges. *China Commun.* **2022**, *19*, 90–108. [CrossRef]
18. Wang, X.; Sun, T.; Duan, X.; Wang, D.; Li, Y.; Zhao, M.; Tian, Z. Holistic service-based architecture for space-air-ground integrated network for 5G-advanced and beyond. *China Commun.* **2022**, *19*, 14–28. [CrossRef]
19. Li, Y.; Li, H.; Liu, W.; Liu, L.; Chen, Y.; Wu, J.; Wu, Q.; Liu, J.; Lai, Z. A case for stateless mobile core network functions in space. In Proceedings of the ACM SIGCOMM 2022 Conference, Amsterdam, The Netherlands, 22–26 August 2022; pp. 298–313.
20. Bouridah, A.; Fajjari, I.; Aitsaadi, N.; Belhadef, H. Optimized scalable SFC traffic steering scheme for cloud native based applications. In Proceedings of the 2021 IEEE 18th Annual Consumer Communications & Networking Conference (CCNC), Las Vegas, NV, USA, 9–12 January 2021; pp. 1–6.
21. Dab, B.; Fajjari, I.; Rohon, M.; Auboin, C.; Diquélou, A. Cloud-native service function chaining for 5G based on network service mesh. In Proceedings of the ICC 2020—2020 IEEE International Conference On Communications (ICC), Dublin, Ireland, 7–11 June 2020; pp. 1–7.
22. Alawe, I.; Hadjadj-Aoul, Y.; Ksentini, A.; Bertin, P.; Darche, D. On the scalability of 5G core network: The AMF case. In Proceedings of the 2018 15th IEEE Annual Consumer Communications & Networking Conference (CCNC), Las Vegas, NV, USA, 12–15 January 2018; pp. 1–6.
23. Nguyen, H.T.; Van Do, T.; Rotter, C. Scaling upf instances in 5g/6g core with deep reinforcement learning. *IEEE Access* **2021**, *9*, 165892–165906. [CrossRef]
24. OpenYurt: An Open Platform That Extends Upstream Kubernetes to Edge. Available online: https://openyurt.io/ (accessed on 22 July 2021).
25. EdgeFoundry: The Preferred Open Source Edge Platform. Available online: https://www.edgexfoundry.org/ (accessed on 18 August 2021).
26. Kubeedge: Kubernetes Native Edge Computing Framework. Available online: https://kubeedge.io/ (accessed on 7 January 2022).
27. Chen, Q.; Zheng, Z.; Hu, C.; Wang, D.; Liu, F. Data-driven task allocation for multi-task transfer learning on the edge. In Proceedings of the 2019 IEEE 39th International Conference On Distributed Computing Systems (ICDCS), Dallas, TX, USA, 7–10 July 2019; pp. 1040–1050.

28. Sutton, R.S.; Barto, A.G. *Reinforcement Learning: An Introduction*; MIT Press: Cambridge, MA, USA, 2018.
29. Alibaba Cloud. Alibaba Cloud Pricing. Available online: https://www.aliyun.com/price/ (accessed on 24 May 2022).
30. Harvard Dataverse. A Multi-Source Dataset of Urban Life in the City of Milan and the Province of Trentino Dataverse. Available online: https://dataverse.harvard.edu/dataset.xhtml?persistentId=doi:10.7910/DVN/EGZHFV (accessed on 16 November 2022).
31. Zheng, J.; Tian, C.; Dai, H.; Ma, Q.; Zhang, W.; Chen, G.; Zhang, G. Optimizing NFV chain deployment in software-defined cellular core. *IEEE J. Sel. Areas Commun.* **2019**, *38*, 248–262. [CrossRef]

electronics

MDPI

Article

A Capacity Enhancement Method for Frequency-Hopping Anti-Jamming Communication Systems

Ziting Yu [1,2], Zhisong Hao [1,2,*], Wang Yao [3,4] and Min Jia [5]

1 The 54th Research Institute of China Electronics Technology Group Corporation, Shijiazhuang 050081, China; yuziting@stu.xidian.edu.cn
2 Science and Technology on Communication Networks Laboratory, Shijiazhuang 050081, China
3 Key Laboratory of Mathematics, Information and Behavioral Semantics, Ministry of Education, Beihang University, Beijing 100191, China; yaowang@buaa.edu.cn
4 Institute of Artificial Intelligence, Beihang University, Beijing 100191, China
5 School of Electronics and Information Engineering, Harbin Institute of Technology, Harbin 150080, China; jiamin@hit.edu.cn
* Correspondence: haozhisong@cetc.com.cn

Abstract: In this paper, we study the enhancement of channel transmission information by expanding the transmission channel in the frequency-hopping rate dimension in a communication system. This is achieved using the frequency-hopping spread spectrum (FHSS) without increasing communication resources, such as power and bandwidth. The anti-jamming capability of the original information is maintained during this process. The spectral characteristics of the extended signal for frequency-hopping (FH) transmission are investigated, a demodulation method based on carrier reconstruction is proposed, the bit error rate performance is simulated, and the capacity enhancement and anti-jamming ability of the extended signal for FH transmission are analyzed.

Keywords: anti-interference communication; FHSS; capacity enhancement; carrier wave reconstruction

1. Introduction

Satellite communication systems are an important component of next-generation mobile communication networks. Through the introduction of satellite communication means, the global coverage of mobile communication systems can be enhanced, thereby expanding applications in such areas as automatic flight piloting in remote areas [1–3], maritime logistics monitoring [4,5], global environmental resource monitoring [6,7], and disaster early warning [8,9].

Satellite communication systems have a significantly increased beam coverage compared to terrestrial communication networks. As long as it is within the beam coverage, it is possible to interfere with satellite communication signals. In addition, satellite communication systems have high-gain characteristics, with the S-band synchronous-orbit communication satellite antenna aperture reaching 15 m [10]. The wide-area coverage and high-gain characteristics of satellite communication systems make the risk and probability of interference significantly higher, especially for wide-beam antenna systems used for signaling or IOT message transmission [11].

In order to improve the anti-jamming performance of satellite communication systems to meet the transmission of highly reliable information, such as signaling transmission, autopilot, far-area flight, disaster warning, etc., the spectrum is usually extended in exchange for a certain anti-jamming gain. The higher the spectrum expansion ratio, the greater the anti-jamming gain. Under the premise of an unchanged transmission bandwidth, the lower the information rate, the stronger the anti-jamming ability. Therefore, in the above scenario, the information rates of each user's signaling secret key, driving instructions, warnings, etc., are generally reduced to the kbps scale for transmission [12,13].

According to Shannon's theorem, there is some room for information capacity improvement before the communication link reaches maximum efficiency after spectrum expansion [14].

Spread spectrum is divided into frequency-hopping spread spectrum and direct spread spectrum. There is a contradiction between the anti-interference capability of information transmission and the need for an information transmission rate. To alleviate the contradiction between the two, this paper proposes a capacity enhancement method for the anti-jamming communication system for FII, which aims to increase the information transmission capacity of the anti-jamming communication system by modulating the incremental information in the frequency-hopping frequency point dimension without guaranteeing the anti-jamming capability of the original system or increasing communication resources, such as the power and bandwidth.

1.1. Related Works

Since the late 1990s, scholars have studied and utilized modulation information on the amount of change in frequency-hopping frequency point positions, using techniques such as differential frequency hopping, etc.

The United States-based organization Lockheed Sanders developed CHESS radio based on differential frequency-hopping technology [15]. With a hopping speed of up to 5000 hops/s and an information transmission rate of up to 19.2 kb/s, CHESS radio is equipped to handle tracking interference, and has considerable resistance to fading ability [16,17]. Zhongying Liu et al. proposed a frequency-hopping detection method based on STFT (short-time Fourier transform) [18], which was combined with the G-function to analyze the frequency characteristics of shortwave differential frequency-hopping signals at each hop time interval, enabling the detection of hopping frequency information. Qian Bo et al. [19] proposed a multi-hop autocorrelation-based differential frequency-hopping (DFH) signal detection method for the non-periodic characteristics of DFH signal hopping sequences.

In recent years, both domestic and foreign scholars in this field have focused on further optimizing security performance using algorithms such as chaotic sequence encryption, as well as the design of the principal rate transfer function and the detection of broadband high-speed frequency-hopping signals.

In 2008, Qu Xiaoxu et al. proposed a FH-DFH approach that combines DFH with the traditional FH technique to eliminate the correlation between successive frequency hopping and to improve the anti-interception performance [20]. Several years later, in 2015, Yong Li, Fuqiang Yao et al. proposed a frequency domain equalization algorithm for wideband DFH systems: the frequency domain decision feedback sliding multi-hop equalization algorithm (DF-SMH-FDE) [21], which can improve the BER performance of differential frequency-hopping (DFH) systems. In 2022, Yin Aibing and Li Yi [22] designed an encryption scheme based on pseudo-random sequences of logistic chaotic mapping to randomly scramble frequency-hopping (FH) patterns to address the poor two-dimensional uniformity of FH patterns in differential frequency-hopping (DFH) communications, which are easily intercepted. In doing so, they strengthened the anti-interception capability; however, this negatively affected the systems' performance. In 2022, Xin Liu et al. [23] proposed an enhanced differential frequency-hopping (EDFH) framework, which adds the transmission and processing processes of training signals to the traditional communication signal transmission and designs a mixed-signal matched filter (CMF) for user signals and interference signals at the receiving end.

The research conducted in this field has demonstrated that differential frequency hopping conducts information modulation from the dimension of the amount of change in the frequency-hopping frequency point position, but the hardware complexity that is required for differential frequency hopping is high, and it only analyzes and utilizes the technology from the perspective of a single modulation method. The capacity-boosting modulation method proposed in this paper is based on the phase modulation and frequency-hopping that are already used in anti-jamming communication systems, in which frequency-hopping dimensional modulation is performed to boost the information capacity. The multidimensional modulation system differs from single dimension modulation in terms of the demodulation method and BER performance.

In frequency-hopping immunity papers, chaotic sequences are used to control frequency-hopping patterns to enhance the security of information transmission. Compared with these papers, the novelty of this paper lies in the use of transmitted information control instead of chaotic sequence control of the frequency-hopping patterns, which on the one hand enhances the unpredictability of frequency-hopping patterns and further improves the security of information transmission; on the other hand, it increases the dimension of the transmitted information and enhances the transmission capacity of the anti-jamming system.

1.2. Novelty and Main Contributions

- Firstly, the capacity enhancement method of FH anti-jamming communication systems is proposed, the capacity enhancement model of FH anti-jamming communication systems is established, the expression of FH capacity enhancement signals is given, and the spectral characteristics of the signal are deduced;
- Secondly, the FH capacity-boosting signal demodulation scheme is designed;
- Finally, the effect of the new method on improving capacity, BER performance, and anti-jamming performance is analyzed.

2. Materials and Methods

2.1. FH System Capacity Enhancement Methodology

Previous FH systems selected carriers of different frequencies for modulation through frequency-hopping patterns; FH patterns carry no information [24] (see Figure 1). A_I denotes the I-way of information A and A_Q denotes the Q-way of information A.

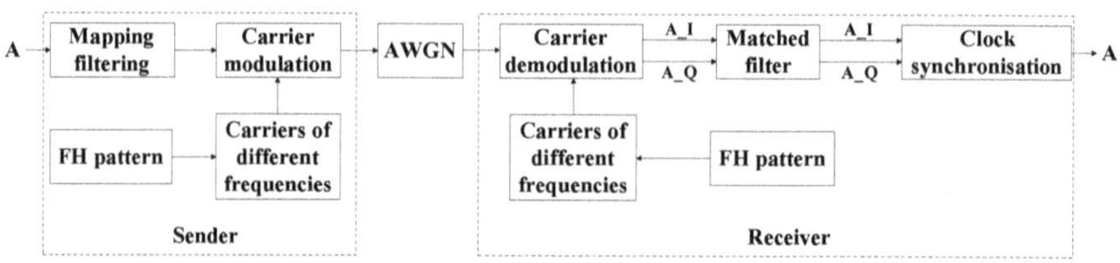

Figure 1. Schematic diagram of information rate enhancement for frequency-hopping systems.

The FH system capacity enhancement method, which is used to enhance the information transmission capacity of the anti-jamming communication system by modulating the information at the frequency-hopping point dimension, is presented in Figure 2. A_I denotes the I-way of information A and A_Q denotes the Q-way of information A. In this method, the information transmitted in the original FH system is referred to as Information A, and the incrementally transmitted information is referred to as Information B.

At the transmitter side, for Information A, BPSK modulation is generally used. After symbol mapping, carrier modulation is performed. The carrier is a frequency-hopping carrier, and the frequency-hopping rate is selected based on Information B. The frequency-hopping period is generally an integer multiple of the period of the code element of Information A. For Information B, at the transmitting end, sequence modulation is performed according to the mapping relationship to transform Information B into control information that does not use frequency carriers. In this case, one code element cycle of Information B corresponds to one or more frequency-hopping cycles. The output at the transmitter side is an FH capacity boost signal carrying both Information A and Information B.

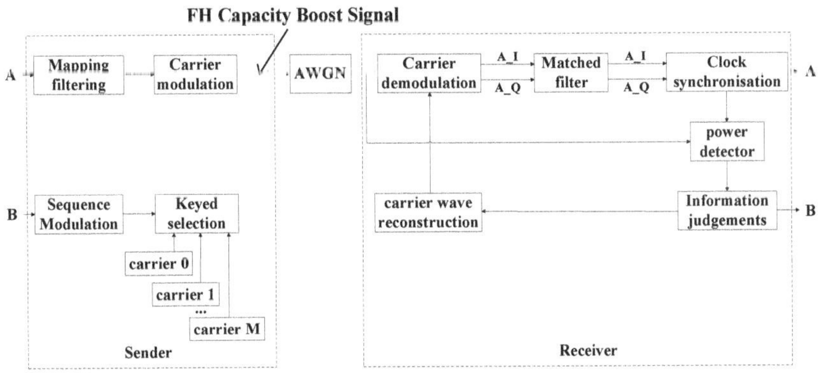

Figure 2. Schematic diagram of information rate enhancement for frequency-hopping systems.

After passing through the AWGN channel, at the receiving end, carrier reconstruction is conducted to form a local carrier using Information B; carrier demodulation is then carried out to form a baseband signal. After the completion of clock synchronization and judgement, Information A is demodulated. At the receiving end, the carrier frequency point is measured using the power detector method, and then the baseband Information B is demodulated according to the modulation mapping relationship.

2.2. FH Capacity Boost Signal Design

2.2.1. FH Capacity Boost Signal Expression

The FH capacity boost signal transmits Information A and Information B in two dimensions: phase and FM point, respectively.

Information A uses BPSK modulation and Information B equivalently uses FSK modulation. Without channel coding, the FH capacity boost signal is denoted as

$$s(t) = \sum_{n=-\infty}^{+\infty} \sum_{m=0}^{M-1} a(n)g_a(t-nT)b_{\omega_m}(n)g_b(t-nT)\cos\omega_m t \tag{1}$$

where the baseband Information A is expressed as $a(n)$. When A = 1, $a(n)$ takes 1, indicating that the carrier phase is 0. When A = 0, $a(n)$ takes -1, which indicates that the carrier phase is π. Information A is an equal probability source.

$$a(n) = \begin{cases} 1 & A=1 \;\; p_A(1)=1/2 \\ -1 & A=0 \;\; p_A(0)=1/2 \end{cases} \tag{2}$$

where the baseband Information B is expressed as $b_{\omega_m}(n)$. The $b_{\omega_m}(n)$ vector is obtained from the modulation mapping relation. $b_{\omega_m}(n)$ is the coefficient before the carrier at each frequency point. When $b_{\omega_m}(n) = 1$, it indicates that this carrier frequency is used for transmission. When $b_{\omega_m}(n) = 0$, it indicates that this carrier frequency is not selected for transmission. $b_{\omega_1}(n)$ to $b_{\omega_m}(n)$ correspond to $\omega_0, \omega_1 \ldots \omega_{M-1}$. In each frequency-hopping cycle, only one $b_{\omega_m}(n)$ is 1 and the others are 0.

When modulating a carrier with different frequency points using Information B, the M frequency points are divided into I groups of J frequency points each, which can modulate k bits of information.

$$I = \frac{M}{J} = 2^k \tag{3}$$

The group label is i (the group label corresponds to the code element label), and $0 \le i \le (I-1)$. The frequency points within each group are labeled j, and $0 \le j \le (J-1)$. J is the number of frequency points in each group.

Putting Information B in terms of one code element per k bits, the corresponding decimal value is i_B. Each Information B code element cycle corresponds to J frequency-hopping cycles. The jth frequency-hopping cycle of the current code element corresponding to $b_{\omega_m}(n)$ is

$$b_{\omega_m}(n) = b_{\omega_{iJ+j}}(n) = \begin{cases} 1 & (i = i_B) \cap (j = j_B) \quad 1/M \\ 0 & (i \neq i_B) \cap (j \neq j_B) \quad 1 - 1/M \end{cases} \tag{4}$$

where $0 \leq m \leq (M-1)$, $0 \leq i \leq (I-1)$, and $0 \leq j \leq (J-1)$.

This demonstrates the correspondence between each k bits and code elements in Information B. The grouping of code elements can undergo real-time chaotic processing to improve the system's anti-interception performance.

2.2.2. FH Capacity Boost Signal Generation

The FH capacity boost signal needs to meet the requirements of frequency-hopping point orthogonality, phase continuity, and code element alignment, as shown in Figure 3. In Figure 3, A denotes information A, B denotes information B.

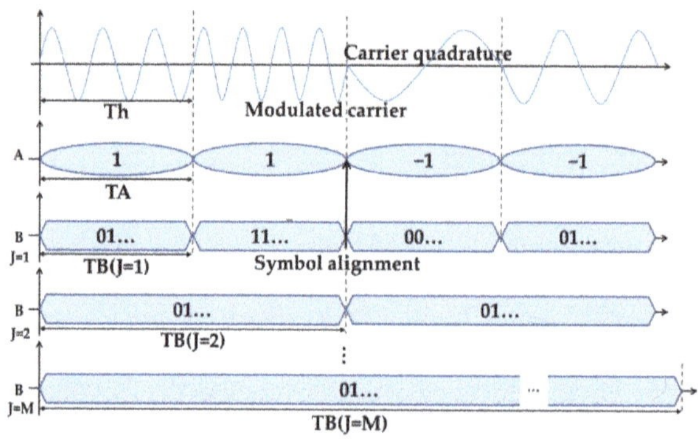

Figure 3. FH capacity boost signal generation.

The frequency-hopping points are orthogonal and the carrier frequency points are ω_0, $\omega_1 = 2\omega_0, \ldots, \omega_{M-1} = M\omega_0$. The carrier signal $\cos\omega_0, \cos\omega_1, \ldots, \cos\omega_{M-1}$ is orthogonal in the period $\cos\omega$. Quadrature frequency hopping is implemented in the baseband first, and then the frequency-hopped signal is up-converted to RF.

In multi-symbol joint detection, to improve the BER performance of Information B without affecting the original anti-jamming performance of Information A, i.e., to maintain the frequency-hopping speed of Information A, multi-symbol joint detection is performed so that a single Information B code element controls one or more frequency-hopping points. At the same time, each set of frequency points is orthogonal to one code element period in the time domain.

To achieve phase continuity—that is, to facilitate the demodulation of Information A at the receiving end—the phase of the signal at different frequency-hopping points should change continuously, to ensure that a symbol period is an integer multiple of the carrier period. To provide the end of a symbol period for the end of the carrier period, the next symbol starts just for the beginning of the carrier period, so that the phase changes continuously.

The start and end moments of the code elements of the originating Information A and Information B are aligned with each other; the code element period of Information A corresponds to an integer number of frequency-hopping cycles; the code element period of Information B corresponds to an integer multiple of the code element period of Information A; and there are an integer number of carrier cycles within each frequency-hopping cycle.

For baseband Information A, the number of frequency-hopping points is M. The frequency-hopping rate is r_H. The channel coding efficiency is μ_A. The information rate is r_A. The code element rate is R_A. In general, the frequency-hopping period T_H is equal to the code word period T_A, or an integer multiple of the code word period T_A.

For baseband Information B, the code element cycle T_B corresponds to j frequency-hopping cycles, T_H, where $j \geq 1$.

2.2.3. FH Capacity Boost Signal Spectral Analysis

To derive the FH capacity boost signal spectrum, s(t) is divided into steady state and alternating waves [25].

- Steady State Wave Analysis

Here, the steady state wave is the statistically averaged component of the random sequence s(t) and a weighted average of the probability of occurrence of each code element.

When both Information A and Information B are equal-probability sources, the number of frequency-hopping points is M, and Information A adopts BPSK modulation; there are a total of 2M possible scenarios for the waveform of s(t) in each frequency-hopping cycle, and the probability of occurrence of each scenario is 1/2M. The statistical average of each frequency-hopping symbol is $v(t) = 0$. Therefore, the s(t) signal contains no steady-state waves, and the s(t) signal spectrum contains no discrete spectral components.

- Alternating Wave Analysis

Alternating waves are defined according to the power spectral density Equation:

$$P_S(\omega) = \lim_{T\to\infty} \frac{E[|S_T(\omega)|^2]}{T} \tag{5}$$

where E denotes the statistical average and T is the interception time, which is set equal to the length of (2N + 1) code elements, $T = (2N+1)T_s$.

First, we find the spectral function, $S_T(t)$, of the function (1) of the random signal s(t). The truncated function, $s_T(t)$, of the random signal, s(t), can be expressed as

$$\begin{aligned} s_T(t) = & \sum_{n=-N}^{+N} a(n)g_a(t-nT)b_{\omega_0}(n)g_b(t-nT)\cos\omega_0 t \\ & + \sum_{n=-N}^{+N} a(n)g_a(t-nT)b_{\omega_1}(n)g_b(t-nT)\cos\omega_1 t \\ & + \ldots + \sum_{n=-N}^{+N} a(n)g_a(t-nT)b_{\omega_M}(n)g_b(t-nT)\cos\omega_M t \end{aligned} \tag{6}$$

For simplicity, the carrier term is initially ignored and the first term of the truncation function is taken to compute the spectrum, $S_{T_0}(\omega)$.

$$s_{T_{\omega_0}}(t) = s_{T_0}(t)\cdot\cos\omega_0 t \tag{7}$$

$$s_{T_0}(t) = \sum_{n=-N}^{+N} a(n)g_a(t-nT)b_{\omega_0}(n)g_b(t-nT) \tag{8}$$

$$S_{T_0}(\omega) = \int_{-\infty}^{\infty} s_{T_0}(t)e^{-j\omega t}dt = \sum_{n=-N}^{+N} a(n)b_{\omega_0}(n)\cdot e^{-j\omega nT}G_{ag}(\omega) \tag{9}$$

where $G_{ab}(\omega) = \int_{-\infty}^{\infty} g_a(t)g_b(t)e^{-j\omega t}dt$.

$$|S_{T_0}(\omega)|^2 = \sum_{k=-N}^{+N}\sum_{n=-N}^{+N} a(n)b(n)\cdot e^{-j\omega(n-k)T}G_{ab}(\omega)G_{ab}{}^*(\omega) \tag{10}$$

$$E[|S_T(\omega)|^2] = \sum_{k=-N}^{+N}\sum_{n=-N}^{+N} E[a(n)b(n)] \cdot e^{-j\omega(n-k)T} G_{ab}(\omega) G_{ab}{}^*(\omega) \tag{11}$$

① When $k \neq n, b(n) \cdot b(k) = 0, a(n) \cdot a(k) = \begin{cases} 1 & 0.5 \\ -1 & 0.5 \end{cases}, E[a(n)^2] = 0, E[b(n)^2] = 0.$

② When $k = n$, $a(n) \cdot a(k) = a(n)^2 = 1$, $b(n) \cdot b(k) = b(n)^2 = \begin{cases} 1 & p = \frac{1}{M} \\ 0 & 1-p \end{cases}$, and $E[a(n)^2] = 1, E[b(n)^2] = \frac{1}{M}$.

$E[a(n)^2], E[b(n)^2]$ has a non-zero value only at $k = n$.

$$E[|S_T(\omega)|^2] = \sum_{n=-N}^{+N} \frac{1}{M} |G_{ab}(\omega)|^2 = \frac{(2N+1)}{M} |G_{ab}(\omega)|^2 \tag{12}$$

T is the interception time; let it be equal to (2N + 1) code element lengths, $T = (2N+1)T_s$.

$$P_S(\omega) = \lim_{N\to\infty} \frac{(2N+1)}{(2N+1)T_s M} |G_{ab}(\omega)|^2 = \frac{\omega_S}{2\pi M} |G_{ab}(\omega)|^2 \tag{13}$$

When the carrier term is added, the power spectral density function of $\cos\omega_0 t$ is $\frac{\pi}{2}[\delta(\omega-\omega_0) + \delta(\omega+\omega_0)]$.

$$P_{S_{\omega_0}}(\omega) = \frac{\omega_S}{4M}[|G_{ab}(\omega-\omega_0)|^2 + |G_{ab}(\omega+\omega_0)|^2] \tag{14}$$

$$P_{S_{\omega M}}(\omega) = \frac{\omega_S}{4M}[|G_{ab}(\omega-\omega_M)|^2 + |G_{ab}(\omega+\omega_M)|^2] \tag{15}$$

Thus, the power spectral density of $s(t)$ is

$$P_S(\omega) = \sum_{m=0}^{M} \frac{\omega_S}{4M}[|G_{ab}(\omega-\omega_m)|^2 + |G_{ab}(\omega+\omega_m)|^2] \tag{16}$$

Taking eight frequency points as an example, when M = 8, the spectrum of the positive frequency point of the signal is shown in Figure 4. Each frequency point has equal power and the positive and negative frequency points are symmetrical about the position of the zero frequency point, which is consistent with Equation (16).

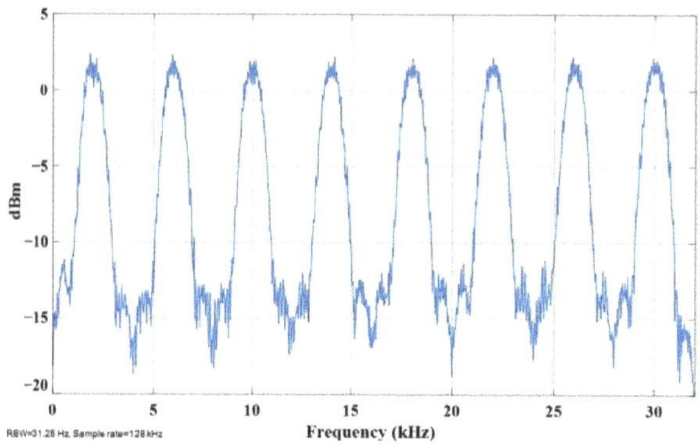

Figure 4. Schematic diagram of the frequency-hopping spectrum of frequency points.

The spectrum is characterized by the following three points:

(1) No steady state waves in the spectrum and no discrete carrier components, unlike FM signals;

(2) The average power at each frequency point is 1/M of the total power, and as the number of frequency points increases, the average power of the signal at each frequency point decreases;
(3) The signal power spectrum at each frequency point is the Fourier transform of the product of the time domain baseband waveforms of signals A and B.

3. Results

3.1. Design of Demodulation Algorithm for FH Capacity Boost Signal

Demodulation Algorithm Flow

As shown in Figure 2, for Information A, coherent demodulation is used to ensure BER performance. Before demodulating Information A, the carrier frequency of the FH capacity boost signal must be determined. After the demodulation of Information B, the carrier reconstruction is performed according to the correspondence between Information B and the carrier frequency, and then Information A is demodulated using the reconstructed carrier of Information B. For Information B, the demodulation performance of both the envelope detection and coherent demodulation converge as the modulation order increases; for simplicity, the demodulation method of envelope detection is used for Information B.

The specific flow of the demodulation algorithm is as follows:

The received fixed-frequency signal is input into the Gardner loop for timing synchronization after passing through the matched filter, after which the reconstructed carrier of Information B, demodulated using the power detector method, is used to perform carrier recovery out of Information A. First, it is determined whether the timing synchronization is completed or not, and the frequency hopping begins once the timing synchronization is completed, as shown in Figure 5.

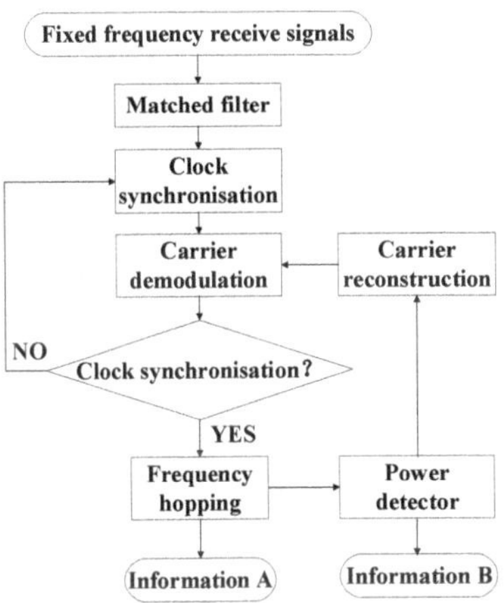

Figure 5. Demodulation flowchart.

Based on the proposed signal generation and demodulation methods, a simulation verification system is established to verify the functional performance parameters.

In the simulation verification model, Gardner's algorithm is used for the timing error detection of Information A. The carrier recovery of Information A is performed using a Costas loop. Before the information transmission link is established, a certain synchronization time is reserved, and the frequency is fixed during the synchronization time without the transmission of Information B.

- Power Detection Method for Demodulation of Information B

The demodulation of Information B is equivalent to the demodulation of FSK signals. There are two different methods that can be used to demodulate FSK signals: the coherent demodulation method and the power detector method. When the number of frequency points gradually increases, the demodulation performance of these two methods is similar. However, the power detector method is relatively simple, so this paper adopts the envelope detector method. The realization block diagram is shown in Figure 6.

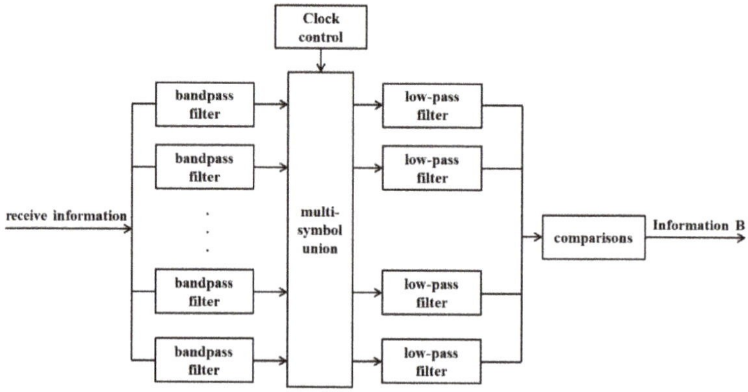

Figure 6. Block diagram of the power detection method.

In order to improve the received signal-to-noise ratio of Information B, the BER can be reduced at the expense of the information rate by using a joint detector judgement of multiple symbols, in which the J frequency-hopping symbols are jointly estimated as one Information B symbol.

When J = 16, the Information B symbol keying selects the carrier frequency group of 16 frequency points, and Information B is recovered using the envelope detection method according to the correspondence between different frequency groups and Information B, as shown in Figure 7.

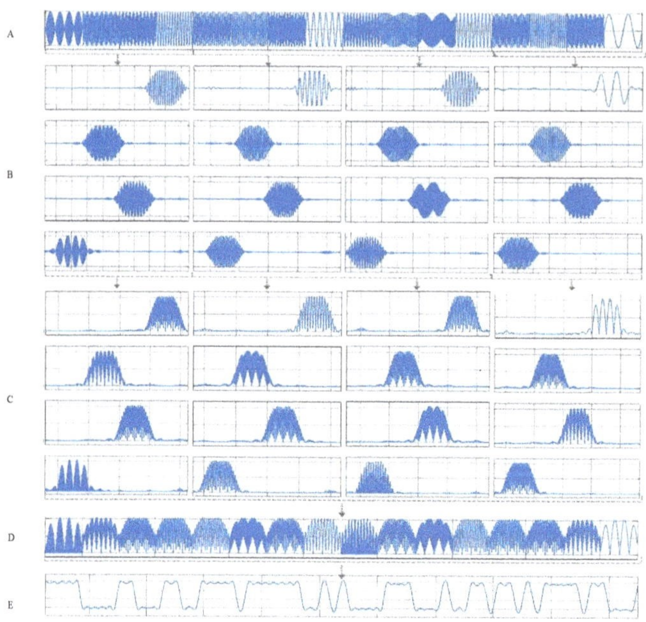

Figure 7. Schematic diagram of joint envelope detection for J = 16 symbols.

Figure 7A shows a received signal containing multiple carrier frequencies, and Figure 7B shows the received signal passing through bandpass filters at different frequencies to separate the different frequency carriers, respectively. As shown in Figure 7C, the signals are integrated to take the absolute value, and Figure 7D,E show the combined signals and the correspondence with Information B in accordance with Information B, respectively. Finally, the baseband Information B is recovered by the low-pass filter.

Through simulation, the spectra after the multi-symbol joint detector for J of one, two, four, eight, and sixteen are shown in Figure 8. The signal power spectral densities are all enhanced with respect to the J = 1 signal, and the power spectral density values at the highest points are enhanced by 4 dB, 9 dB, 12 dB, and 15 dB, respectively.

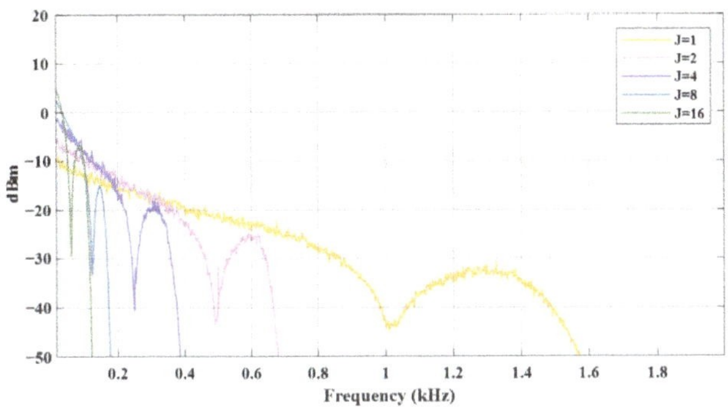

Figure 8. Spectrum after joint multi-symbol wave detection.

- Clock Synchronization

For the Gardner algorithm [26], each code element needs two sampling points, and each code element cycle only needs to calculate one timing error. The Gardner algorithm structure is shown in Figure 9.

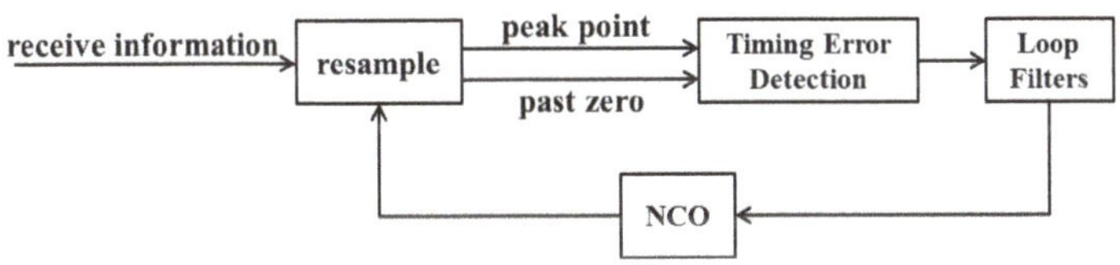

Figure 9. Gardner structure.

At the receiving end of Figure 2, the I and Q path baseband signals of Information A are sampled at twice the code word rate to generate two paths: the peak point and the over-zero point. After timing error detection is performed, the timing error information is fed into the loop filter and the phase of the resampled clock signal is controlled by the NCO. The NCO output is also used for envelope detection and carrier reconstruction for Information B.

When Information B is not transmitted, the system receiver is equivalent to a general BPSK demodulator, and the clock synchronization link is normally locked. In this paper, we focus on the work of the clock loop after the normal transmission of Information A, the moment at which Information B begins to be transmitted (the moment when the carrier begins to hop), and the impact of Information B on the clock loop in the presence of different J values of the error code. The clock loop locking condition is represented by the clock phase offset statistics' mean square difference value and the clock phase maximum offset value.

The quality of clock recovery varies for different Eb/N0 values for Information A and different J values for Information B. The quality of clock recovery is expressed as the average frequency shift difference of the clock-phase error signal and the maximum offset value of the clock phase. The clock recovery quality is expressed in terms of the average frequency shift difference, σ_{t_mean}, of the clock-phase error signal and the maximum shift value, e_{t_max}, of the clock phase, where e_t is the error value between the clock recovered from the timing loop and the standard clock, and $\overline{e}_t$ is the average value of e_t.

$$\sigma_{t\_mean} = \frac{\sum_{n} |e_t(n) - \overline{e}_t|}{n} \tag{17}$$

e_{t_max} denotes the maximum offset of the clock from the start of transmission of Information B

$$\sigma_{t\_max} = \max|e_t(n)| \tag{18}$$

Tables 1–3 show the standard deviation values of the clock-phase shift and the maximum shift values of the clock phase within 10 s after the start of frequency hopping, obtained via simulation for different values of J at different SNRs, where the information rate of Information A is 1 kbps, the frequency-hopping points are 64, and the Eb/N0 of Information A is 4 dB, 9 dB, and 14 dB. The offset value is normalized by the hopping period T_H; a maximum offset value of 0.11 is offset by 0.11 times the hopping period.

Table 1. Clock-locked data for Eb/N0 = 4 dB with different J values.

J	Clock-Phase Shift Standard Deviation	Maximum Clock-Phase Offset
1	0.06	0.234375
2	0.054688	0.234375
4	0.05	0.22
8	0.05	0.2
16	0.04	0.193
32	0.0351	0.185

Table 2. Clock-locked data for Eb/N0 = 9 dB with different J values.

J	Clock-Phase Shift Standard Deviation	Maximum Clock-Phase Offset
1	0.032656	0.103
2	0.033563	0.109375
4	0.034609	0.12
8	0.036	0.12
16	0.035	0.11
32	0.031	0.1

Table 3. Clock-locked data for Eb/N0 = 14 dB with different J values.

J	Clock-Phase Shift Standard Deviation	Maximum Clock-Phase Offset
1	0.023438	0.085938
2	0.020781	0.039063
4	0.016406	0.054688
8	0.0186	0.04
16	0.0218	0.041
32	0.023	0.04

Figures 10 and 11 present the clock locking for J values of 1, 2, 4, 8, 16, and 32 for the Eb/N0 values of Information A of 4 dB,9 dB, and 14 dB, respectively. Figure 10 shows the standard deviation of the clock-phase offset and Figure 11 shows the maximum clock-phase offset. For the same J value, the clock-locking quality becomes progressively better with the increase in the Eb/N0 values; the effect of the J value on the clock-locking quality is weaker than that of Eb/N0.

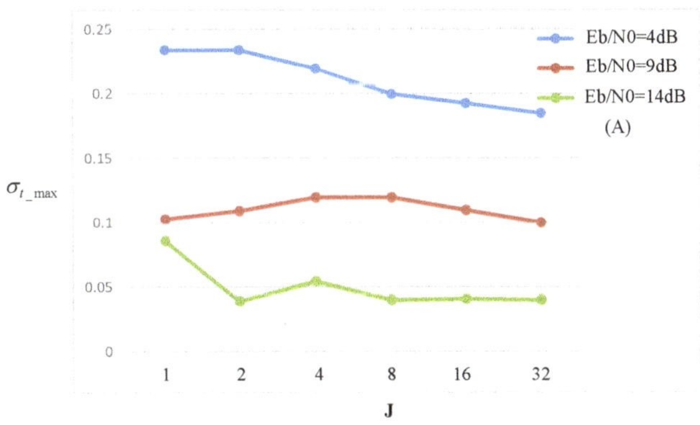

Figure 10. Standard deviation of clock-phase shift for different Eb/N0 values and different values of J.

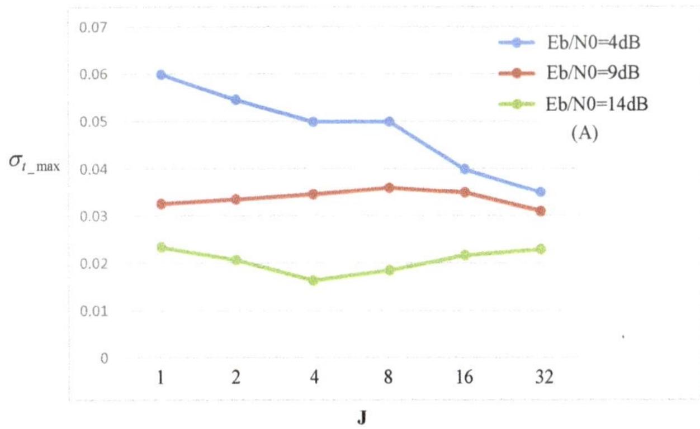

Figure 11. Maximum shift of clock-phase for different Eb/N0 values and different values of J.

- Carrier Synchronization

The carrier recovery uses the Costas ring [27], which has a block diagram, as shown in Figure 12. It mainly consists of a phase detector, a numerically controlled oscillator, and a loop filter. The phase detector is used to identify the phase, i.e., to track the phase information, which is generally realized with a multiplier. The quadrature frequency converter module multiplies the received IF signal and the local carrier to form I- and Q-way baseband signals. The matched filter module filters out the high harmonics of the baseband signal. The carrier phase error extraction and loop filter cooperate to form the carrier-phase error signal, which controls the carrier signal in the carrier reconstruction module for phase adjustment; the carrier reconstruction module carries out carrier selection and reconstruction according to Information B, as well as the mapping relationship between Information B and the carriers of different frequency-hopping points.

The quality of carrier recovery varies for different Eb/N0 values for Information A and different J values for Information B. The quality of carrier recovery is expressed as the average offset difference, σ_c, of the carrier-phase error signal, where e_c is the error between the carrier recovered from the carrier loop and the carrier at the transmitter, and $\overline{e}_c$ is the average value of e_c.

$$\sigma_c = \frac{\sum_{n} |e_c(n) - \overline{e}_c|}{n} \quad (19)$$

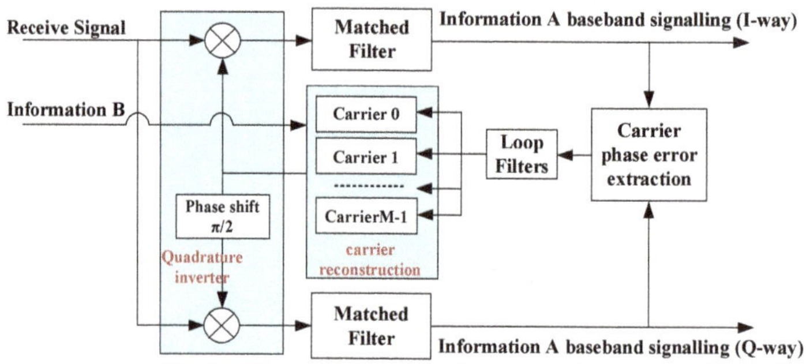

Figure 12. Costas ring structure diagram.

Figure 13 shows the variation in the standard deviation of the carrier-phase error signal for the J values of 1, 2, 4, 8, 16, and 32 for the Eb/N0 values of Information A of 4, 9, and 14 dB, respectively. In addition, since the error code of Information B creates errors in the reconstructed carrier, which also leads to degradation in the quality of the carrier recovery, the error case of the reconstructed carrier is also presented in Figure 13, denoted as cw_e, where cw_s denotes the carrier used for modulation at the transmitter and cw_r denotes the carrier recovered from the loop to be used for carrier reconstruction.

$$cw_e = cw_r - cw_s \tag{20}$$

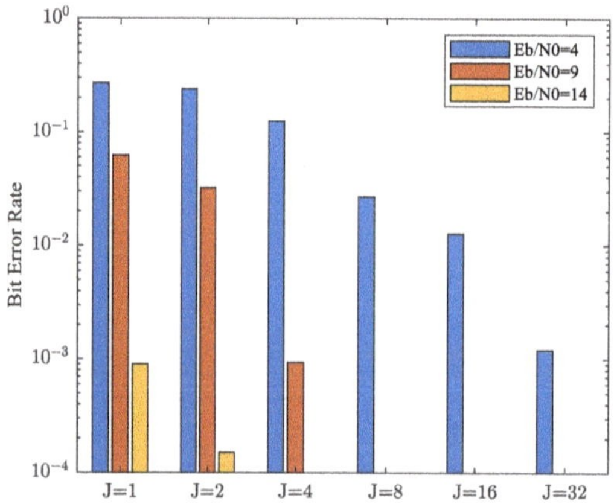

Figure 13. Quality of carrier recovery under different conditions.

Figure 13 is a histogram of the error probability of the reconstructed carrier, and it can be seen that the error probability of the reconstructed carrier gradually decreases as the Eb/N0 value of Information A gradually increases with the same J value. At the same Eb/N0 value, the error probability of the reconstructed carrier decreases as the value of J increases.

Figure 14 shows the variation in the standard deviation of the carrier phase error signal for the J values of 1, 2, 4, 8, 16, and 32 for the Eb/N0 values of Information A of 4, 9, and 14 dB, respectively. As shown in Figure 14, the carrier recovery quality varies with the Eb/N0 value of Information A. At an Eb/N0 value of 14 dB for Information A, the value of σ_c varies between 0.025 and 0.028. At an Eb/N0 value of 9 dB for Information A, σ_c varies between 0.046 and 0.05, and the carrier jitter worsens by a factor of two, relative to the previous condition. At an Eb/N0 value of 14 dB for Information A, σ_c varies between 0.07 and 0.08, and the carrier jitter deteriorates further. Additionally, the carrier recovery quality is not sensitive to the value of J or the number of hopping cycles occupied by the Information B code element, and the carrier quality, σ_c, varies within 0.01 between different values of J at the same signal-to-noise ratio, which suggests that carrier recovery can be carried out normally on the basis of the reconstructed carrier.

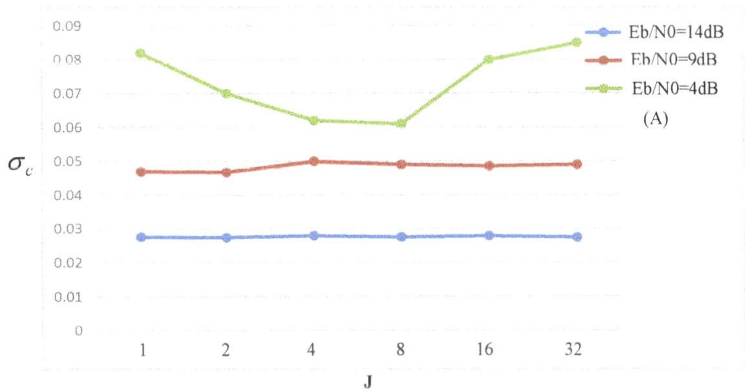

Figure 14. Variation in carrier recovery quality with J for different signal-to-noise ratios.

3.2. *BER Performance Analysis*

3.2.1. Information A BER Analysis

Without using channel coding, an information rate of 1 kbps for the Information A signal is generated; this signal is mapped using the BPSK modulation method and then subjected to eightfold sampling square-root ascending cosine-forming filtering to form the Information A baseband signal. The Information A baseband signal and the frequency-hopping carrier signal are multiplied to form the FH capacity boost signal. The FH carrier signal is created using Information B by selecting different frequency carriers according to the keying rule. The simulation specific parameters are shown in Table 4.

Table 4. BER performance simulation parameters.

Parameter Name	Parameter Value
Information A modulation	BPSK
Information A information rate	1 kbps
Information A channel coding	unencoded
Carrier hopping speed	1k hops/s
Number of carrier hopping points	8
Value of J	1, 2, 4, 8, 16, 32

After passing through a Gaussian white noise channel, Information A is demodulated and then adjudicated bit by bit. After that, the BER is calculated.

Since the demodulation performance of Information A depends on Information B for carrier reconstruction, the BER of Information B influences the carrier reconstruction effect, which in turn affects the demodulation performance of Information A.

The BER performance of Information A is simulated under seven cases of J values such as 1, 2, 4, 8, 16, 32, and 32 encoded, as well as coding with different Eb/N0 conditions for Information A. The simulation results are shown in Figure 15. To simplify the simulation algorithm, eight frequency-hopping points are selected, and to ensure signal orthogonality in the Information B code element cycle, the same frequency-hopping points are used in each Information B code element for J values of 8, 16, and 32.

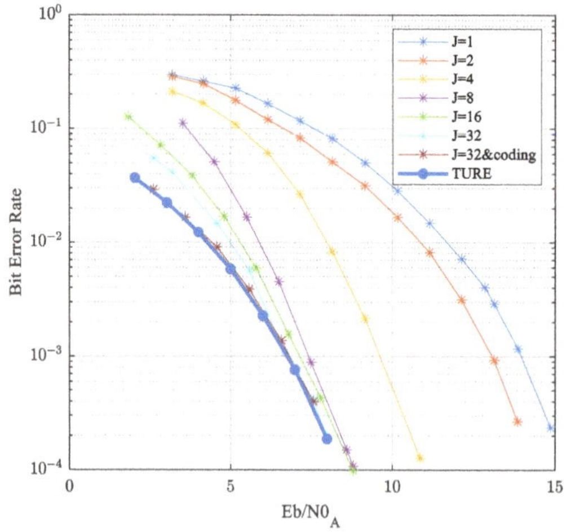

Figure 15. Information A BER test data for different J values.

As shown in Figure 15, the BER curve of Information A gradually converges to the BPSK theoretical value as the value of J increases. At J = 32 plus coding, the theoretical value of Information A is consistent when the BER of Information A at an order of −2 magnitude (the operating point of Information A after adding channel coding).

3.2.2. Information B BER Analysis

After the FH capacity boost signal passes through the Gaussian white noise channel, the baseband Information B is recovered using envelope detection.

To reduce the BER of Information B, two methods are used to reduce the information rate (joint multi-symbol judgement) in addition to channel coding.

Multi-symbol joint judgement reduces the information rate to achieve a lower BER by sacrificing the transmission rate of Information B and estimating multiple symbols as one symbol. The Information B code element rates are 1000, 500, 250, 125, 62.5, and 31.25 bps for the code elements of joint judgement of 1, 2, 4, 8, 16, and 32, respectively. The BER improvement of Information B is shown in Figure 16. The more symbols that are united, the better the BER performance of Information B.

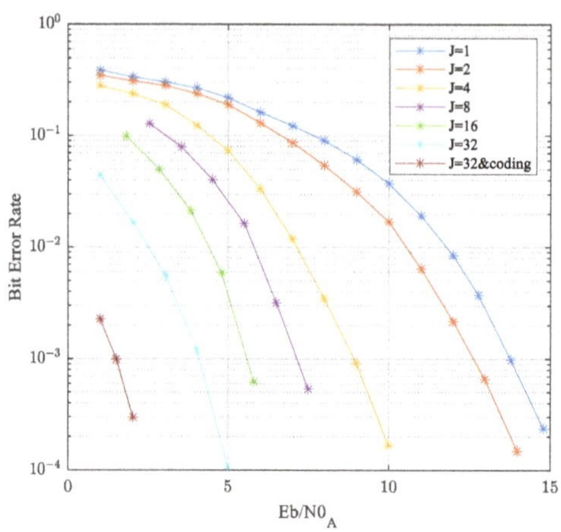

Figure 16. Information B BER test data for different rates.

To further reduce the bit error rate, the (2, 1) convolutional code is added at the Information B code element rate of 31.25 bps. A simulation verifies that the bit error rate reduces further as a result. When J = 32 and (2, 1) convolutional coding is used, the Information B BER is less than three orders of magnitude lower than the Information A BER.

3.3. Transmission Upgrade Analysis

The information rate of Information B represents the system transmission capacity enhancement capability.

To ensure the BER performance of Information B, one Information B code element cycle, T_B, corresponds to j frequency-hopping cycles, T_H, in which $j \geq 1$. The code elements of each Information B are orthogonal within the code element cycle. The M frequency-hopping points are divided into 2^k groups, with each group including j frequency points. Each Information B code element carries k bits of information.

Together, the number of code elements, the code element rate, and the coding efficiency of Information B determine the information rate of Information B.

$$r_B = \frac{k\mu_B}{jT_H} = \frac{kh\mu_B}{j} = \frac{h\mu_B \log(M/j)}{j} \tag{21}$$

where h is the hopping speed, which is the reciprocal of the hopping period, T_H; μ_B is the coding efficiency of Information B.

Equation (21) shows that the larger the number of frequency-hopping points M, the higher the frequency-hopping rate and the higher the information enhancement capacity.

If $M = 1024$, the rate of frequency hopping is $r_H = 10k$ and the information rate of Information A is $r_A = 5$ kbps. To enable Information B to obtain sufficient BER to support Information A for carrier reconstruction, every four frequency-hopping cycles correspond to one code word cycle of Information B, in which $j = 4$. Then, the equivalent Information B code element rate is 2.5 kbps; each code element carries four bits of information (1/2 channel coding). At this stage, the information rate of Information B is $r_B = 10$ kbps.

3.4. Analysis of Anti-Interference Capability

Resistance to Broadband Interference

1. Information A Broadband-Jamming Capability

The anti-interference gain is generally used to reflect the signal's ability to resist broadband interference. When the anti-jamming gain is enhanced by expanding the bandwidth, the anti-jamming gain is the expanded bandwidth compared with the bandwidth of the original information modulation signal, as shown in the following equation:

$$G_{PA} = \frac{P_i}{P_A} = \frac{W_s N_0}{E_b r_A} = \frac{W_s}{r_A Th_A} \tag{22}$$

G_{PA} indicates Information A immunity gain, W_s denotes the total spreading bandwidth, r_A is the information rate of A, and Th_A is the demodulation BER threshold of Information A.

Anti-jamming gain is essentially the value of the power that the jamming source needs to prioritize relative to the jammed link in order to achieve the jamming effect when the jamming source is in the same location as the transmitter terminals of the jammed system with the same aperture antennae, and the jammed system has an infinite power link margin.

Since there is no change in W and R before and after the capacity enhancement method is applied to Information A, the incremental Information B also barely affects the demodulation BER threshold, Th_A, of A. Thus, the signal A anti-interference gain is unchanged before and after the use of the capacity-boosting method, i.e., the addition of the transmitted Information B in the channel has little effect on signal A's ability to resist broadband interference.

2. Information B Broadband Interference Resistance

Similarly, the immunity gain of Information B can be expressed as follows:

$$G_{PB} = \frac{P_i}{P_B} = \frac{W_s N_0}{E_b r_B} = \frac{W_s}{r_B Th_B} \quad (23)$$

Since the expanded bandwidths of Information B and Information A are the same, the anti-interference gain of Information B depends on the product of the information rate of B and the demodulation threshold of B.

In general, as long as the number of frequency-hopping points and the frequency-hopping speed reach a certain value, when the information rate of Information B is equal to that of Information A, and if the BER demodulation threshold is also the same, the anti-jamming ability of Information B and Information A is the same.

3. Resistance to Narrowband Tracking Interference for Information A

Information A's ability to resist narrowband tracking interference depends on factors such as the frequency-hopping speed and frequency-hopping pattern. The addition of incremental Information B does not reduce the frequency-hopping speed. The frequency-hopping pattern changes from being driven by sequence polynomials or functions to being driven by Information B, which drastically reduces its predictability. The ability of Information A to resist narrowband tracking interference is strengthened accordingly.

4. Resistance to Narrowband Tracking Interference for Information B

The demodulation of Information B uses power detection. When the narrowband interference signal exists on the same frequency as the signal, the power of Information B is strengthened, which is conducive to the judgment of Information B. Similarly, the frequency-hopping pattern of the signal is driven by Information B, which is unpredictable and decipherable. Therefore, Information B has a strong anti-narrowband interference capability.

3.5. Comparison with OFDM Signals

OFDM is a multicarrier modulation with high frequency utilization and good resistance to multipath fading [28]. It is generally used in scenarios where frequency resources are limited and frequency resource efficiency is prioritized, such as terrestrial mobile communications. In mobile communication 5G systems, OFDM signals are divided into two categories: OFDM access signals and OFDM transmission signals.

FH capacity-boosting signals with high power utilization are generally used in scenarios where power is limited and transmission reliability is a priority, such as anti-jamming communications, satellite signaling beams, and private network communications.

3.5.1. Comparison of FH Capacity Enhancement Signals with OFDM Access Signals

OFDM access signals, typically used in multi-user access scenarios, transmit narrowband information such as access signaling, with the sub-bandwidth typically on the order of kHz. Within the communication bandwidth, W_s, the frequency resource is divided into a number of sub-bandwidths, and each sub-bandwidth is divided into a number of time slots. At the time of access, each user is assigned to a time slot within a sub-bandwidth to communicate.

The spectral structure of the OFDM access signal is shown in Figure 17.

With the same bandwidth resource-limiting parameter (bandwidth W_s) and power resource-limiting parameter (BER threshold Th_A), according to Equation (22), FH capacity-boosting signal A is able to obtain the same anti-jamming gain when the information rate of the FH capacity-boosting signal A is the same as the information rate of the OFDM access signal. At this time, the FH capacity enhancement signal is able to transmit Information B in addition to Information A. Therefore, the FH capacity enhancement signal transmission rate is higher than the OFDM access signal.

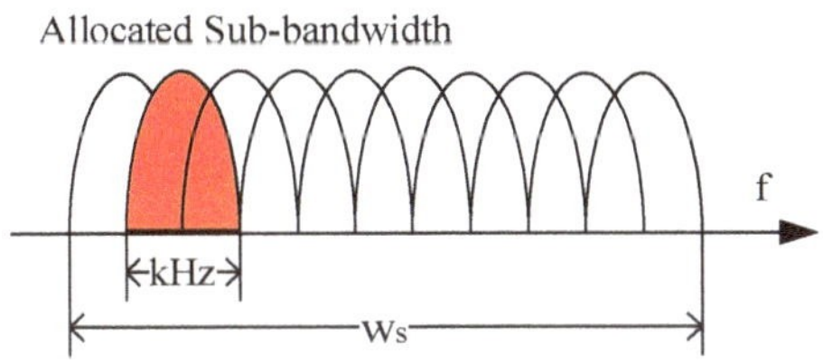

Figure 17. OFDM access signal spectrum structure.

FH capacity-boosting signals and OFDM access signals both suffer from random variations in frequency parameters. The FH capacity-boosting signals use the frequency parameter variations for transmitting Information B, and the OFDM access signals are used to indicate multiple access information.

3.5.2. Comparison of FH Capacity Enhancement Signals with OFDM Transmission Signals

OFDM transmission signals, generally used to transmit broadband information, occupy one or more sub-bandwidths, up to all sub-bandwidths within the bandwidth, when the transmission rate is at its highest. OFDM transmission signals give priority to the utilization of frequency resources, so higher-order modulation is generally used.

OFDM transmits the signal using a coding efficiency of η and a modulation order of λ, when the demodulation threshold of OFDM is Th_{OFDM}, with an occupied bandwidth of W_s and a transmission information rate of $r = W_s\lambda/\eta$. For ease of calculation, the bandwidth of the 1/2 sub-bandwidth is ignored, as shown in Figure 18.

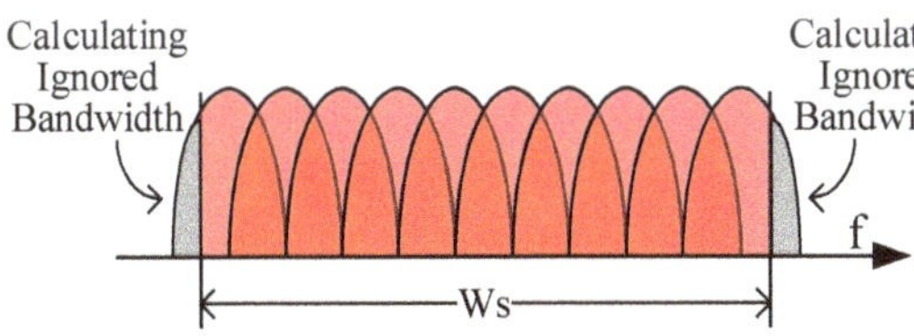

Figure 18. Spectrogram of OFDM transmitted signals.

The information transmission rate is $2r_{FH-A}$ and the demodulation threshold is Th_A for the case of doubling the signal rate with FH capacity enhancement.

According to Equation (22), the ratio of the immunity of the two types of signals depends on the following equation:

$$\gamma = \frac{r_{FH-A}(Th_A)}{r_{OFDM}(Th_{OFDM})} \tag{24}$$

According to Equation (24), for a certain occupied bandwidth, when the BPSK regime is used for FH capacity enhancement signals and the 16QAM regime is used for OFDM signals. The demodulation SNR requirement for OFDM high-order modulated signals is four times that of FH capacity-boosted signals (6 dB). At this point, when r_{OFDM} is 1000 (30 dB) times as large as r_{FH-A}, the FH capacity-boosting signal is 36 dB more resistant to interference than the OFDM transmission signal. When r_{OFDM} and r_{FH} (the total rate of messages A and B) are equal, FH capacity-boosting signal A is more resistant to interference than the OFDM transmission signal by 9 dB.

4. Discussion

Interference signals can be carried to the internal communication system through two ways. One way is through the coupling of the internal transmission line of the equipment; this type of interference increases the electromagnetic interference shielding performance of the equipment chassis materials for suppression [29–32]. The other is spatially coupled and enters the interior of the system through the receiving antenna, where the signal is usually suppressed through the means of spectrum expansion. After the signal spectrum is expanded, lower-rate signals are generally transmitted.

By expanding the information transmission channel in the frequency dimension of frequency-hopping signals, we increased the amount of transmitted information and enhanced the information transmission efficiency of the anti-jamming satellite communication system.

In this paper, a simulation model of a frequency-hopping spread-spectrum communication system is established, the signal generation method and spectral characteristics are presented, and the BER rate performance of the system is simulated.

The transmission of Information B does not change the frequency-hopping rate or bandwidth of Information A, so it does not affect the anti-jamming performance of Information A. Due to the unpredictability of Information B, different carrier frequencies are driven for frequency hopping. This further enhances the unpredictability of Information A while increasing the communication capacity, increasing the anti-interception capability of Information A.

In future work, we will continue to study the optimization of demodulation algorithms to enhance the information capacity of FH signals and DS signals, and we will continue to investigate the impact of broadband channel characteristics and digital compensation [33,34].

5. Patents

Transmission capacity enhancement method for spread spectrum anti-jamming communication system: China, ZL202310279789.8 [P]. 2023-3-22.

Author Contributions: Conceptualization, Z.H. and Z.Y.; methodology, Z.Y. and Z.H.; software, Z.Y., W.Y. and M.J.; validation, Z.Y., W.Y. and M.J.; formal analysis, M.J.; investigation, W.Y.; resources, M.J.; data curation, Z.Y. and M.J.; writing—original draft preparation, Z.Y. and Z.H.; writing—review and editing, W.Y. and M.J.; visualization, W.Y.; supervision, W.Y.; project administration, M.J. and W.Y.; funding acquisition, M.J. All authors have read and agreed to the published version of the manuscript.

Funding: This research was funded by the National Natural Science Foundation of China [62231012] and the Natural Science Foundation for Outstanding Young Scholars of Heilongjiang Province [YQ2020F001].

Data Availability Statement: The simulation of the FH capacity boost signal can be provided: https://blog.csdn.net/yuziting/article/details/133881184?csdn_share_tail=%7B%22type%22%3A%22blog%22%2C%22rType%22%3A%22article%22%2C%22rId%22%3A%22133881184%22%2C%22source%22%3A%22yuziting%22%7D; https://download.csdn.net/download/yuziting/88437737. Accessed from 17 October 2023 to the present. The data presented in this study are available on request from the corresponding author due to ethical and privacy restrictions.

Conflicts of Interest: The authors declare no conflict of interest.

References

1. Chen, Z.; Scott, T.R.; Bearman, S.; Anand, H.; Keating, D.; Scott, C.; Arrowsmith, J.R.; Das, J. Geomorphological Analysis Using Unpiloted Aircraft Systems, Structure from Motion, and Deep Learning. In Proceedings of the 2020 IEEE/RSJ International Conference on Intelligent Robots and Systems (IROS), Las Vegas, NV, USA, 25–29 October 2020; pp. 1276–1283. [CrossRef]
2. Hook, L.R.; Skoog, M.; Garland, M.; Ryan, W.; Sizoo, D.; VanHoudt, J. Initial considerations of a multi-layered run time assurance approach to enable unpiloted aircraft. In Proceedings of the 2018 IEEE Aerospace Conference, Big Sky, MT, USA, 3–10 March 2018; pp. 1–11. [CrossRef]
3. Jasiunas, M.; Chakraborty, A.; Kearney, D. A distributed operating system supporting strong mobility of reconfigurable computing applications in a swarm of unpiloted airborne vehicles. In Proceedings of the 2009 International Conference on Field-Programmable Technology, Sydney, Australia, 9–11 December 2009; pp. 392–398. [CrossRef]
4. Kim, K.H.; Hong, B.H. Maritime logistics and applications of information technologies. In Proceedings of the 40th International Conference on Computers & Indutrial Engineering, Awaji, Japan, 25–28 July 2010; pp. 1–6. [CrossRef]
5. Zhu, S.; Du, G. Evaluation of the Service Capability of Maritime Logistics Enterprises Based on the Big Data of the Internet of Things Supply Chain System. *IEEE Consum. Electron. Mag.* **2023**, *12*, 100–108. [CrossRef]

6. Ditsariyakul, P.; Sutheparuks, L.; Limlamai, N. Role of Thailand Remote Sensing Center in data service for global environment monitoring. In Proceedings of the IGARSS'93—IEEE International Geoscience and Remote Sensing Symposium, Tokyo, Japan, 18–21 August 1993; Volume 4, pp. 1841–1842. [CrossRef]
7. Chang, Y.-G.; Chen, C.-H.; Huang, H.-L.; Miao, H.-H. Geographic information systems and global positioning systems for watershed management. IGARSS 2001. Scanning the Present and Resolving the Future. In Proceedings of the IEEE 2001 International Geoscience and Remote Sensing Symposium (Cat. No.01CH37217), Sydney, Australia, 9–13 July 2001; Volume 5, pp. 2112–2114. [CrossRef]
8. Rajarapollu, P.; Sharma, V.K. Design and analysis of satellite based disaster warning system—II. In Proceedings of the TENCON 2007—2007 IEEE Region 10 Conference, Taipei, Taiwan, 30 October–2 November 2007; pp. 1–4. [CrossRef]
9. Fan, Q.; Huang, H.; Wu, J.; Niu, W.; Zhang, Q.; Zhang, Y. Design and Construction of Multi-Source Satellite High-Frequency Power Grid Fire Risk Monitoring and Early Warning System. In Proceedings of the 2023 5th International Conference on Electronic Engineering and Informatics (EEI), Wuhan, China, 23–25 June 2023; pp. 262–268. [CrossRef]
10. Tiezzi, F.; Padilla, J.; Vigano, C. S-band transmit/receive antenna with electronically switched beams for mobile satellite systems. In Proceedings of the 2012 6th Advanced Satellite Multimedia Systems Conference (ASMS) and 12th Signal Processing for Space Communications Workshop (SPSC), Vigo, Spain, 5–7 September 2012; pp. 62–67. [CrossRef]
11. Jin, L.; Wang, L.; Jin, X.; Zhu, J.; Duan, K.; Li, Z. Research on the Application of LEO Satellite in IOT. In Proceedings of the 2022 IEEE 2nd International Conference on Electronic Technology, Communication and Information (ICETCI), Changchun, China, 27–29 May 2022; pp. 739–741. [CrossRef]
12. Ban, Y.; Li, B.; Le, Q. Optimization and simulation of network management signaling transmission in satellite communication systems. *Radio Eng.* **2017**, *47*, 5–9.
13. Chang, D.-C.; Cheng, P.-W.; Lee, C.-H.; Wu, C.-T. Broadband wide-beam circular polarization antenna for global navigation satellite systems application. In Proceedings of the 2015 Asia-Pacific Symposium on Electromagnetic Compatibility (APEMC), Taipei, Taiwan, 26–29 May 2015; pp. 44–46. [CrossRef]
14. Shannon, C.E. A mathematical theory of communication. *Bell Syst. Tech. J.* **1948**, *27*, 379–423. [CrossRef]
15. Herrick, D.L.; Lee, P.K. CHESS a New Reliable High Speed HF Radio. In Proceedings of the MILCOM '96 IEEE Military Communications Conference, McLean, VA, USA, 24–24 October 1996; Volume 3, pp. 684–690. [CrossRef]
16. Herrick, D.L.; Lee, P.K.; Ledlow, L.L. Correlated Frequency Hopping: An Improved Approach to HF Spread Spectrum Communications. In Proceedings of the 1996 Tactical Communication Conference, Fort Wayne, IN, USA, 30 April–2 May 1996; pp. 320–324.
17. Mills, D.G.; Edelson, G.S.; Egnor, D.E. A multiple access differential frequency hopping system. In Proceedings of the IEEE MILCOM'03, Bostom, MA, CSA, 13–16 October 2003; pp. 1184–1189.
18. Liu, Z.; Zhang, Y.; Yao, F. A shortwave DFH hop detection method based on the combination of STFT and G-function. *J. Electron.* **2003**, *1*, 13–16.
19. Qian, B.; Feng, Y.; Pan, C.; Tian, B. A differential frequency hopping signal detection method based on multi-hop autocorrelation. *J. Aeronaut.* **2011**, *32*, 2268–2276.
20. Xiaoxu, Q.; Jingyi, L.; An, L. Improving Anti-interception Performance of Differential Frequency Hopping Communication. In Proceedings of the 2008 International Seminar on Future BioMedical Information Engineering, Wuhan, China, 18 December 2008; pp. 295–299. [CrossRef]
21. Li, Y.; Yao, F.; Xu, B.; Zhao, L. Equalization Algorithm in Frequency Domain for Broadband DFH Systems. In Proceedings of the 2015 4th International Conference on Computer Science and Network Technology (ICCSNT), Harbin, China, 19–20 December 2015; pp. 1154–1159. [CrossRef]
22. Yin, A.; Li, Y. Design of a differential frequency hopping scheme based on chaotic encryption algorithm. *Opt. Commun. Res.* **2022**, *4*, 69–72. [CrossRef]
23. Liu, X.; Chen, L.; Yao, C.; Wang, M.; Song, X. A novel enhanced anti-jamming system for differential frequency hopping communication. *Comput. Appl. Res.* **2022**, *39*, 1820–1824. [CrossRef]
24. Yu, J.; Wang, C.; Shen, L. Principles of Spread Spectrum Communication Technology and its Applications. *China Wirel.* **2010**, *03*, 44–47.
25. Fan, C.; Cao, L. *Principles of Communication*, 6th ed.; National Defence Industry Press: Beijing, China, 2006; pp. 230–231.
26. Gardner, F. A BPSK/QPSK Timing-Error Detector For Sampled Receivers. *IEEE Trans. Commun.* **1986**, *34*, 423–429. [CrossRef]
27. Best, R.E.; Kuznetsov, N.V.; Leonov, G.A. Tutorial on dynamic analysis of the Costas loop. *Annu. Rev. Control* **2016**, *42*, 27–49. [CrossRef]
28. Schmidl, T.M.; Cox, D.C. Robust frequency and timing synchronization for OFDM. *IEEE Trans. Commun.* **1997**, *45*, 1613–1621. [CrossRef]
29. Amini, M.; Kamkar, M.; Rahmani, F.; Ghaffarkhah, A.; Ahmadijokani, F.; Arjmand, M. Multilayer Structures of a $Zn_{0.5}Ni_{0.5}Fe_2O_4$-Reduced Graphene Oxide/PVDF Nanocomposite for Tunable and Highly Efficient Microwave Absorbers. *ACS Appl. Electron. Mater.* **2021**, *3*, 5514–5527. [CrossRef]
30. Zhou, M.; Gu, W.; Wang, G.; Zheng, J.; Pei, C.; Fan, F.; Ji, G. Sustainable wood-based composites for microwave absorption and electromagnetic interference shielding. *J. Mater. Chem. A* **2020**, *8*, 24267–24283. [CrossRef]

31. Ryu, S.H.; Han, Y.K.; Kwon, S.J.; Kim, T.; Jung, B.M.; Lee, S.B.; Park, B. Absorption-dominant, low reflection EMI shielding materials with integrated metal mesh/TPU/CIP composite. *Chem. Eng. J.* **2022**, *428*, 131167. [CrossRef]
32. Stupar, S.L.; Vuksanović, M.M.; Mijin, D.Ž.; Milanović, B.C.; Joksimović, V.J.; Barudžija, T.S.; Knežević, M.R. Multispectral electromagnetic shielding and mechanical properties of carbon fabrics reinforced by silver deposition. *Mater. Chem. Phys.* **2022**, *289*, 126495. [CrossRef]
33. Zhisong, H.; Zhiming, Z. High symbol-rate polarization interference cancelation for satellite-to-ground remote sensing data transmission system. *EURASIP J. Wirel. Commun. Netw.* **2018**, *286*, 1–10.
34. Hao, Z.; He, C.; Jia, M.; Wu, L. Design of Multichannel Adaptive Filter by Constructing Multidimensional Wiener-Hopf Equation. *China Commun.* **2022**, *19*, 124–134. [CrossRef]

electronics

Article

Optimal Beamwidth for Maximizing Uplink Coverage Probability in Quasi Earth-Fixed LEO Satellite Communication System

Chun-Tai Liu [1,2,*] and Jen-Yi Pan [1,*]

1 Department of Communications Engineering, National Chung Cheng University, Chiayi 621301, Taiwan
2 Advanced Institute of Manufacturing with High-Tech Innovations, National Chung Cheng University, Chiayi 621301, Taiwan
* Correspondence: chuntailiu@alum.ccu.edu.tw (C.-T.L.); jypan@ccu.edu.tw (J.-Y.P.)

Abstract: Satellite communication is proposed to fulfill the ubiquitous coverage for next-generation wireless networks. Considering the propagation delay and path loss, low-earth orbit (LEO) satellites are widely adopted. However, since the beam boresight directions become close in quasi-earth-fixed cells (QEFC) scenarios at low elevation angles, the interference increases and causes low communication quality. This paper introduces the optimal beamwidth maximizing uplink coverage probability scheme for quasi-earth-fixed cells in LEO satellite communication systems. The proposed scheme dynamically adjusts the beamwidth to achieve max uplink coverage probability at different elevation angles. The simulation results show that the proposed scheme matches the exhaustive search method in different scenarios and target signal-to-interference-plus-noise ratios. Furthermore, the proposed scheme significantly mitigates interference and improves the uplink coverage probability. Compared with the 3GPP setting, the proposed scheme improves the uplink coverage probability by 0.93 at time 100 s, and compared with the three-color frequency reuse, the proposed scheme improves the uplink coverage probability by 0.195 at time 100 s. The dynamic beamwidth and boresight direction adjustments enable the satellite to maintain seamless and reliable communication services across a wide range of operating conditions, ultimately realizing the goal of ubiquitous communications in the QEFC LEO satellite system.

Keywords: satellite communications; quasi-earth-fixed cells; low-earth orbit satellite; optimal beamwidth; uplink coverage enhancement

1. Introduction

The fifth-generation mobile network (5G) is expected to provide high-quality communications, such as a high sum rate, low latency, extended coverage, and low power consumption [1]. However, ubiquitous coverage remains a significant challenge to solve. According to [2], an estimated 34 percent of the global population cannot use the Internet. To ensure the anytime and anywhere 5G service, satellite communication, also called non-terrestrial networks (NTNs), was proposed and is becoming increasingly popular. The satellite communication system provides a comprehensive coverage area, an attractive solution for ubiquitous wireless coverage for next-generation wireless networks, especially in areas where terrestrial networks are difficult to deploy or are cost-prohibitive [3]. The Third-Generation Partnership Project (3GPP), an international standardization organization for mobile telecommunications, started studying and standardizing how to implement satellite communication on the 5G New Radio (NR) system in Release 15 and 17. Compared with medium-earth and geostationary orbit, low-earth orbit (LEO) satellites, with altitudes less than 2000 km, they are an attractive solution to achieving ubiquitous wireless coverage and high spectral efficiency by considering propagation delay and channel attenuation.

Two types of LEO service links are introduced in the 3GPP standard (as shown in Figure 1):

- Quasi earth-fixed cells (QEFC): satellites can compensate and steer the beams at the same position on the earth while satellites move.
- Earth-moving cells (EMC): satellites cannot provide steerable beams. The position of beams changes while satellites move.

Since the earth-moving LEO satellite moves very fast and each provides many beams, user equipment (UE) can stay inside the beam footprint for only a few seconds [4]. The rapid footprint location change creates problems for paging and handover [5]. Although 3GPP reduces the neighbor cell measurements for handover procedures with predictable satellite locations using ephemeris information in Release 17, the network still needs to spend a lot of resources on random access procedures to handle the frequent handovers of numerous UEs.

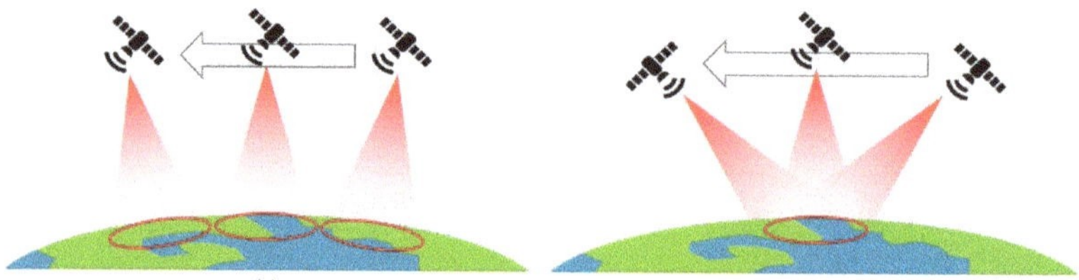

Figure 1. Service link types (**a**) earth-moving cells and (**b**) quasi earth-fixed cells.

The QEFC LEO satellites are designed to steer beams to specific areas on Earth within configurable time frames, solving frequent handover issues. Over the years, 3GPP has developed various handover strategies for the QEFC scenario, such as conditional handover in Release 17 and satellite switch with re-sync in Release 18 [6]. The QEFC scenario holds significance for the cellular satellite communications network. However, despite this importance, only a few studies have discussed and analyzed the performance of QEFC satellites. The authors of [7] evaluated the throughput and reference signal received power with different beam-fixed time durations, and the authors of [8] evaluated different elevation angles' downlink signal-to-noise-plus-interference ratio (SINR). The evaluation results show that UEs have low communication quality at low elevation angles. However, they did not delve into the reasons behind this performance drop or propose any solutions. Hence, addressing the issue of low SINR to enhance signal quality for mobile satellite communications in the QEFC scenario is essential. This paper aims to fill this gap by evaluating and analyzing the SINR of QEFC LEO satellites and proposing solutions to mitigate the low SINR problem, ultimately striving towards achieving ubiquitous coverage.

Since the max transmission power of a handheld user devices is limited to 0.2 watts, it is hard for the receiver to decode the uplink signal successfully with the long-distance communication system. Some uplink coverage enhancements were proposed to strengthen the receiving power in the 3GPP NR standard, including physical uplink shared channel (PUSCH) repetition [9], joint channel estimation [9], and peak-to-average-power ratio reduction [10]. To confirm whether the satellite communication system can provide 5G service, 3GPP analyzed the uplink link budget for NTNs. Unfortunately, the analysis showed that the carrier-to-noise ratio is insufficient to provide voice-over NR (VoNR) service (4.75 kbps) at an elevation angle of 30 degrees for LEO 1200 km satellites with the existing PUSCH repetition method [11,12]. Thus, 3GPP discussed and supported joint channel estimation for NR NTN in Release-18. Moreover, 3GPP did not consider interference; interference leads to worse coverage performance.

The interference mitigation scheme must be studied to provide ubiquitous coverage in the QEFC satellite communication system. One of the well-known interference mitigation solutions for satellite communication is frequency reuse. The frequency reuse scheme divides the frequency band into sub-bands so that the adjacent beams have disjoint frequency bands [13]. Although frequency reuse significantly mitigates the interference, it reduces the available bandwidth for each beam. Furthermore, frequency reuse does not solve the severe path loss problem; the performance still degrades at low elevation angles due to high path loss. Another interference mitigation scheme is beamwidth control. SpaceX in [14] mentions that the increasing size of the spot on the ground increases the potential for interference with other beams using the same frequencies. SpaceX controls the half-power beamwidth between 2.15° and 2.45° in the Ku-band to mitigate interference by controlling the active antenna elements at various steering angles [14]. The authors of [15] controlled the beamwidth and altitude of the satellite to maximize uplink coverage probability. The controlled beamwidth mitigates interference by limiting the interference region and improves the uplink coverage probability. However, the authors of [15] did not consider the antenna pattern effect, an essential issue in directional antenna systems.

Based on the above discussion, this paper evaluates the performance of QEFC LEO satellites, an area that has received relatively less attention. We analyze serving and interference power for each elevation angle by considering the antenna gain, antenna pattern, and large-scale fading. To improve the uplink coverage probability and solve the low SINR problem highlighted in [7,8], especially at low elevation angles, we propose the optimal beamwidth maximizing uplink coverage probability scheme. The proposed scheme maximizes uplink coverage probability with target SINR for QEFC LEO satellites by tuning beamwidth for different elevation angles.

The remainder of the paper is organized as follows. The primary notations in this paper are listed in Table 1. Section 2 introduces the system model, which is the large-scale channel model described in [16]. Section 3 introduces the proposed optimal beamwidth maximizing uplink coverage probability. Three parts are included in the proposed scheme: beam boresight direction, uplink coverage probability analysis model, and beamwidth optimization. The numerical results and discussion of the proposed scheme are shown in Section 4. Finally, the concluding remarks are in Section 5.

Table 1. The description of notations.

Notation	Distriptions
$\mathcal{F}, \mathcal{T}$	The set of the simulated frequency band and time. $\lvert\mathcal{F}\rvert = 3$ for FRF3 and $\lvert\mathcal{F}\rvert = 1$ for non-frequency reuse.
$\mathcal{L}_f, \mathcal{S}_l$	The set of beams in frequency band f ($f \in \mathcal{F}$), UE serving by beam l ($l \in \mathcal{L}_f$).
$\epsilon_l, \epsilon_{s,l}$	Elevation angle of beam l, elevation angle of user s served by beam l.
ζ_ϵ	Clutter loss. Clutter loss is 0 for the LOS state. The values for the (NLOS state at elevation angle ϵ are defined in [16].
κ_ϵ	Shadow fading at elevation angle ϵ. Shadow fading is zero mean Gaussian variable.
$\sigma_{LOS,\epsilon}, \sigma_{NLOS,\epsilon}$	The variation in shadow fading for lien-of-sight (LOS) and non line-of-sight (NLOS) states at elevation angle ϵ [16].
$p_{LOS,\epsilon}, p_{NLOS,\epsilon}$	LOS probability and NLOS probability at elevation angle ϵ [16].
$\varphi_{k,l}$	Angle difference of boresight direction between beam l and beam k ($l, k \in \mathcal{L}$ and $l \neq k$).
φ_l^s	Angle difference of UE s and boresight direction of beam l.
$\mathbb{P}(x), \mathbb{P}_y(x)$	The probability of the condition x to be satisfied, $\mathbb{P}_y(x) = \mathbb{P}(x\mid y)$.

2. System Model

This paper evaluates and analyzes the multi-beam QEFC LEO satellite system with inter-beam interference. We use the NTN large-scale channel model defined in [16]. We assume all users are transmitting with the same power P_t. The receiving power $P_{s,l}(t,m)$ from user s to beam l in satellite with equivalent antenna aperture m at time t is given by

$$P_{s,l}(t,m) = P_t - L_s(t) - \kappa_{\epsilon_{s,l}(t)} - \zeta_{\epsilon_{s,l}(t)} + G(\varphi_l^s(t), m). \quad (1)$$

Path loss L_s of user s consists of free-space path loss (FSPL), ionospheric loss of 2.2 dB [17], atmospheric loss, building entry loss, and polarization loss of 3 dB [17]. This paper assumes all users are outdoors, and atmospheric loss is negligible in the S-band. Atmospheric loss and building entry loss are equal to zero. The path loss is obtained as follows [16]:

$$L_s(t) = FSPL_s(t) + 2.2 + 3 \tag{2}$$

$$FSPL_s(t) = 32.45 + 20\log_{10} f_c + 20\log_{10} d_s(t) \tag{3}$$

$$d_s(t) = ||\mathbf{x}_{sat}(t) - \mathbf{x}_s(t)|| \tag{4}$$

where $\mathbf{x}_{sat}$ is the location vector of satellite, $\mathbf{x}_s$ is the location vector of UE s, d_s is the distance between user s and the satellite, h is the satellite height, R_E is the earth radius, and f_c is carrier frequency.

In the 3GPP satellite communication system, the equivalent antenna aperture is used to model the satellite antenna. The larger equivalent antenna aperture represents more active antenna elements, which form a narrower and higher power gain beam. The antenna gain is calculated as follows [16,18]:

$$G(\phi, m) = 10\log_{10} \eta\pi^2 m^2 \lambda^{-2} g(\phi, m) \tag{5}$$

$$g(\phi, m) = \begin{cases} 1, & \text{if } \phi = 0 \\ 4\left|\frac{J_1(\pi m \lambda^{-1} \sin\phi)}{\pi m \lambda^{-1} \sin\phi}\right|^2, & \text{otherwise} \end{cases} \tag{6}$$

where m is equivalent antenna aperture, ϕ is the angle between the boresight of the beam and UE, $J_1(.)$ is the first-kind and first-order Bessel function, λ is the wave length, and η is the antenna aperture efficiency 0.57. The maximum antenna gain is $10\log_{10} \eta\pi^2 m^2 \lambda^{-2}$ dBi. Finally, the SINR $\xi_{s,l}$ from user $s \in \mathcal{S}_l$ to the serving beam $l \in \mathcal{L}_f$ in the satellite expressed as linear form can be derived as follows

$$\xi_{s,l}(t, m) = \frac{P_{s,l}(t, m)}{\sum_{k \in \mathcal{L}_f \backslash l} \sum_{s' \in \mathcal{S}_k} \rho_{s',s} P_{s',l}(t, m) + N} \tag{7}$$

$$\text{s.t.} \sum_{s' \in \mathcal{S}_k} \rho_{s',s} = 1 \tag{8}$$

where N is noise power, and $\rho_{s',s} \in \{0, 1\}$ is the scheduling factor that captures the interference from other beams in shared resources. While $\rho_{s',s} = 1$ represents that user s' transmits in the same radio resource with user s and interferes with user s; while $\rho_{s',s} = 0$ represents that user s' does not interfere user s. In this paper, we assume the number of scheduled users in one resource block of each beam is limited to one, as described in (8).

3. Optimal Beamwidth Maximizing Uplink Coverage Probability Scheme

The proposed optimal beamwidth maximizing the uplink coverage probability scheme aims to address the increasing FSPL and mitigate inter-beam interference simultaneously by tuning the active antenna element number to control the beamwidth and antenna gain. This scheme utilizes a satellite equipped with a phased array antenna to facilitate adjustable beamwidth. The phased array antenna allows beam direction management as the satellite moves, meeting the demand for steerable beams in the QEFC system. This section introduces the management of the beam boresight direction, the uplink coverage probability analysis model, and the optimization function of the proposed scheme.

3.1. Beam Boresight Direction

In the QEFC satellite system, the network and satellite need to calculate and decide the beam boresight direction during the beam-fixed time. This section introduces the beam topology defined in [18] and formulates the beam boresight direction. The beam position is expressed as Earth-Centered Earth-Fixed (ECEF) coordinates $[x_l\ y_l\ z_l]^{\mathrm{T}}$. Our beam topology follows [18] with the adjacent beam spacing is 3.82°, half-power beamwidth is 4.4127°, and the distance of the adjacent beam is around 40.2 km. With the pre-defined beam topology and the satellite orbit, the beam boresight direction $[\phi_l(t)\ \theta_l(t)]$ can be calculated based on the satellite angular speed ω and relative beam position $[x_l(t)\ y_l(t)\ z_l(t)]^{\mathrm{T}}$ at time t as follows:

$$\omega = \sqrt{\frac{G_E \cdot M_E}{(R_E+h)^3}} = \sqrt{\frac{3.98\times 10^{14}}{(R_E+h)^3}} \tag{9}$$

$$\mathbf{v}_l(t) = \begin{bmatrix} \hat{x}_l(t) \\ \hat{y}_l(t) \\ \hat{z}_l(t) \end{bmatrix} = \begin{bmatrix} \cos(\omega t) & 0 & -\sin(\omega t) \\ 0 & 1 & 0 \\ \sin(\omega t) & 0 & \cos(\omega t) \end{bmatrix} \begin{bmatrix} x_l \\ y_l \\ z_l \end{bmatrix} \tag{10}$$

$$\begin{bmatrix} \hat{x}_0(0) \\ \hat{y}_0(0) \\ \hat{z}_0(0) \end{bmatrix} = \begin{bmatrix} x_0 \\ y_0 \\ z_0 \end{bmatrix} = \begin{bmatrix} 0 \\ 0 \\ R_E \end{bmatrix} \tag{11}$$

$$\delta_l(t) = \arccos \frac{\hat{z}_l(t)}{R_E} \tag{12}$$

$$\phi_l(t) = \arccos \frac{R_E + h - R_E \cos\delta_l(t)}{\sqrt{R_E^2 + (R_E+h)^2 - 2R_E(R_E+h)\cos\delta_l(t)}} \tag{13}$$

$$\theta_l(t) = \arctan \frac{\hat{y}_l(t)}{\hat{x}_l(t)} \tag{14}$$

$$\epsilon_l(t) = \frac{\pi}{2} - \lambda_l(t) - \phi_l(t) \tag{15}$$

where G_E is the earth's gravitational constant, M_E is the earth's mass, $\phi_l(t)$ and $\theta_l(t)$ are the zenith and azimuth angle of departure, $l = 0$ denotes the center beam, and $t = 0$ denotes the time when $\epsilon_0(0) = \frac{\pi}{2}$. The example for (12) to (15) is shown in Figure 2. With the defined beam boresight direction, the angular difference between beam l and user s can be calculated as follows:

$$\varphi_l^s(t) = \arccos \frac{||\mathbf{x}_{sat}(t) - \mathbf{x}_s(t)||^2 + ||\mathbf{x}_{sat}(t) - \mathbf{v}_l(t)||^2 - ||\mathbf{x}_u(t) - \mathbf{v}_l(t)||^2}{2||\mathbf{x}_{sat}(t) - \mathbf{x}_u(t)|| \cdot ||\mathbf{x}_{sat}(t) - \mathbf{v}_l(t)||}. \tag{16}$$

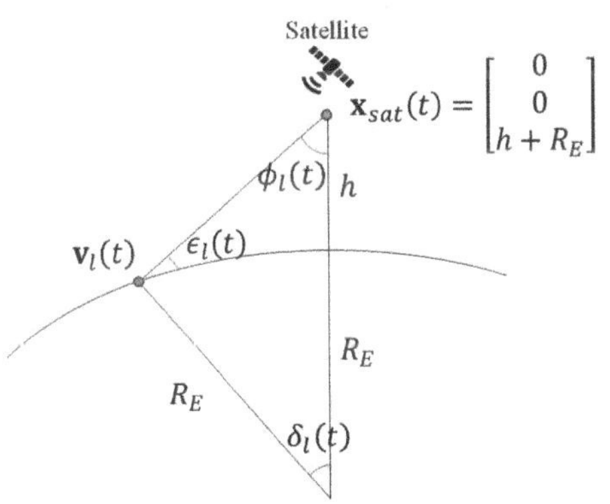

Figure 2. The illustration of Equations (12)–(15).

3.2. *Uplink Coverage Probability Analysis Model*

This section uses mathematical reasoning to build the uplink coverage probability analysis model. The uplink coverage probability analysis model calculates the uplink coverage probability with different time t (different elevation angle), equivalent antenna aperture m, and target SINR γ. Since the elevation angle and footprint are different within each beam, the uplink coverage probability of the satellite should be calculated beam by beam. We calculate the uplink coverage probability $p(t, m, \gamma)$ of the satellite by the uplink coverage probability of each beam with weighting as follows:

$$p(t, m, \gamma) = \sum_{f \in \mathcal{F}} \sum_{l \in \mathcal{L}_f} w_l \mathbb{P}(\zeta_{s,l}(t, m) \geq \gamma) \tag{17}$$

$$\text{s.t. } |\mathcal{F}| = 1,\ t \in \{\mathcal{T} \mid \frac{\pi}{6} \leq \epsilon_l(t) \leq \frac{\pi}{2}\} \tag{18}$$

where w_l is the ratio of the coverage area of beam l. Since the FSPL and elevation angle result in minor changes within a beam, our model uses the location of the beam center to present the FSPL and elevation angle to reduce the computation complexity as follows:

$$\begin{aligned} &\mathbb{P}(\zeta_{s,l}(t, m) \geq \gamma) \\ &= \mathbb{P}(-\kappa_{\epsilon_{s,l}} - \zeta_{\epsilon_{s,l}} + G(\varphi_l^s(t), m) \geq \alpha_{s,l}(t)) \\ &\simeq \mathbb{P}(-\kappa_{\epsilon_l} - \zeta_{\epsilon_l} + G(\varphi_l^s(t), m) \geq \alpha_l(t)) \end{aligned} \tag{19}$$

$$\alpha_l(t) = \gamma - P_t + L_l(t) + W_l(t) \tag{20}$$

$$W_l(t) = 10 \log_{10}(10^{N/10} + \bar{I}_l(t)) \tag{21}$$

where $W_l(t)$ is the interference-plus-noise power of beam l at time t, and $\bar{I}_l(t)$ is the expect interference power received in beam l at time t. Similarly, our model uses the location of the interference beam center to present the FSPL and elevation angle. Since shadow fading follows zero-mean Gaussian distribution, it is neglectable for the expected interference. Moreover, since the clutter is zero for LOS users, the LOS probability should be considered. Considering the LOS probability and received antenna gain, the expected interference power in linear form is calculated as

$$\bar{I}_l(t) = \sum_{k \in \mathcal{L}_f \setminus l} p_{LOS,\epsilon_k} 10^{(P_t - L_k + G(\varphi_{k,l}(t), m))/10} + p_{NLOS,\epsilon_k} 10^{(P_t - L_k + G(\varphi_{k,l}(t), m))/10} \tag{22}$$

where $\varphi_{k,l}(t)$ denotes the angular difference between two beams. The angular difference can be derived similarly to (16) as follows:

$$\varphi_{k,l}(t) = \arccos \frac{||\mathbf{x}_{sat}(t) - \mathbf{v}_k(t)||^2 + ||\mathbf{x}_{sat}(t) - \mathbf{v}_l(t)||^2 - ||\mathbf{v}_k(t) - \mathbf{v}_l(t)||^2}{2||\mathbf{x}_{sat}(t) - \mathbf{v}_k(t)|| \cdot ||\mathbf{x}_{sat}(t) - \mathbf{v}_l(t)||}. \tag{23}$$

To calculate the uplink coverage probability in (19), the cumulative distribution function (CDF) for clutter loss, shadow fading, and antenna gain needs to be discussed. Clutter loss and shadow fading distributions follow LOS probability and Gaussian distribution, which can be calculated with low complexity. However, the adjacent beam spacing for each beam is different, so it is complex to calculate the probability of serving antenna gain. Thus, we simplify the model by assuming the adjacent beam spacings of each beam are the same value $\bar{\varphi}_t$. $\bar{\varphi}_t$ equals to the average of adjacent beam spacings. With the assumed adjacent beam spacing $\bar{\varphi}_t$ in the hexagonal layout, the receive antenna gain $G(\varphi_l^s(t), m)$ from user s to serving beam l is calculated by assuming $0 \leq \varphi_l^s(t) \leq \frac{\bar{\varphi}_t}{\sqrt{3}}$, and the probability of receive antenna gain is determined as follows (as shown in Figure 3):

$$p_{\bar{\varphi}_t}(\varphi) = \begin{cases} \frac{2\pi\varphi}{A}, & \text{if } 0 \leq \varphi \leq \frac{\bar{\varphi}_t}{2} \\ \frac{12\varphi}{A}[\frac{\pi}{6} - \arccos(\frac{\bar{\varphi}_t}{2\varphi})], & \text{if } \frac{\bar{\varphi}_t}{2} < \varphi \leq \frac{\bar{\varphi}_t}{\sqrt{3}}. \end{cases} \tag{24}$$

$$A = \frac{\sqrt{3}\bar{\varphi}_t^2}{2}. \tag{25}$$

Finally, the (19) can be driven to joint CDF form as follows:

$$
\begin{aligned}
&\mathbb{P}(-\kappa_{\epsilon_l} - \zeta_{\epsilon_l} + G(\varphi_{s,l}(t), m) > \alpha_l(t)) \\
&= \mathbb{P}(G(\varphi_{s,l}(t), m) - \zeta_{\epsilon_l} - \alpha_l(t) \geq \kappa_{\epsilon_l}) \\
&= \int_0^{\frac{\bar{\varphi}_t}{\sqrt{3}}} p_{\bar{\varphi}_t}(\varphi)\mathbb{P}(G(\varphi, m) - \zeta_{\epsilon_l} - \alpha_l(t) \geq \kappa_{\epsilon_l})\, d\varphi \\
&= p_{LOS,\epsilon_l} \int_0^{\frac{\bar{\varphi}_t}{\sqrt{3}}} p_{\bar{\varphi}_t}(\varphi)\mathbb{P}_{LOS}(\kappa_{\epsilon_l} \leq G(\varphi, m) - \alpha_l(t))\, d\varphi + p_{NLOS,\epsilon_l} \int_0^{\frac{\bar{\varphi}_t}{\sqrt{3}}} p_{\bar{\varphi}_t}(\varphi)\mathbb{P}_{NLOS}(\kappa_{\epsilon_l} \leq G(\varphi, m) - \alpha_l(t) - \zeta_{\epsilon_l})\, d\varphi \\
&= p_{LOS,\epsilon_l} \int_0^{\frac{\bar{\varphi}_t}{\sqrt{3}}} p_{\bar{\varphi}_t}(\varphi)\frac{1}{2}[1 + erf(\frac{G(\varphi, m) - \alpha_l(t)}{\sigma_{LOS,\epsilon_l}\sqrt{2}})]\, d\varphi + p_{NLOS,\epsilon_l} \int_0^{\frac{\bar{\varphi}_t}{\sqrt{3}}} p_{\bar{\varphi}_t}(\varphi)\frac{1}{2}[1 + erf(\frac{G(\varphi, m) - \alpha_l(t) - \zeta_{\epsilon_l}}{\sigma_{NLOS,\epsilon_l}\sqrt{2}})]\, d\varphi
\end{aligned}
\quad (26)
$$

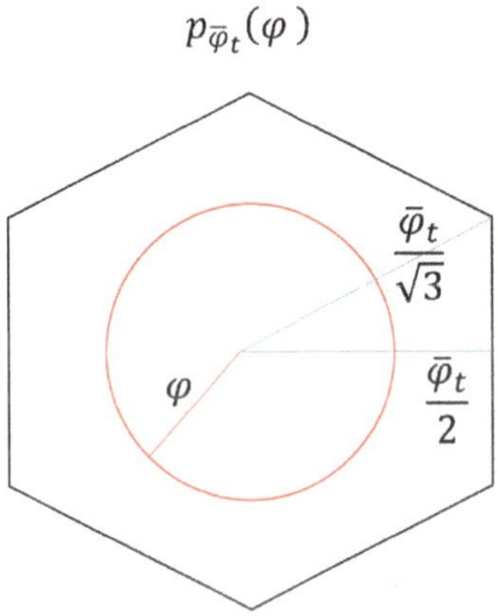

Figure 3. The illustration of probability for receiving antenna gain.

3.3. Beamwidth Optimization

Controlling the beamwidth through the number of active antenna elements, also known as the equivalent antenna aperture, is crucial for optimizing coverage in NTNs. Increasing the number of active antenna elements results in a narrower beam with higher antenna gain and vice versa. Although a narrow beam provides higher antenna gain, leading to improved signal strength and quality for users within the beam, it also results in a smaller footprint. This limited coverage area may allow users to move out of the serving area, potentially causing service interruptions or handover issues. To address this challenge, the optimal equivalent antenna aperture at different times (elevation angles) and service demands (target SINRs) can be determined using the uplink coverage probability function as follows:

$$m_t^* = \underset{m}{\operatorname{argmax}}\; p(t, m, \gamma) \quad (27)$$

$$\text{s.t. } (17), (18), (26). \quad (28)$$

Using the optimization function, satellites can dynamically adjust the number of active antenna elements during the beam-fixed time to achieve optimal performance with the service demand.

Finally, we summarize the proposed optimal beamwidth maximizing the uplink coverage probability scheme in Algorithm 1, which consists of two parts. In the first part, the satellite steers the beam to the specified position, where the departure angle follows $\phi_l(t)$ and $\theta_l(t)$ as calculated in (13) and (14). In the second part, the optimal equivalent antenna aperture m_t^* is calculated, and the antenna configuration is adjusted accordingly based on this optimal equivalent antenna aperture.

Algorithm 1 Optimal Beamwidth Maximizing Uplink Coverage Probability Scheme

Input: Beam position on Earth $\mathbf{v}_l(0)$, satellite hight h, and time t
Output: Beam boresight direction $\phi_l(t)$, $\theta_l(t)$, and optimal equivalent antenna aperture m_t^*

1: **for all** Beam $l \in \mathcal{L}_f$ **do**
2: Calculate relative beam position $\mathbf{v}_l(t)$ from (9) to (11)
3: Calculate beam boresight direction $\phi_l(t)$ and $\theta_l(t)$ from (13) and (14)
4: Satellite steers the beam l following the beam boresight direction $\phi_l(t)$ and $\theta_l(t)$
5: **end for**
6: Calculate the optimal equivalent antenna aperture m_t^* from (27) and (28)
7: Satellite tunes the beamwidth based on the optimal equivalent antenna aperture m_t^*

4. Numerical Results and Discussion

The previous section introduces the proposed optimal beamwidth maximizing uplink coverage probability. In this section, we compare the proposed scheme with the following:

1. 3GPP setting: the equivalent antenna aperture is 2 m, the max receive antenna gain is 30 dBi, and the half-power beamwidth is 4.4127° [18].
2. 3GPP setting with three-color frequency reuse (FRF3): The frequency band is divided into three sub-bands, and the adjacent beams are allocated to different sub-bands. Other assumptions follow the 3GPP setting. The three-color frequency reuse is also known as frequency reuse factor three.
3. Exhaustive antenna aperture search: exhaustively search the equivalent antenna aperture with the highest uplink coverage probability in a predefined target SINR for each simulated time. The exhaustive antenna aperture search denotes the upper bound of the uplink coverage probability for non-frequency reuse.

We calculate the uplink coverage probability using the Monte Carlo method. The evaluated scenarios are the NTN rural and NTN urban [16]. To fully address the interference from other satellites, we simulate a 19-beam satellite scenario with the wrap-around mechanism. Other simulation assumptions are summarized in Table 2. We evaluate the performance with target SINR −5.6 dB and −8.61 dB. The target SINR −5.6 dB and −8.61 dB denote the required SINR to satisfy 0.02 block error rate for VoNR with 10 and 20 times PUSCH repetition (transport block size is 184 bits and modulation order is quadrature phase shift keying).

Table 2. Simulation assumptions.

Assumption	Value
Satellite height	600 km
Satellite beam pattern	19 spot beams with wrap-around
Max antenna gain for 3GPP setting and FRF3	30 dBi
Carrier frequency	2 GHz
UE-type	Handheld
UE antenna configuration	Omni-directional, −5.5 dBi antenna gain [19]
UE transmit power	23 dBm
UE attachment	RSRP
Scheduled bandwidth	180 kHz (1 resource block)
Noise power	−147 dBW

4.1. Signal Quality

In this section, we show and discuss the results of signal quality. Firstly, we show the simulation result of the proposed scheme for the equivalent antenna aperture, antenna gain, half-power beamwidth, and elevation angle of the center beam in Figure 4. The results show that the optimal equivalent antenna aperture increases while the elevation angle decreases. The increasing equivalent antenna aperture provides a higher antenna gain and narrower beamwidth. Figure 5 shows the CDF of SINR for different settings in the NTN rural scenario. With the significant interference mitigation, the FRF3 performs best on SINR at $t = 0$, as shown in Figure 5a. Compared to the 3GPP setting, the proposed scheme improves the SINR by 8.15 dB and 7.4 dB at the 50-percentile point and 5-percentile point at $t = 60$ and by 12.25 dB and 12.23 dB at the 50-percentile point and 5-percentile point at $t = 100$. Compared to the FRF3, the proposed scheme improves the SINR by 2.29 dB and 1.63 dB at the 50-percentile point and 5-percentile point at $t = 60$ and by 7.68 dB and 8.53 dB at the 50-percentile point and 5-percentile point at $t = 100$. The SINR results for the NTN urban scenario are shown in Figure 6. The figure also shows that the proposed scheme significantly improves the SINR at low elevation angles.

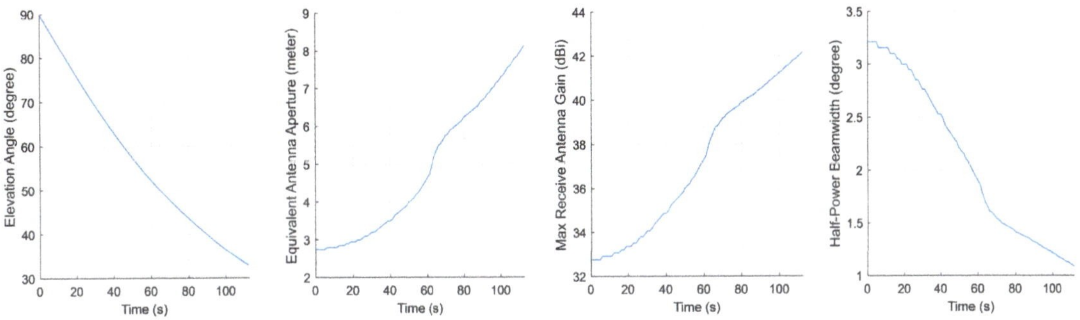

Figure 4. Elevation angle of the center beam $\epsilon_0(t)$ and optimal solutions of equivalent antenna aperture, max receive antenna gain, and half-power beamwidth with target SINR −5.6 dB in NTN rural scenario.

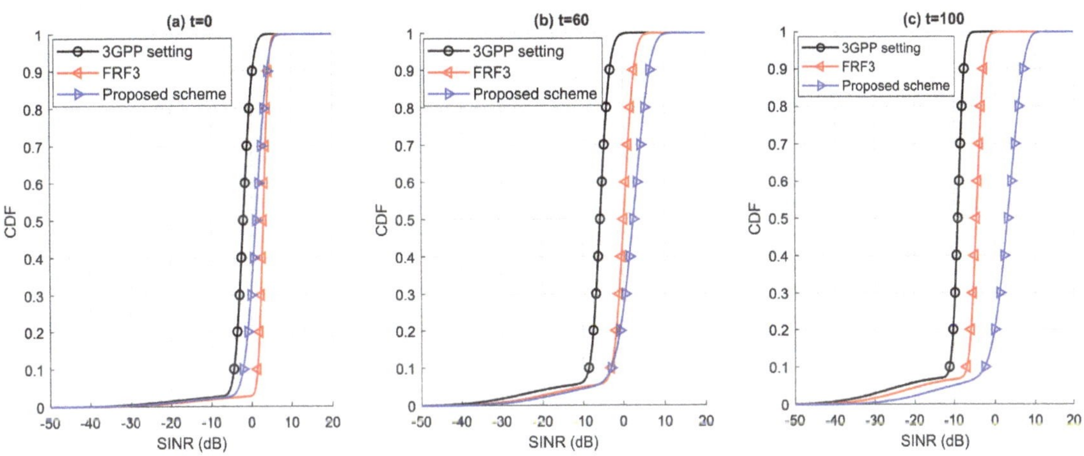

Figure 5. The SINR results for 3GPP setting, FRF3, and the proposed scheme (target SINR −5.6 dB) in NTN Rural scenario at (**a**) t = 0, (**b**) t = 60, (**c**) t = 100.

Figure 7 shows the CDF of interference power for different settings in the NTN rural scenario. As shown in Figure 7, the interference power increases while time increases (elevation angle decreases) for both the 3GPP setting, FRF3, and the proposed scheme. In Figure 7a, compared to the 3GPP setting, the proposed scheme and the FRF3 decrease interference power by 1.3 dBW and 11.6 dBW at the 95-percentile point. In Figure 7b, compared to the 3GPP setting, the proposed scheme decreases interference power by 1.8 dBW and 5.2 dBW at the 95-percentile point and the 50-percentile point, and the FRF3 decreases interference power by 6.6 and 8.7 dB at the 95-percentile point and the

50-percentile point. In Figure 7c, compared to the 3GPP setting, the proposed scheme decreases interference power by 1.2 dBW and 5.5 dBW at the 95-percentile point and the 50-percentile point, and the FRF3 decreases interference power by 4.9 dBW and 5.2 dB at the 95-percentile point and the 50-percentile point.

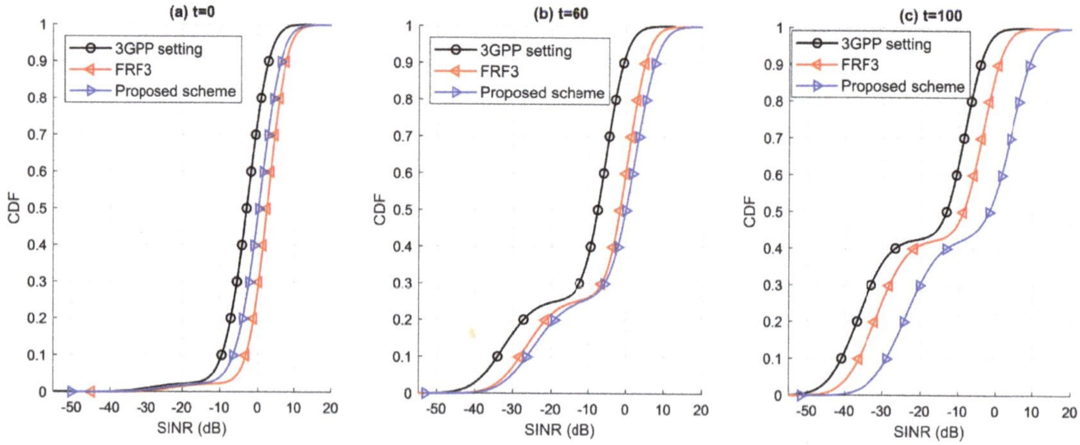

Figure 6. The SINR results for 3GPP setting, FRF3, and the proposed scheme (target SINR −5.6 dB) in NTN Urban scenario at (**a**) t = 0, (**b**) t = 60, (**c**) t = 100.

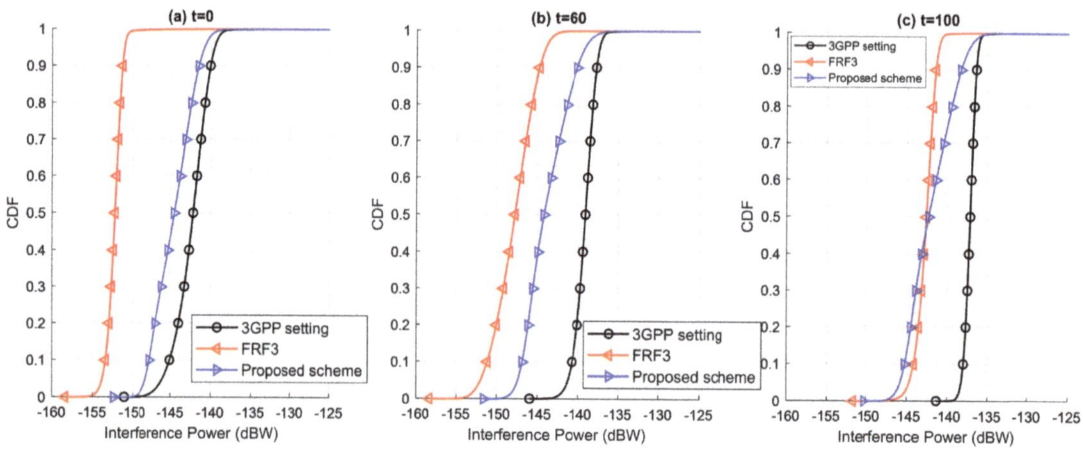

Figure 7. The interference results for 3GPP setting, FRF3, and the proposed scheme (target SINR −5.6 dB) in NTN Rural scenario at (**a**) t = 0, (**b**) t = 60, (**c**) t = 100.

The results presented in Figure 7 highlight a notable increase in interference over time. This trend primarily arises due to the significant overlap of beam footprints. Figure 8 visually represents the beam topology and footprints at $t = 0$ and $t = 100$ for the 3GPP setting and FRF3. It illustrates how the beam footprints expand, particularly at low elevation angles, thereby reducing the angular difference between the serving beam and interfering UE, consequently leading to increased interference. Furthermore, Figure 9 shows the angular difference and antenna gain from interfering UEs to the center beam. The interfering UEs in the figure are at the center of their serving beams. Figure 9 demonstrates a decrease in the angular difference between interfering UEs and the center beam over time, accompanied by increased received antenna gain. Particularly for the UE located at $(x, y) = (70, 0)$ kilometers, the angular difference decreases from 6.65° to 2.3°, increasing antenna gain from 12 dBi to 26.7 dBi. This increase in antenna gain ultimately contributes to the escalating interference levels. As shown in Figure 10, the average interference rises from 0.64×10^{-14} to 2.08×10^{-14} watts for the 3GPP setting. Despite the FRF3 aims to mitigate interference, the average interference still rises from 0.06×10^{-14} to 0.66×10^{-14} watts.

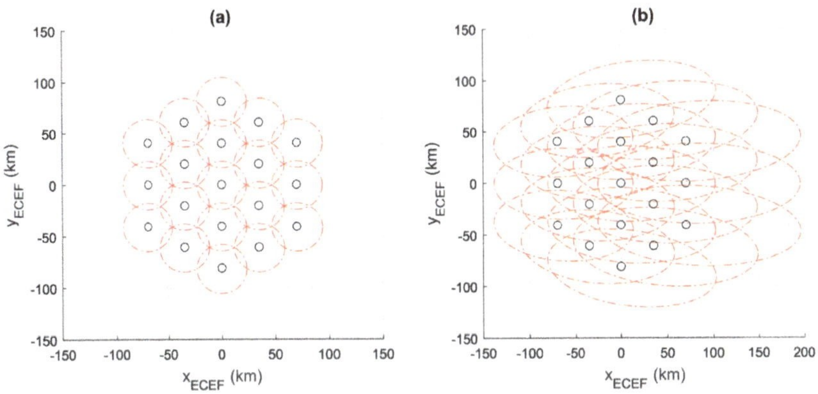

Figure 8. Beam topology and footprints with elevation angle (**a**) $\epsilon_0(0) = 90°$ and (**b**) $\epsilon_0(100) = 36.53°$ for the 3GPP setting and FRF3. Black dots represent the position of the beam center, and red circles represent the footprints of half-power beamwidth (4.4127°).

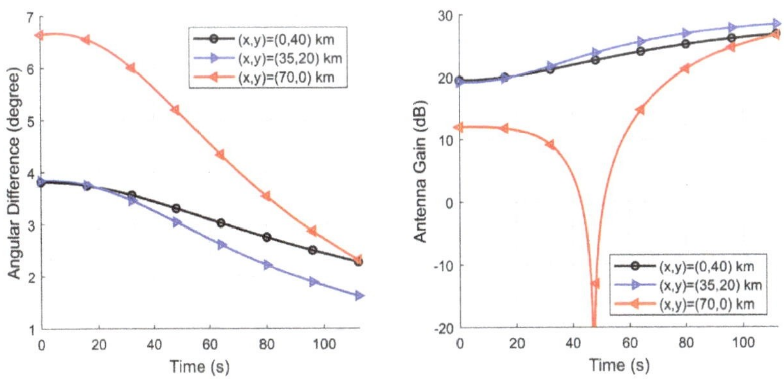

Figure 9. Angular difference $\varphi_0^s(t)$ and received antenna gain $G(\varphi_0^s(t), 2)$ between interfering UE and the center beam.

Figure 4 illustrates that the satellite forms a narrower beam with the proposed scheme to mitigate interference as time progresses. Furthermore, as shown in Figure 11, the proposed scheme effectively reduces the overlap of beam footprints by adjusting the beamwidth. Specifically, the half-power beamwidth is adjusted to only 1.212° at $t = 100$, significantly reducing the overlap of the beam footprints. Although the average interference power for all three schemes increases as the elevation angle decreases and FRF3 exhibits lower interference power than the proposed scheme, the proposed scheme demonstrates a minor increment owing to its narrower beamwidth. Specifically, the 3GPP setting and FRF3 experience an increase in average interference power of 1.44×10^{-14} (225%) and 0.6×10^{-14} (1000%) watts, respectively, while the proposed scheme only sees an increase of 0.5×10^{-14} (123%) watts. This difference can be attributed to the proposed scheme's narrower beamwidth, effectively reducing interference overlap. The proposed scheme also has a significant impact on serving power. As explained in [16,18], FSPL increases with decreasing elevation angle, leading to a decline in the average serving power for both the

3GPP setting and FRF3, as shown in Figure 10. Conversely, the average serving power for the proposed scheme increases from 0.77×10^{-14} to 2.16×10^{-14} watts due to higher antenna gain. This higher antenna gain compensates for the growing FSPL and mitigates the impact of growing interference.

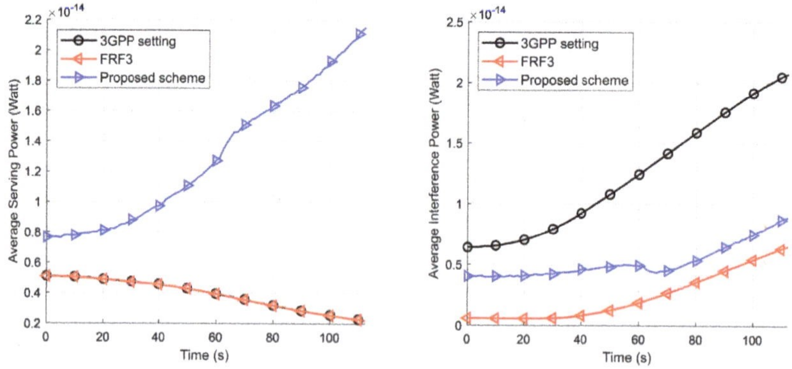

Figure 10. Average serving power and average interference power for QEFC LEO satellite in NTN rural scenario. The target SINR for the proposed scheme is −5.6 dB.

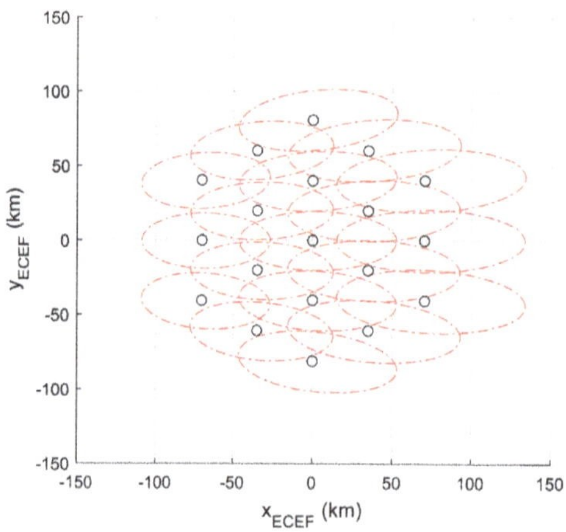

Figure 11. Beam topology and footprints with elevation angle $\epsilon_0(100) = 36.53°$ for the proposed optimal beamwidth scheme. Black dots represent the position of the beam center, and red circles represent the footprints of half-power beamwidth (1.212°).

4.2. Uplink Coverage Probability

Figures 12 and 13 show the uplink coverage probability for different target SINR in NTN rural and NTN urban scenarios. The simulations presented in Figures 12 and 13 show the proposed scheme's effectiveness in maximizing uplink coverage probability. The results demonstrate that the scheme's performance closely approximates the upper bound established by exhaustive antenna aperture search methods. Specifically, the figures illustrate that the proposed scheme achieves notable proximity to the upper bound, with a maximum deviation of around 0.9% in the NTN rural scenario and around 5% in the NTN urban scenario. This indicates the robustness and efficacy of the proposed scheme in optimizing beamwidth.

Furthermore, the proposed optimal beamwidth maximizing uplink coverage probability scheme is advantageous over the 3GPP setting. As shown in Figures 7 and 10, the interference increases at low elevation angles, which degrades the uplink coverage probability. In the NTN rural scenario, with a target SINR of −8.61 dB, the uplink coverage probability for the 3GPP setting starts above 0.95 but sharply declines to less than 0.36 by $t = 100$. Similarly, in the scenario with a target SINR of −5.6 dB, the uplink coverage probability for the 3GPP setting nearly diminishes to zero by $t = 100$. In contrast, the proposed scheme maintains an uplink coverage probability of at least 0.93 throughout all simulated times in the NTN rural scenario, irrespective of the target SINR. In the NTN urban scenario, where the low LOS probability impacts the uplink coverage probability, the proposed scheme still significantly improves compared to the 3GPP setting.

Despite FRF3 being considered a helpful interference mitigation scheme, the proposed scheme demonstrates improved uplink coverage probability through increased serving power, as shown in Figure 10. In the NTN rural scenario, the proposed scheme and FRF3 exhibit similar uplink coverage probability performance within $t = 0$ and $t = 80$. Beyond $t = 80$, the proposed scheme outperforms FRF3. In the NTN urban scenario, the uplink coverage probability of FRF3 surpasses that of the proposed scheme by approximately 0.085 and 0.03 at the beginning, with target SINR values of −5.6 and −8.61, respectively. However, as time progresses, the uplink coverage probability of FRF3 decreases, eventually falling below that of the proposed scheme around $t = 55$ and falling below by approximately 0.176 and 0.08 at $t = 100$, with target SINR values of −5.6 and −8.61, respectively.

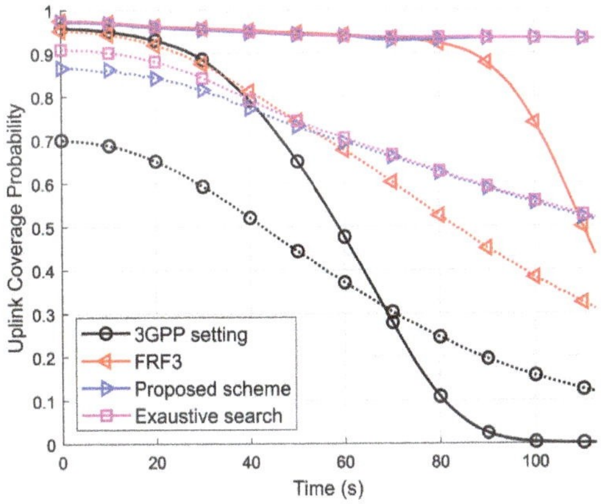

Figure 12. Simulation results of uplink coverage probability with target SINR −5.6 dB in rural (dashed) and urban (solid) scenario.

The simulation results and analysis show the proposed scheme's advantages. The proposed scheme significantly enhances uplink coverage probability, SINR, and interference mitigation. Furthermore, it closely matches the results obtained from exhaustive search methods. These improvements indicate that the proposed scheme effectively addresses the low SINR problem in the QEFC scenario. However, it is essential to acknowledge its limitations. Typically, satellites are equipped with two types of antennas: phased array antennas or parabolic antennas. Since the proposed scheme focuses on adjustable beamwidth, it may not be suitable for satellites equipped with parabolic antennas. Moreover, as illustrated in Figure 4, the optimal equivalent antenna aperture is substantial, three times larger than that of the 3GPP setting. Achieving such high antenna gain and narrow beam design requires a significant number of antenna elements. This affects the satellite's weight and, consequently, the deployment cost. Hence, a trade-off exists between price and performance in the QEFC scenario.

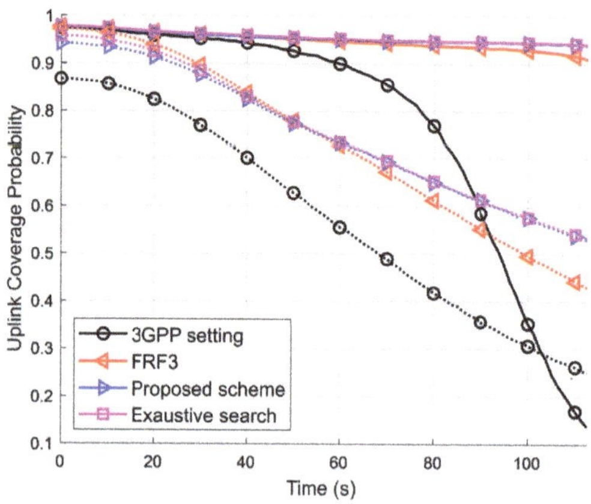

Figure 13. Simulation results of uplink coverage probability with target SINR −8.61 dB in rural (dashed) and urban (solid) scenario.

5. Conclusions

In this paper, we undertake a comprehensive performance analysis of the QEFC LEO satellite system. Our simulations show that QEFC LEO satellites encounter significant challenges, including high interference levels and low SINR. To address these issues and achieve ubiquitous coverage, we introduce an approach to optimizing beamwidth to maximize uplink coverage probability. Our proposed scheme aims to enhance SINR by determining the optimal beamwidth configuration that maximizes uplink coverage probability. We compare the performance of our proposed scheme against two existing configurations: the 3GPP setting and the three-color frequency reuse scheme.

Simulation results demonstrate that our proposed scheme yields notable improvements in serving power and effectively mitigates interference power compared to conventional configurations. Moreover, our scheme achieves the highest uplink coverage probability, closely approaching the upper bound of achievable uplink coverage probability. In the NTN rural scenario, the proposed scheme maintains the uplink coverage probability above 0.93. By leveraging optimized beamwidths, our proposed scheme offers a promising solution to the challenges faced by QEFC LEO satellite systems, ultimately enhancing network performance and facilitating ubiquitous coverage for users. However, considering the cost implications of the proposed scheme, future studies should explore strategies to maximize performance while operating within budget constraints.

Author Contributions: Conceptualization, C.-T.L. and J.-Y.P.; methodology, C.-T.L.; software, C.-T.L.; validation, C.-T.L.; formal analysis, C.-T.L.; investigation, C.-T.L.; resources, C.-T.L.; data curation, C.-T.L.; writing—original draft preparation, C.-T.L.; writing—review and editing, C.-T.L. and J.-Y.P.; visualization, C.-T.L.; supervision, C.-T.L. and J.-Y.P.; project administration, J.-Y.P.; funding acquisition, J.-Y.P. All authors have read and agreed to the published version of the manuscript.

Funding: This research was funded by the National Science and Technology Council, Taiwan, R.O.C., under Grants MOST 112-2218-E-305-001 and NSTC 112-2218-E-194-004.

Data Availability Statement: Data are available in a publicly accessible repository. The data presented in this study are openly available in FigShare at https://doi.org/10.6084/m9.figshare.25497694.v2, accessed on 31 March 2024.

Acknowledgments: This research was partly supported by the Advanced Institute of Manufacturing with High-Tech Innovations from the Featured Areas Research Center Program within the framework of the Higher Education Sprout Project by the Ministry of Education (MOE) in Taiwan.

Conflicts of Interest: The authors declare no conflict of interest.

Abbreviations

The following abbreviations are used in this manuscript:

3GPP	Third-Generation Partnership Project
5G	Fifth-Generation mobile network
ECEF	Earth-Centered Earth-Fixed
EMC	Earth-moving cells
FRF3	Three-color frequency reuse
FSPL	Free-space path loss
LEO	Low earth orbit
LOS	Line-of-sight
NLOS	Non-line-of-sight
NR	New Radio
NTN	Non-terrestrial networks
PUSCH	Physical uplink shared channel
QEFC	Quasi earth-fixed cells
SINR	Signal-to-interference-plus-noise ratio
UE	User equipment
VoNR	Voice-over New Radio

References

1. Liao, Q.; Kaneko, M. Global Energy Efficiency Optimization of a Ka-Band Multi-Beam LEO Satellite Communication System. *IEEE Access* **2021**, *9*, 55232–55243. [CrossRef]
2. Measuring Digital Development: Facts and Figures 2022. Available online: https://www.itu.int/hub/publication/d-ind-ict_mdd-2022/ (accessed on 1 April 2024).
3. Xia, S.; Jiang, Q.; Zou, C.; Li, G. Beam Coverage Comparison of LEO Satellite Systems Based on User Diversification. *IEEE Access* **2019**, *7*, 181656–181667. . [CrossRef]
4. Juan, E.; Lauridsen, M.; Wigard, J.; Mogensen, P. Performance Evaluation of the 5G NR Conditional Handover in LEO-based Non-Terrestrial Networks. In Proceedings of the 2022 IEEE Wireless Communications and Networking Conference (WCNC), Austin, TX, USA, 10–13 April 2022. . [CrossRef]
5. Darwish, T.; Kurt, G. K.; Yanikomeroglu, H.; Bellemare, M.; Lamontagne, G. LEO Satellites in 5G and Beyond Networks: A Review From a Standardization Perspective. *IEEE Access* **2022**, *10*, 35040–35060. . [CrossRef]
6. 3GPP R2-2313877. RACH-less satellite switch procedure. In Proceedings of the 3GPP TSG RAN WG2 Meeting #124, Chicago, IL, USA, 13–17 November 2023.
7. Yu, J.; Kim, J. H. Analysis of SINR according to Elevation Angle in Earth Fixed Beam. In Proceedings of the 2022 27th Asia Pacific Conference on Communications (APCC), Jeju Island, Republic of Korea, 19–21 October 2022. . [CrossRef]
8. Baik, J. S.; Kim, J. H. Analysis of the Earth Fixed Beam Duration in the LEO. In Proceedings of the 2021 International Conference on Information Networking (ICOIN), Jeju Island, Republic of Korea, 13–16 January 2021. . [CrossRef]
9. Yi, J.; Kim, Y.; Ryu, H. Performance of Uplink Coverage Enhancement Schemes for 5G NR in 3GPP. In Proceedings of the 2022 IEEE 95th Vehicular Technology Conference: (VTC2022-Spring), Helsinki, Finland, 19–22 June 2022. . [CrossRef]
10. Nasarre, I.P.; Levanen, T.; Pajukoski, K.; Lehti, A.; Tiirola, E.; Valkama, M. Enhanced Uplink Coverage for 5G NR: Frequency-Domain Spectral Shaping With Spectral Extension. *IEEE Open J. Commun. Soc.* **2021**, 2, 1188–1204. . [CrossRef]
11. 3GPP R1-2208321. Final Report of 3GPP TSG RAN WG1 #110 v1.0.0. In Proceedings of the 3GPP TSG RAN WG1 Meeting #110bis-e, Online Meeting, 10–19 October 2022.
12. 3GPP R1-2207811. Summary #4 on 9.12.1 Coverage enhancement for NR NTN. In Proceedings of the 3GPP TSG RAN WG1 Meeting #110, Toulouse, France, 22–26 August 2023.
13. Vázquez, M.Á.; Pérez-Neira, A.; Christopoulos, D.; Chatzinotas, S.; Ottersten, B.; Arapoglou, P.; Ginesi, A.; Taricco, G. Precoding in Multibeam Satellite Communications: Present and Future Challenges. *IEEE Wirel. Commun.* **2016**, *23*, 88–95. . [CrossRef]
14. Spacex Non-Geostationary Satellite System Attachment a Technical Information to Supplement Schedule S. November 2016. Available online: https://fcc.report/IBFS/SAT-LOA-20161115-00118/1158350 (accessed on 1 April 2024).
15. Al Homssi, B.; Al-Hourani, A. Optimal Beamwidth and Altitude for Maximal Uplink Coverage in Satellite Networks. *IEEE Wirel. Commun. Lett.* **2022**, *11*, 771–775. [CrossRef]
16. 3GPP TR 38.811. Study on New Radio (NR) to Support Non-Terrestrial Networks (Release 15). V15.4.0, September 2020. Available online: https://portal.3gpp.org/desktopmodules/Specifications/SpecificationDetails.aspx?specificationId=3234 (accessed on 1 April 2024).
17. 3GPP R1-2208321. Discussion on NR NTN coverage enhancement. In Proceedings of the 3GPP TSG RAN WG1 Meeting #110, Toulouse, France, 22–26 August 2023.

18. 3GPP TR 38.821. Solutions for NR to Support Non-Terrestrial Networks (NTN) (Release 16). V16.1.0, May 2021. Available online: https://portal.3gpp.org/desktopmodules/Specifications/SpecificationDetails.aspx?specificationId=3525 (accessed on 1 April 2024).
19. 3GPP RP-222654. Revised WID: NR NTN (Non-Terrestrial Networks) enhancements. In Proceedings of the 3GPP TSG RAN Meeting #97-e, Online Meeting, 12–16 September 2022.

electronics

MDPI

Article

Deep Reinforcement Learning-Based Multipath Routing for LEO Megaconstellation Networks

Chi Han [1], Wei Xiong [1,2,*] and Ronghuan Yu [1,2]

1 National Key Laboratory of Space Target Awareness, Space Engineering University, Beijing 101400, China; hanchi@hgd.edu.cn (C.H.); yrh1983@163.com (R.Y.)
2 School of Space Information, Space Engineering University, Beijing 101400, China
* Correspondence: xiongwei@hgd.edu.cn

Abstract: The expansion of megaconstellation networks (MCNs) represents a promising solution for achieving global Internet coverage. To meet the growing demand for satellite services, multipath routing allows the simultaneous establishment of multiple transmission paths, enabling the transmission of flows in parallel. Nevertheless, the mobility of satellites and time-varying link states presents a challenge for the discovery of optimal paths and traffic scheduling in multipath routing. Given the inflexibility of traditional static deep reinforcement learning (DRL)-based routing algorithms in dealing with time-varying constellation topologies, DRL-based multipath routing (DMR) enabled by a graph neural network (GNN) is proposed as a means of enhancing the transmission performance of MCNs. DMR decouples the stochastic optimization problem of multipath routing under traffic and bandwidth constraints into two subproblems: multipath routing discovery and multipath traffic scheduling. Firstly, the minimum hop count-based multipath route discovery algorithm (MHMRD) is proposed for the computation of multiple available paths between all source and destination nodes. Secondly, the GNN-based multipath traffic scheduling scheme (GMTS) is proposed as a means of dynamically scheduling the traffic on each available path for each data stream, based on the state information of ISLs and traffic demand. Simulation results demonstrate that the proposed scheme can be scaled to constellations with different configurations without the necessity for repeated training and enhance the throughput, completion ratio, and delay by 42.64%, 17.39%, and 3.66% in comparison with the shortest path first algorithm (SPF), respectively.

Keywords: satellite network; multipath routing; deep reinforcement learning; traffic scheduling; hop count

1. Introduction

The characteristics of megaconstellation networks (MCNs), such as wide area coverage, low latency, and high bandwidth, confer upon them unique advantages in remote area communications and delay-sensitive services (e.g., emergency communications and disaster monitoring) [1]. The aforementioned advantages have resulted in the accelerated development of MCNs, which are anticipated to fulfill a multitude of services, including communications, remote sensing, navigation, and positioning. Furthermore, MCNs are poised to become a pivotal component of the infrastructure of the space-air-ground integrated network (SAGIN) [2–4].

In order to meet the growing demand for satellite communication services, multipath routing is a promising approach. Multipath routing allows multiple transmission paths to be established simultaneously between the source and destination nodes for each data stream, which can increase network throughput [5]. In contrast to single-path routing, multipath routing allows for the optimal utilization of available bandwidth, thereby alleviating congestion, improving throughput, and enhancing fault tolerance. This is achieved by enabling a single service to be transmitted over multiple paths in parallel. Consequently, multipath routing facilitates the resilience of resource-constrained satellite networks to withstand node or link failure scenarios. However, the high dynamics of satellite network topologies and transmission link states presents challenges for multipath routing discovery and multipath traffic scheduling [6–8].

Multipath routing discovery is the first challenge posed by the time-varying nature of satellite topology and traffic. Factors such as changes in satellite attitude and complex electromagnetic environments can cause laser link transmission rates to decrease or even fail. As a result, link failures may occur frequently in MCNs rather than being an exceptional event, as in terrestrial networks. The frequent occurrence of link passes or failures impedes the ability of the broadcast mechanism to discover and maintain routes in a timely manner, as is the case in terrestrial networks. This results in an increase in the overheads associated with broadcast and route control. Consequently, topology-based ad hoc routing protocols, such as ADOV [9], are suboptimal in MCNs. Furthermore, due to the mobility and energy constraints of LEO satellites, established links are often not stably maintained. Topological changes in the path discovery process may result in the rapid expiration of discovered paths, with the corresponding paths no longer existing at the time of data transfer. The above factors require multipath routing discovery to be able to sense link failures in a timely manner and reasonably set multiple alternative paths. Some schemes currently deploy path computation in the ground network control center (NCC) [10]. The NCC improves network transmission performance by comprehensively considering the distribution of terrestrial user traffic and link utilization and making full use of the link redundancy brought about by the mesh topology of the satellite network to plan multiple parallel transmission paths for each flow at the same time.

Multipath traffic scheduling in MCNs is the second challenge. Complex electromagnetic environments, satellite attitude changes, and other factors result in fluctuations in link quality. As a consequence, traditional multipath traffic scheduling based on static routing is no longer applicable in MCNs. Furthermore, the time-varying nature of terrestrial user traffic and the fluctuation in link quality render the satellite network incapable of accurately identifying traffic patterns, and fine-grained traffic scheduling between multiple paths frequently lacks effective information. Deep reinforcement learning (DRL) has demonstrated significant advantages in the description of spatiotemporal features, flow decisions, and other related areas. A considerable number of scholars have opted to adopt DRL for the purpose of adaptive multipath flow scheduling, as evidenced by [11–14]. However, the time-varying satellite networks and fluctuating quality of inter-satellite links (ISLs) necessitate the aggregation of multiple time slices for decision making in models employing DRL. As the constellation size increases, the model's dimension expands dramatically, resulting in a significant increase in computational effort and a reduction in model scalability. Graph neural networks (GNNs) are capable of identifying features and patterns in graph-structured data, performing relational reasoning and combinatorial generalization on graph-structured data [15,16], and they can be applied without additional model tuning to constellations of different topologies and sizes [17,18].

Recently, DRL has demonstrated considerable potential in learning temporal and spatial traffic characteristics for the purpose of making routing decisions. Researchers have employed DRL for the purpose of adaptive traffic assignment, with the aim of replacing static traffic segmentation. However, DRL-based models require the ability to summarize continuous topological snapshots in order to make decisions in a dynamic satellite environment. Satellite networks have high-dimensional and sparse state spaces, and the dimensionality of DRL models increases dramatically with the size of the constellation and the level of traffic demand, which in turn leads to poor model scalability. Graph neural networks (GNN) facilitate the inference of relational data and the generalization of combinatorial structures within graph-structured information. This enables application of the DRL model to satellite constellations of varying sizes without the need for additional modifications. In light of the aforementioned analysis, we propose DRL-based multipath routing (DMR) embedded by a GNN to improve the transmission performance of MCNs.

The multipath routing problem is initially modeled as a stochastic optimization problem, with the objective of maximising network efficiency under traffic and bandwidth constraints. In order to obtain the optimal solution, the original problem is decoupled into two subproblems: multipath routing discovery and multipath traffic scheduling. With regard to the initial subproblem, it is proposed that the redundant topology of the satellite network be fully exploited, with the introduction of a minimum hop count-based multipath routing discovery (MHMRD) algorithm. MHMRD is designed to compute a number of available paths between all source and destination nodes. On this basis, the second subproblem is addressed by the design of GNN-based multipath traffic scheduling (GMTS). GMTS models the multipath traffic scheduling process as a Markov decision process (MDP), which dynamically schedules traffic on each available path for each data stream based on satellite network link state information and traffic demand. Simulation results on Iridium and OneWeb demonstrate that the proposed scheme is not reliant on retraining for different-sized constellations. Furthermore, the network throughput, average delay, and flow completion rate exceeded those of the baseline scheme.

The main contributions of this paper are summarized as follows:

- The multipath routing problem in satellite networks is modeled as a stochastic optimization problem under traffic and bandwidth constraints. The objective is to maximize network efficiency. The problem is decoupled into a multipath routing discovery subproblem and a multipath traffic scheduling subproblem.
- In order to address the multipath routing discovery subproblem, the MHMRD algorithm has been proposed as a means of planning multiple minimum hop count paths for each flow between any pair of nodes in the network based on the link state information.
- The multipath traffic scheduling subproblem is modeled as an MDP, with the GMTS scheme proposed to dynamically schedule the proportion of traffic on each available path for each data stream. GMTS is a scalable solution which can be applied to different constellations.

The rest of this paper is organized as follows. Section 2 provides some related work. In Section 3, we provide the system model and problem formulation. In Section 4, DRL-based multipath routing is described. Section 5 provides the simulation and main results. Finally, we summarize the results in Section 6.

2. Related Work

2.1. Multipath Routing

In satellite networks, there are multiple paths with an equal number of hops from end to end. In order to improve the reliability of data transmission and enhance the network load balancing capability, researchers have proposed the use of multipath routing in satellite networks, where each flow is divided into multiple subflows for parallel transmission along different paths. In order to distribute traffic over multiple paths, the traditional equal-cost multipath (ECMP) method performs static traffic splitting based on message information. However, the ECMP approach does not consider the limitations of network parameters such as the bandwidth and delay and is prone to congestion when the network load is high. To achieve dynamic flow management for multipath routing, a software-defined network (SDN)-based satellite network architecture has been proposed for central topology control and traffic control [18–21]. In the SDN-based network architecture, the data plane and control plane are separated, the SDN controller determines the routing based on the network parameters and QoS requirements, and the satellite is responsible for data forwarding. In [22], the authors attempted to integrate network coding and multipath routing in order to enhance the data transmission efficiency of satellite networks. However, the centralized control approach is susceptible to high latency issues. In [23], in order to reduce the control latency and improve the transmission stability, the authors employed a distributed approach for routing control. In [24], the authors proposed network coding-based multipath cooperative routing (NCMCR). In order to optimize the transmission capacity, each flow in NCMCR is transmitted along multiple disjoint links. However, the aforementioned multipath routing schemes do not consider the link parameters of the various forwarding paths and thus are unable to adaptively perform reasonable traffic splitting. In [1], the authors designed an adaptive traffic balancing scheme among multiple temporal paths by combining sparse and redundant network coding mechanisms. This approach enables the achievement of deterministic delay guarantees in limited node resources and multi-user competition scenarios. Nevertheless, all of the aforementioned routing schemes utilize static topologies, which precludes them from adapting to satellite networks with dynamically changing topologies.

2.2. Intelligent Routing

The time-varying satellite network environment and dynamic interstellar links bring challenges to satellite network routing. In order to ensure the network's quality of service under the conditions of proliferating user and constellation sizes, multipath routing requires more efficient path planning and finer traffic control strategies. Reinforcement learning is an intelligent tool which supports decision making by interacting with the environment [25–27], and scholars have proposed numerous intelligent routing schemes based on reinforcement learning. In [25], the authors developed a supervised deep learning system for the construction of routing tables. In [28], the authors employed graph-based deep learning in satellite networks through a neural network architecture called teh graph-query neural network. Nevertheless, such supervised learning-based routing schemes are constrained by their inability to generalize and adapt across different constellations. To address this issue, the authors of [29] proposed a Markov decision process (MDP) model for satellite network routing, employing multi-agent deep reinforcement learning (DRL) to satisfy diverse quality of service (QoS) requirements. However, the fixed goal policy renders DRL-based routing algorithms inflexible, as they are unable to adapt to dynamically changing networks. It is evident that traditional static objective configurations are unable to reflect the varying importance of different metrics in dynamic network environments. In order to address this issue, the authors of [17,30] integrated a graph neural network (GNN) into DRL intelligences, enabling the timely adjustment of network performance metrics by predicting the trend of the optimization objectives of routing algorithms. This enables the intelligences to learn optimal paths which can adapt to different environmental changes. In [31,32], the authors proposed different DRL architectures, all of which dynamically adjust the link weights according to the load of the key nodes or the link load. In [33], the authors extended the TCP options to piggyback the relevant control information and flexibly support communication between the subflows of the transport layer and the SDN controller. An SDN cooperated MPTCP (scMPTCP) architecture was proposed, which selects routes for new subflows based on the available bandwidth of each route and avoids the bottleneck of other subflows. It also can adapt to changes in network load. In [34], the authors proposed a DQN controller considering a path loss model based on the Markov decision process model for network selection and adaptive resource allocation in heterogeneous networks.

Nevertheless, an examination of the aforementioned research reveals that existing DRL-based studies are not readily scalable due to their inherent coupling to the input states and limited topology, which is consistent with the training data. To resolve the aforementioned contradiction, the proposed DRL-based multipath routing embedded with a GNN can readily identify the optimal action within a continuous action space. DMR decouples the multipath routing problem into distinct components—multipath discovery and multipath traffic scheduling—to reduce the solution's complexity. Concurrently, the target policy adopted by DMR can be modified in accordance with the prevailing environmental conditions, thereby enhancing the flexibility of the system with respect to different topologies.

3. System Model and Problem Formulation

3.1. Multipath Scenario

The multipath routing scenario in the MCN is shown in Figure 1, where the satellite network is divided into mutually independent data planes and control planes via the SDN. In the terrestrial network, data in the send buffer are transmitted to the sender through routers over multiple hops and aggregated into multiple subflows. Each subflow can select the optimal transmission path according to the current network condition and then converge at the receiver.

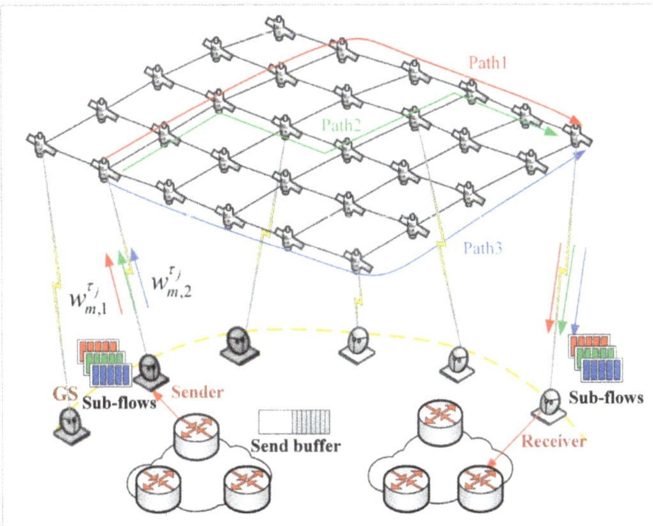

Figure 1. Multipath routing in the megaconstellation network.

The data plane comprises the Walker Delta constellation $N_S N_P / N_P / F$, which is distributed in a uniform and symmetrical manner, and the ground station. The N_P orbital planes are distributed uniformly along the equator, while the N_S satellites are distributed uniformly on each plane. Each satellite is capable of being connected to four inter-satellite links (ISLs), comprising two intra-plane ISLs and two inter-plane ISLs. In the control plane, the network control center assumes the functions of unified topology management, traffic scheduling, and path control. Due to the mobility of LEO megaconstellation networks, inter-plane ISLs are subject to frequent disruptions and rebuilds, and the topology is in a state of constant flux. Consequently, the NCC must be continuously updated to accommodate the evolving topology, which presents a significant challenge to multipath routing planning and traffic scheduling.

3.2. Multipath Routing and Traffic Model

The MCN is modeled as a spatiotemporal graph $G = \{(V, E, T, N)\}$, where $V = S \cup G$ is the node consisting of all satellites and terrestrial gateways and $E = E_S \cup E_G$ is the set of ISLs E_S and the set of terrestrial links E_G, while T is the time slot vector of a length N. As illustrated in Figure 1, the mesh topology of the MCN allows for the existence of multiple potential routes between any given pair of nodes. Assume that the number of packet flows at time slot t is M (i.e., $P_t = \{ (s_i, d_i) | i = 1, 2, \cdots, M\}$). In other words, the variable M represents the number of subflows, as shown in Figure 1, where three subflows in the sender are awaiting transmission via multiple paths. The number of available paths between the source node s_i and destination node d_i is L_{s_i, d_i}, denoted as $R_t = \left\{ p^t_{1,1}, \cdots, p^t_{1,L}, \cdots, p^t_{M,L} \right\}$. Here, $p^t_{m,l}$ is the lth path between the mth pair of source and destination nodes, consisting of all ISLs on that path (i.e., $p_{m,l}^t = \left\{\varepsilon_1, \varepsilon_2, \cdots, \varepsilon_j, \cdots, \varepsilon_{n_{m,l}}\right\}$, where $n_{m,l}$ is the total hop count of the current link).

Let $q_{\varepsilon_j}^t$ be the queuing delay of the link ε_j, $r_{\varepsilon_j}^t$ be the packet processing delay, and $\gamma_{\varepsilon_j}^t$ be the link distance. Then, the delay of the lth transmission path of the mth source-destination node pair can be expressed as follows:

$$\eta_{m,x}^t = \sum_{j=1}^{n_{m,x}} \left(q_{\varepsilon_j}^t + r_{\varepsilon_j}^t + \frac{\gamma_{\varepsilon_j}^t}{c} \right) \tag{1}$$

where c is the vacuum speed of light and $n_{m,l}$ denotes the hop count of the current link. Each source node can divide the traffic into L_{s_i,d_i} subflows to pass through L_{s_i,d_i} paths in parallel. Let $\omega_{m,l}^t$ be the traffic ratio of each subflow. Then, the sum of L_{s_i,d_i} subflows between any pair of source and destination nodes should be one, which can be expressed as follows:

$$\sum_{l=1}^{L} \omega_{m,l}^t = 1, \forall m = 1, 2, \cdots, M \tag{2}$$

Therefore, the traffic of link ε_j at time t can be expressed by

$$f_{\varepsilon_j}^t = \sum_{m=1}^{M} \sum_{l=1}^{L} b_m^t \omega_{m,l}^t \delta_{m,l}^t \tag{3}$$

where b_m^t is the bandwidth requirement of the mth pair to the source and destination nodes and $\delta_{m,l}^t$ is the path packet loss rate. Noting that the maximum bandwidth of the interstellar link is BW, the ISL traffic should not exceed the maximum bandwidth limit, which can be expressed as follows:

$$f_{\varepsilon_j}^t \leq BW, \forall \varepsilon_j \in p_{m,l}^t \tag{4}$$

Consequently, the throughput between the mth node pair can be expressed by

$$f_{m,l}^t = \max\left\{f_{\varepsilon_j}^t\right\}, \forall \varepsilon_j \in p_{m,l}^t \tag{5}$$

3.3. Problem Formulation

In order to measure the delay and throughput of ISLs, the utility function of the MCN is defined as

$$U(\bar{f}_t, \bar{d}_t) = \beta_1 \log(\bar{f}_t) - \beta_2 \log(\bar{d}_t) \tag{6}$$

where β_1 and β_2 are the importance coefficients of the throughput and delay, respectively, β_1 and β_2 satisfy $\beta_1 + \beta_2 = 1$, and $\bar{f}_t$ and $\bar{d}_t$ are the average throughput and delay of the current time slot, respectively, which can be expressed as follows:

$$\bar{f}_t = \sum_{m=1}^{M} \sum_{l=1}^{|f_m|} \sum_{y=1}^{|p_k|} \frac{\nabla_{pak}}{t}, \forall t \in [1, T] \tag{7}$$

$$\bar{d}_t = \frac{\sum_{m=1}^{M} \sum_{l=1}^{|f_m|} \sum_{y=1}^{|p_k|} d_{m,l}^t \kappa_y}{\sum_{m=1}^{M} \sum_{l=1}^{|f_m|} \sum_{y=1}^{|p_k|} \kappa_y}, \forall t \in [1, T] \tag{8}$$

where $|f_m|$ is the number of data flows between the mth node pair, $|p_k|$ is the number of packets for the kth data flow, ∇_{pak} is the packet size, and $\kappa_y \in \{0, 1\}$ is a binary variable which measures whether the yth packet was successfully sent.

Consequently, the satellite network multipath routing optimization can be expressed as the following optimization problem:

$$
\begin{aligned}
P0: & \max U(\bar{f}, \bar{d}), \forall t \in [1, T] \\
& C1: \bar{d} = \frac{1}{N} \sum_{i=1}^{N} \bar{d}_t \\
& C2: \bar{f} = \frac{1}{N} \sum_{i=1}^{N} \bar{f}_t \\
s.t. \quad & C3: f^t_{\varepsilon_j} \leq BW, \forall \varepsilon_j \in p^t_{m,l} \\
& C4: \sum_{l=1}^{L} \omega^t_{m,l} = 1, \forall m \in [1, M] \\
& C5: d^t_{m,l} \leq d^t_{m,y}, \forall 1 \leq l \leq y \leq L, \forall m \in [1, M]
\end{aligned}
\tag{9}
$$

where constraints C1 and C2 denote the average delay and throughput of the satellite network, respectively, constraint C3 stipulates that the the throughput of link ε_j at any given moment t cannot exceed the bandwidth limit, constraint C4 denotes that the total communication between any pair of nodes is the cumulative result of multiple sub-streams traversing multiple paths, and finally, constraint C5 describes the delay constraints of all candidate paths between each pair of nodes. It can be seen that P0 is an NP-hard problem with a rugged solution space. P0 consists of two parts—multipath routing discovery and multipath traffic scheduling—which converge slowly and may fall into local optima if solved by stochastic optimization. In order to reduce the solution's complexity, this paper decouples the problem into two subproblems, routing discovery and traffic scheduling, to improve the efficiency of the algorithm.

4. DRL-Based Multipath Routing

In this paper, we propose DRL-based multipath routing (DMR), as shown in Figure 2, which first uses the minimum hop count based-multipath discovery algorithm to solve for multiple paths with acceptable delays and then utilizes the GMTS algorithm to determine the traffic scheduling scheme based on the state of the constellation network.

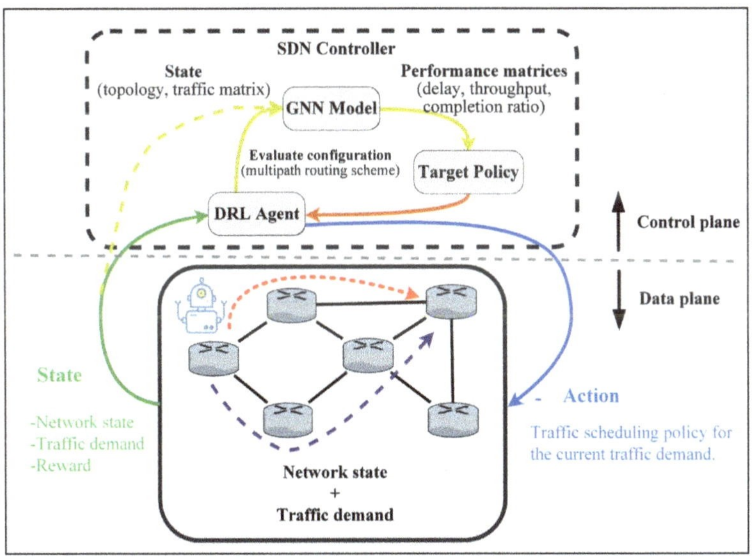

Figure 2. The overall structure of DRL-based multipath routing.

4.1. Multipath Routing Discovery

The objective of the multipath routing discovery subproblem is to identify the optimal delay path and the suboptimal delay path. In MCNs, the propagation delay, which depends on the number of path hops, represents a significant component of the end-to-end delay. Consequently, multipath paths are primarily considered for the hop-optimal and suboptimal paths within the network. The multipath routing discovery subproblem can be expressed as follows:

$$\begin{aligned} P1:\ & \min d^t_{m,l}, \forall t \in T, m \in [1,M], l \in [1,L] \\ s.t.\ \ & d^t_{m,l} \le d^t_{m,y}, \forall 1 \le l \le y \le L \end{aligned} \tag{10}$$

The minimum hop count path between any pair of nodes is initially determined. On this basis, the NCC determines the suboptimal transmission path (i.e., the backup path) for each flow in descending order of traffic. This is because centralized traffic is more likely to cause bottleneck link congestion than decentralized tiny traffic. In the Wakler Delta constellation $N_S N_P / N_P / F$, as shown in Figure 3, the inclination is α, and the right ascension of the ascending node (RAAN) difference of adjacent orbits is $\Delta\Omega = 2\pi/N_P$. The phase difference between neighboring satellites in the same orbit is $\Delta\Phi = 2\pi/N_S$, while the phase difference between neighboring satellites in adjacent orbits is $\Delta v = 2\pi F/N_S N_P$, while u_1 and u_2 are the argument of latitude (or phase angle) values of the source and destination satellites, respectively. The argument of latitude is the angle between the ascending node and the satellite and basically defines the position of the satellite in the orbit, where $\zeta(u_1)$ and $\zeta(u_2)$ indicate the longitude difference of the source and destination satellite to the corresponding ascending nodes, respectively. In the Walker Delta constellation, the end-to-end hop count includes both transverse inter-plane hops H_h and intra-plane hops H_v.

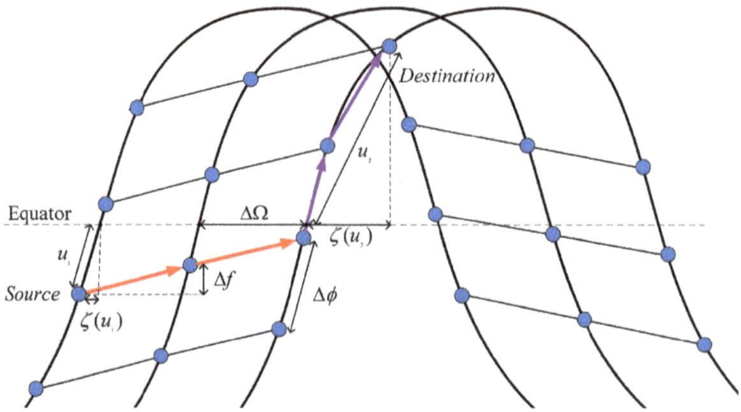

Figure 3. Track of sub-satellite point of megaconstellation networks.

4.1.1. Inter-Plane Hops H_h

The number of inter-plane hops depends on the RAAN difference $\Delta\Upsilon_0$ of the orbits where the source and destination nodes are located. Given a source and destination satellite pair (s,d), $\Delta\Upsilon_0$ is calculated as follows:

$$\Delta\Upsilon_0 = (\Upsilon_2 - \Upsilon_1) \bmod 2\pi \in [0, 2\pi] \tag{11}$$

where Υ_1 and Υ_2 are the RAAN values of s and d, respectively. If the destination satellite is to the west of the source satellite, then the RAAN difference is $2\pi - \Delta\Upsilon_0$. Since the RAAN difference between adjacent orbits is constant (i.e., $\Delta\Omega = 2\pi/N_P$), the inter-plane hops in the west and east directions can be expressed as follows:

$$H_h^{\leftarrow} = \left\langle \frac{2\pi - \Delta\Upsilon_0}{\Delta\Omega} \right\rangle \tag{12}$$

$$H_h^{\rightarrow} = \left\langle \frac{\Delta\Upsilon_0}{\Delta\Omega} \right\rangle \tag{13}$$

where $\langle x \rangle = \text{sgn}(x)\lfloor x + 1/2 \rfloor$ denotes the integer closest to x, $H_h^{\leftarrow}$ is the hop count between the orbital planes to the west, and $H_h^{\rightarrow}$ is the eastward inter-plane hop count.

4.1.2. Intra-Plane Hops H_v

The intra-plane hops depend on the phase angle difference Δu of the satellite. Each intra-plane hop increases the phase angle by $\Delta\Phi$, and each inter-plane hop results in a phase angle increment of Δf. Therefore, the phase angle of the destination satellite can be expressed as follows:

$$u_2 = u_1 + \Delta f \cdot H_h^{\rightarrow} + \underbrace{\Delta\Phi \cdot H_v^{\nearrow}}_{\Delta\vec{u}} \tag{14}$$

where $H_v^{\nearrow}$ is the eastward intra-plane hop count. In order to compute H_v, it is necessary to first eliminate the phase difference $\Delta f \cdot H_h^{\rightarrow}$ caused by H_h from the phase difference $\Delta u = u_2 - u_1$, where u_1 and u_2 are the argument of latitude values of the source and destination satellites, respectively. To distinguish between the phase difference due to eastward and westward propagation, Δu can be expressed separately as follows:

$$\Delta\vec{u} = (u_2 - u_1 - H_h^{\rightarrow} \cdot \Delta f) \bmod 2\pi \tag{15}$$

$$\Delta\overleftarrow{u} = (u_2 - u_1 + H_h^{\leftarrow} \cdot \Delta f) \bmod 2\pi \tag{16}$$

where Δf is the phase angle change due to inter-plane hops. As illustrated in Figure 3, the track of the sub-satellite point is divided into an ascending segment (from southwest to northeast) and a descending segment (from northwest to southeast), allowing for the propagation of data in both directions. Consequently, in this paper, the intra-plane hops in the four directions are calculated, which can be expressed as follows:

$$H_v^{\nwarrow} = \left| \frac{\Delta\overleftarrow{u}}{\Delta\Phi} \right| \tag{17}$$

$$H_v^{\nearrow} = \left| \frac{\Delta\vec{u}}{\Delta\Phi} \right| \tag{18}$$

$$H_v^{\swarrow} = \left| \frac{2\pi - \Delta\overleftarrow{u}}{\Delta\Phi} \right| \tag{19}$$

$$H_v^{\searrow} = \left| \frac{2\pi - \Delta\vec{u}}{\Delta\Phi} \right| \tag{20}$$

where $\Delta\overleftarrow{u}$ and $\Delta\vec{u}$ denote the phase difference due to westward and eastward propagation between adjacent orbits, respectively. Meanwhile, $H_v^{\nwarrow}$, $H_v^{\nearrow}$, $H_v^{\swarrow}$, and $H_v^{\searrow}$ denote the phase difference due to hops in the orbit toward the northwest, northeast, southwest, and southeast, respectively.

The inter-plane hops H_h and the intra-plane hops, H_v were calculated earlier. Consequently, the end-to-end minimum hop count in the inclined orbit constellation can be expressed as follows:

$$H = \min \begin{Bmatrix} H_h^{\leftarrow} + H_v^{\nwarrow} \\ H_h^{\leftarrow} + H_v^{\swarrow} \\ H_h^{\rightarrow} + H_v^{\nearrow} \\ H_h^{\rightarrow} + H_v^{\searrow} \end{Bmatrix} \tag{21}$$

On the basis of obtaining the shortest hop count path, multiple available paths are generated. Taking the suboptimal path computation between node pairs (s,d) as an example, firstly, the occupancy frequency $F_{a,b}^t$ of each link and the minimum hop count path set $R_{a,b}^t$ are computed. Meanwhile, the link occupancy frequency threshold ξ_ε is defined. The weight $m_{a,b}^t \in [r_1, r_2]$ of link (a,b) is set if $F_{a,b}^t \leq \xi_\varepsilon$; otherwise, $m_{a,b}^t \in [r_2, r_3]$, where $r_1 < r_2 < r_3$ are random numbers. In order to prevent the interconnection of disparate paths, the NCC will eliminate the links in $R_{a,b}^t$ from the network. Subsequently, the NCC will identify the optimal feasible paths for (s,d) among the remaining links, utilising the principle of the shortest end-to-end hop count. The aforementioned process is repeated continuously until L possible paths between any two nodes in the network have been identified. The specific process of MHMRD is illustrated in Algorithm 1. The computational complexity of Algorithm 1 is $O(N \log N)$.

Algorithm 1: Minimum hop count-based multipath route discovery (MHMRD).

Input: $\{(V, E, T, N)\}$, source-destination node pair P
Output: Candidate paths $\mathbf{R}$
Initialization: $\mathbf{R} = \emptyset$
for $i = 1 : N$ **do**
  Compute the minimum hop count path $\mathbf{R}^t$ via Equation (21);
  Compute other available paths;
  for (s,d) *in* P **do**
    for $l = 1 : L-1$ **do**
      Compute the occupancy frequency $F_{a,b}^t$ of each ISL;
      for (a,b) *in* $\mathbf{R}^t$ **do**
        $F_{a,b}^t \leftarrow F_{a,b}^t + 1$;
      end
      for (a,b) *in* $\mathbf{E}^t$ **do**
        if $F_{a,b}^t \leq \xi_\varepsilon$ **then**
          $m_{a,b}^t \in [r_1, r_2]$;
        end
        if $F_{a,b}^t > \xi_\varepsilon$ **then**
          $m_{a,b}^t \in [r_2, r_3]$;
        end
      end
      for e *in* $\mathbf{R}_{s,d}^t$ **do**
        Remove the edge from $\mathbf{E}^t$;
      end
      Compute the feasible paths $\mathbf{R}$ via minimum hop count path;
    end
  end
end

4.2. Multipath Traffic Scheduling

Based on the completion of multipath path discovery, the second subproblem of multipath routing for MCNs is multipath traffic scheduling, which can be expressed as follows

$$\begin{aligned} P2: & \max U(\vec{f},\vec{d}), \forall t \in T \\ s.t. & \sum_{l=1}^{L} \omega_{m,l}^{t} = 1, \forall m = 1, \cdots, M \\ & f_{\varepsilon_j}^{t} \le BW, \forall \varepsilon_j \in p_{m,l}^{t} \end{aligned} \quad (22)$$

Since the traffic assignment decision in satellite networks depends on the current node and surrounding node states in the network, independent of the historical state, the multipath traffic scheduling subproblem can be described as a Markov decision process (MDP). The occurrence of failures in ISLs or nodes has the effect of impairing the functionality of the satellite network, resulting in frequent alterations to the network topology. Given that the GNN has a superior generalization capability for topologies of varying sizes, this paper proposes the adoption of GNN-based multipath traffic scheduling (GMTS), as illustrated in Figure 4. The DRL agent uses a proximal policy optimization (PPO) algorithm, which is an actor-critic algorithm. At each stage of the MDP, the actor model selects a traffic scheduling decision, and the critic model assigns a score to that decision based on the environmental reward feedback. The process is repeated continuously, with the objective of optimizing the flow scheduling task in order to maximize the cumulative reward.

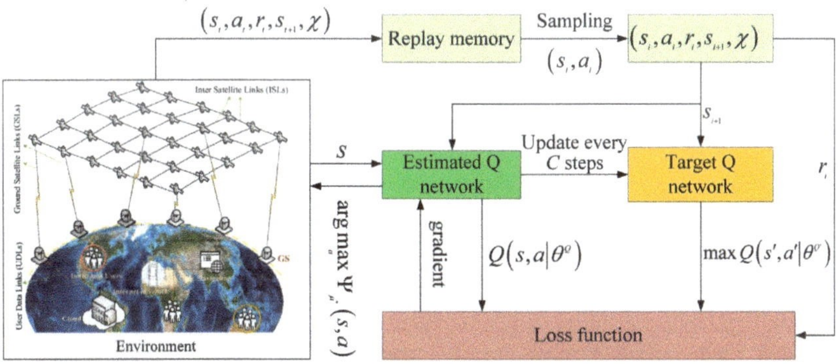

Figure 4. The framework of the proposed GNN-based multipath traffic scheduling (GMTS) algorithm.

4.2.1. State

By dividing the time into multiple time slots, the traffic demand matrix within each time slot can be denoted as $TR = [b_1, b_2, \cdots, b_M]$, where b_i denotes the bandwidth between the ith pair of nodes. According to Equation (5), the throughput of the jth link is f_j. Consequently, the residual bandwidth of the ith link, denoted by c_i, can be expressed in terms of the link bandwidth BW:

$$c_i = BW - f_i, i \in [1, L] \quad (23)$$

Given that the maximum number of hops between pairs of nodes is $H_{\max}$, the jth path between the ith pair of nodes can be designated as $p_{i,j} = \{e_1, e_2, \cdots, e_{H_{\max}}\}$, and the corresponding path matrix can be designated as $P_t = [p_{1,1}, \cdots, p_{1,L}, \cdots, p_{M,L}]$. Therefore, the state vector at time slot t can be expressed as follows:

$$\mathbf{s}_t = [C_t, TR_t, G_t, P_t], \mathbf{s}_t \in S \quad (24)$$

where S is the satellite network state space.

4.2.2. Action

Based on the information in the state space, NCC splits the network traffic between each node pair onto L candidate paths which are precomputed. Therefore, the action is defined as a vector $\mathbf{a}_t$ consisting of the proportion of traffic on the candidate paths, which can be expressed as follows:

$$\mathbf{a}_t = \left[\omega_{1,1}^t, \cdots, \omega_{1,L}^t, \cdots, \omega_{M,L}^t\right] \tag{25}$$

where the proportion of flow on each path should satisfy Equation (2).

4.2.3. Reward

After executing action $\mathbf{a}_t$ under the state $\mathbf{s}_t$ at time t, the agent will gain feedback to evaluate the effectiveness of the action. In multipath traffic scheduling, the reward is the objective function of subproblem P2, which can be expressed as follows:

$$r_t = \sum_{m=1}^{M} \sum_{l=1}^{L} U(\bar{f}_{m,l}, \bar{d}_{m,l}) \tag{26}$$

where $\bar{f}_{m,l}$ is the average throughput on the lth path between the mth pair of nodes and $\bar{d}_{m,l}$ is the average delay on the lth path between the mth pair of nodes. The agent will seek to complete the multipath traffic scheduling with a higher throughput and lower latency. Since the decision making is divided into multiple phases, the cumulative reward R_t, based on a defined single momentary reward, can be expressed as follows:

$$R_t = \sum_{k=0}^{\infty} \rho^k r_{t+k+1} \tag{27}$$

where $\rho \in [0,1]$ is the discount factor. Given that there are multiple possible actions in each state, the average gain for each state and the average gain for each state action can be expressed as follows:

$$\Gamma_\mu(s) = E_\mu(G_t | \mathbf{s}_t = s) \tag{28}$$

$$\Psi_\mu(s,a) = E_\mu(G_t | \mathbf{s}_t = s, \mathbf{a}_t = a) \tag{29}$$

Here, $\Psi_\mu(s,a)$ is maximized when the agent of the DRL finds the optimal policy μ^*, converting the process of updating $\Psi_\mu(s,a)$ with the number of iterations to the Bellman form; in other words, we have

$$\Psi_\mu(s,a) = E_\mu\left[r_t + \rho \Psi_\mu(\mathbf{s}_{t+1}, \mathbf{a}_{t+1}) \middle| \mathbf{s}_t = s, \mathbf{a}_t = a\right] \tag{30}$$

After the DRL converges to a stable level, the corresponding optimal action a_{μ^*} can be expressed as follows:

$$a_{\mu^*} = \arg\max_a \Psi_{\mu^*}(s,a) \tag{31}$$

4.3. *Training Process of GMTS*

The pseudo-code for the GMTS training process is shown in Algorithm 2 with the computational complexity of $O(L)$. The initialization of $\mu(s|\theta^\mu)$ and $Q(s,a|\theta^Q)$ is first required, where θ^μ and θ^Q are hyperparameters of the actor network and critical network, respectively. During the training phase, the experience replay buffer is employed to store environmental interaction data, including states, actions, and rewards. In order to prevent overfitting, the generated samples $(s_t, a_t, r_t, s_{t+1}, \chi)$ are stored in the experience replay buffer at each decision step. At this point, $\chi \in \{0,1\}$ is used to determine whether the training of the model is complete in the current phase. The sample $(s_i, a_i, r_i, s_{i+1}, \chi)$ will be extracted at specified intervals in accordance with the playback buffer threshold B_{max} with the objective of training the network. At this point, the target Q value can be expressed as follows:

$$y_i = r_i + \rho(1-\chi)\left[Q\left(s_{t+1}, a_{t+1}|\theta^Q\right)\right] \tag{32}$$

Algorithm 2: Training process of the GNN-based multipath traffic scheduling (GMTS).

Input: B_{max}, critic network, actor network, and target network
Output: a_{μ^*}
for $i = 1 : I$ **do**
  Reset the initial state as s_0
  for $t = 1 : T$ **do**
    Compute the action a_t, reward r_t, and new state s_{t+1}
    Store $(s_t, a_t, r_t, s_{t+1}, \chi)$ in the replay buffer
    Update state s_t to s_{t+1} **while** $|B| \leq B_{max}$ **do**
      for $j = 1 : N_B$ **do**
        Calculate p_i according to Equation (34)
        Calculate ω_i according to Equation (36)
        Calculate the target Q value y_i according to Equation (32)
        Update the probability of sampling transition according to Equation (35)
      end
    end
    Calculate the loss of the critic network and actor network according to Equations (38) and (39)
    Update the parameter of actor network and critic network based on the gradient descending approach.
  end
end

In the GMTS algorithm, the temporal difference error (TD error) is employed as a means of gauging the significance of transitions. The TD error between the target Q value and the current Q value can be expressed as follows:

$$\delta_i = \left|y_i - Q\left(s_i, a_i|\theta^Q\right)\right| \tag{33}$$

The corresponding proportional prioritization is

$$p_i = \delta_i + \eta \tag{34}$$

where η denotes an extremely small amount and η is added so that the samples can be sampled even when the TD error is zero. The stochastic sampling method is adopted, and the probability of sampling transition i is

$$P(i) = \frac{p_i^\alpha}{\sum_k p_k^\alpha} \tag{35}$$

where α determines how much prioritization is used. When $\alpha = 0$, then the sampling method is degraded to random sampling. The importance sample (IS) weights of $(s_i, a_i, r_i, s_{i+1}, \chi)$ are

$$\omega_i = \left(\frac{P(i)}{P_{min}}\right)^\beta \tag{36}$$

where β is used for biased and unbiased control. Given the learning rate ϑ^Q, the actor network θ^Q can be updated in the following way:

$$\theta^Q \leftarrow \theta^Q + \vartheta^Q \delta_i \nabla_{\theta^Q} Q\left(s, a|\theta^Q\right) \tag{37}$$

Therefore, the loss function of the critic network and actor network can be expressed as follows:

$$L_q = \frac{1}{N_B} \sum_{i=1}^{B_{\max}} \omega_i \delta_i^2 \tag{38}$$

$$L_\mu = -Q\left(s, a|\theta^Q\right) \tag{39}$$

where N_B is the size of the replay buffer B.

4.4. Workflow of DMR

The DRL-based multipath routing (DMR) proposed in this paper can be divided into two phases, minimum hop count-based multipath path discovery (MHMRD) and GNN-based multipath traffic scheduling (GMTS), as shown in Figure 5. In the first stage, the network control center first plans multiple optimal paths between arbitrary node pairs based on the minimum hop count principle according to the network topology. Specifically, based on the periodicity and predictability of the satellite orbits, the network is divided into multiple time slices according to time slot intervals, and the network topology remains stable in each discrete time slice. Within each time slot, MHMRD calculates L minimum hop count available paths for each service based on the priority order defined by the MHMRD based on the end-to-end service volume. After completing the calculation for the current timeslot, it moves to the next timeslot. In the second phase, the network control center generates the split ratios for each service on different paths based on the model obtained from GMTS training and sends them to the corresponding satellites at the appropriate time before the start of each time slot.

On the one hand, this paper adopts priority experience replay to speed up the convergence of the model by sampling the samples in the experience replay pool according to a certain priority. On the other hand, due to the dynamic nature of satellite networks, the network topology and traffic matrices are time-varying, and traditional DNNs cannot effectively handle such changing inputs. Therefore, in this paper, we adopt GNN for variable size graph structures and route configurations accordingly. Since GNNs have better generalization capabilities for dynamic network structures at different scales, they can be used to aggregate elementary features without specifying the input dimensions.

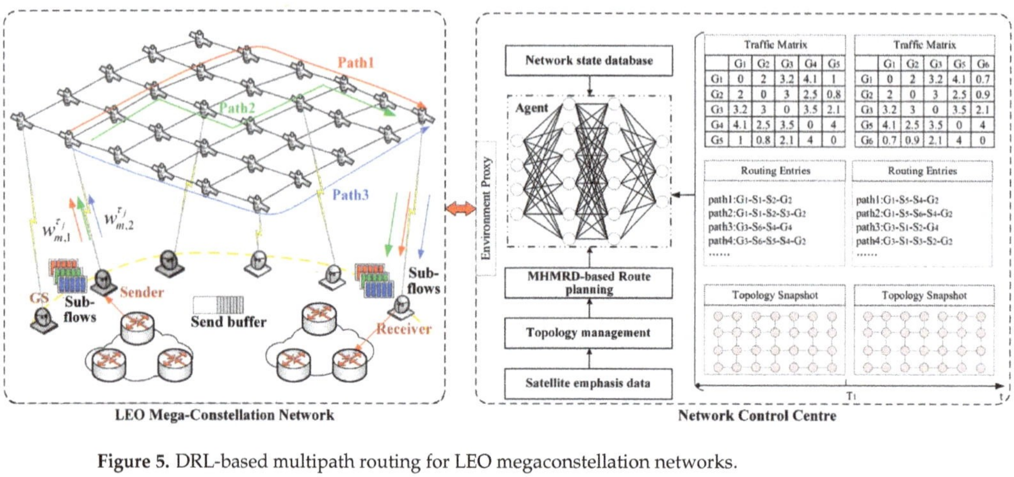

Figure 5. DRL-based multipath routing for LEO megaconstellation networks.

5. Performance Evaluation

5.1. Simulation Set-Up

In this paper, we implement routing computation and multipath traffic scheduling for LEO megaconstellations in NS3 to analyze the performance of the proposed DMR algorithm. In order to analyze the scalability of multipath routing on different constellations, this paper conducts tests on two different-sized constellations. The first constellation was the Iridium constellation [35] with 66 polar orbiting satellites, and the second was the OneWeb constellation [36] with 648 satellites, which contains a total of 18 orbital planes with 36 satellites evenly distributed on each orbital plane. The parameters of the two constellations are shown in Table 1. In order to validate the performance of multipath routing under different sizes of traffic demand, a traffic dataset was generated based on the ground traffic density, where the number of source-destination node pairs was 50. The proposed GMTS algorithm was implemented based on Python 3.9 and Pytorch 1.14, and the corresponding parameters are shown in Table 2.

Table 1. Constellation parameters.

Constellation	N_P	N_S	N_G	ε	h	E
Iridium	6	11	16	90°	780 km	20°
OneWeb	18	36	16	53°	550 km	20°

Table 2. Training parameters.

	Parameter	Value
Network parameter	N_g	24
	L_q	100
	∇_{pak}	1 KB
Learning parameter	N_B	32
	B_{max}	100
	α	0.6
	β	0.5
	ρ	0.8

The performance of the proposed minimum hop count multipath path discovery-based (MHMRD) algorithm, which distributes end-to-end traffic in equal proportions to precomputed multiple paths, was tested first. Based on this, the GMTS algorithm was trained using inclined orbit constellation topology and traffic matrices [37], which were deployed to the NCC after training and tested for key metrics such as the average throughput, latency, and flow completion rate of the satellite network. The following comparison algorithms were used in this paper:

- Shortest path first algorithm (SPF): SPF employs the average transmission delay as a link metric with the objective of minimizing the total delay.
- Network coding-based multipath cooperative routing (NCMCR) [24]: NCMCR is designed to address the routing challenges posed by a frequently changing topology and potentially sparse and intermittent connectivity. The NCMCR algorithm takes advantage of the predictability of the relative motion of the satellites, with the time-varying topology modeled as a spatiotemporal map.
- Ant-based multipath backbone routing for load balancing (AMBRLB) [38]: In order to overcome the issues of traffic overflow and overhead, an ant-based load balancing multipath backbone routing algorithm in MANET was proposed. Upon the initiation of transmission by a source node to a destination node, the ant colony optimization (ACO) algorithm is employed to identify multiple paths with the highest probability of success.

- Deep deterministic policy gradient traffic engineering (DDPG-TE) [39]: This method employs the DDPG algorithm to dynamically allocate the traffic proportions of different transmission paths. It should be noted that the model is not applicable to arbitrary satellite constellations, as the dimensions of the state and action spaces are constrained by the size of the input topology and traffic matrices.

5.2. Results and Analysis

5.2.1. Throughput

The average throughput of the MCN over the Iridium constellation and OneWeb constellation are illustrated in Figure 6, with each data point representing the average value of the throughput under five consecutive topological snapshots. It can be observed that, in the context of multipath routing, as the input traffic volume increased, a greater proportion of traffic was allocated to multiple paths for forwarding, based on different policies. This resulted in an overall increase in the average network throughput. Nevertheless, the increase in throughput gradually slowed down. The proposed DMR algorithm exhibited a superior throughput at varying user sizes. DMR enhanced the average throughput by 30.26% and 8.45% in comparison with SPF and NCMCR, respectively, when the total service volume reached its upper limit. Figure 6b illustrates the throughput of multipath routing on the OneWeb constellation. It can be observed that DMR employed the same model as that used for the Iridium constellation, while the other algorithms were retrained based on the characteristics of the new network. It can be observed that the DMR algorithm continued to demonstrate superior performance relative to the other benchmark algorithms in the context of changes in the constellation topology and size, reflecting the enhanced generalization ability of DMR. When the traffic intensity was 8 Gbps, DMR demonstrated an improvement in the average throughput of 42.64% and 9.55% in comparison with SPF and NCMCR, respectively.

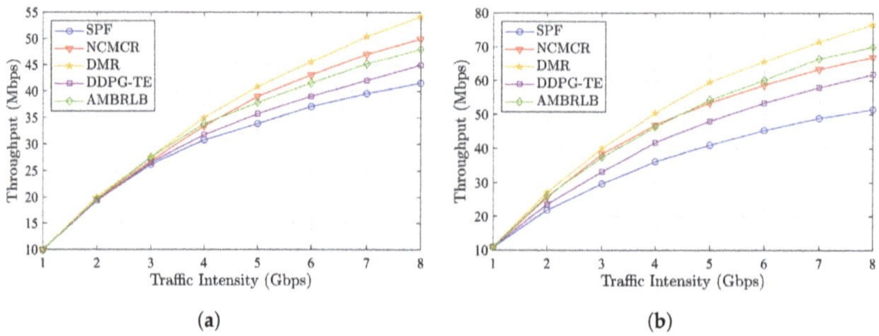

Figure 6. Average throughput on different constellations with various traffic intensity. (**a**) Average throughput for Iridium. (**b**) Average throughput for OneWeb.

5.2.2. Average Flow Completion Ratio

Figure 7 illustrates the average flow completion ratio for varying traffic intensities on the Iridium and OneWeb constellations. As the intensity of traffic increased, some nodes in the network gradually became congested. This is evidenced by a decreasing trend in the average flow completion ratios of all five routing schemes. At a traffic intensity of 8 Gbps, the average flow completion ratio of the proposed DMR scheme was enhanced by 17.39% and 11.52% on the Iridium and OneWeb constellations, respectively, in comparison with SPF. Meanwhile, on the OneWeb constellation, the average stream completion rate of the DMR algorithm improved by 1.98%, 2.58%, and 6.49% compared with NCMCR, AMBRLB, and DDPG-TE, respectively. Consequently, it can be seen that the average flow completion ratio on the larger OneWeb constellation was consistently higher than that on the smaller Iridium constellation, even under varying traffic intensities. This demonstrates the advantage of large-scale constellations in handling high loads. It can therefore be concluded that the proposed DMR scheme demonstrates superior generalization ability across different scale constellations, with a higher flow completion ratio than the baseline scheme.

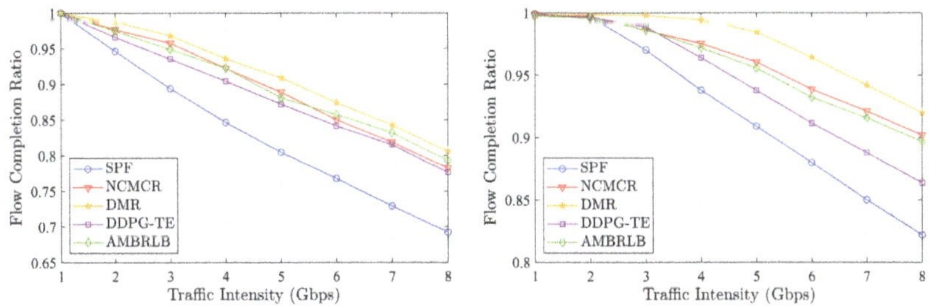

(a) (b)

Figure 7. Average flow completion ratio of different constellations with various traffic intensity. (**a**) Average flow completion ratio for Iridium constellation. (**b**) Average flow completion ratio for OneWeb constellation.

5.2.3. Average End-to-End Delay

Figure 8 illustrates the average end-to-end delay as a function of the traffic intensity on the Iridium and OneWeb constellations. As the network load increased, the queuing delay also rose, with more flows transmitted along the candidate paths with higher hop counts. Consequently, the average end-to-end delay gradually increased for all of the compared schemes. In the Iridium constellation, SPF achieved the lowest end-to-end delay when the network load was low. This is due to the fact that the scheme always transmits the packets along the shortest path when all the links in the network are in a normal state. As the network load increased, when the total service volume reaches the upper bound, SPF is unable to dynamically change the path to avoid congested paths caused by the increased load. Consequently, on the Iridium constellation, the DMR scheme reduced the average end-to-end delay by 3.67%, 1.78%, 1.86%, and 0.86% compared with SPF, NCMCR, DDPG-TE, and AMBRLB, respectively. Upon extending multipath routing to larger OneWeb constellations, the delay for DMR approached that of NCMCR, AMBRLB, and DDPG-TE when the traffic intensity reached its upper limit. This is due to the fact that both the DMR scheme and the DDPG-TE scheme take into account the dynamic distribution of traffic across different paths. In general, the mean end-to-end delay of DMR at a high load was situated between those of SPF and DDPG-TE. The primary objective of multipath routing is to enhance the network's throughput and resilience when the topology is disrupted. In small-scale constellations, multipath routing offers additional routing options which can be employed to circumvent congested nodes in a timely manner, thereby reducing the delay. However, in large-scale constellations, there are always sufficient alternative routes, and the improvement in end-to-end delay metrics is not substantial.

The experimental results demonstrate that the SPF scheme with the shortest path exhibited a lower average delay in low-load scenarios. However, SPF exhibited the lowest average throughput and average flow completion ratio, and the performance of SPF deteriorated rapidly under high-load conditions. In contrast, the proposed DMR scheme exhibited a better average throughput and flow completion ratio under different loads, and it exhibited a better generalization ability over constellations of different sizes and topologies without repeated training.

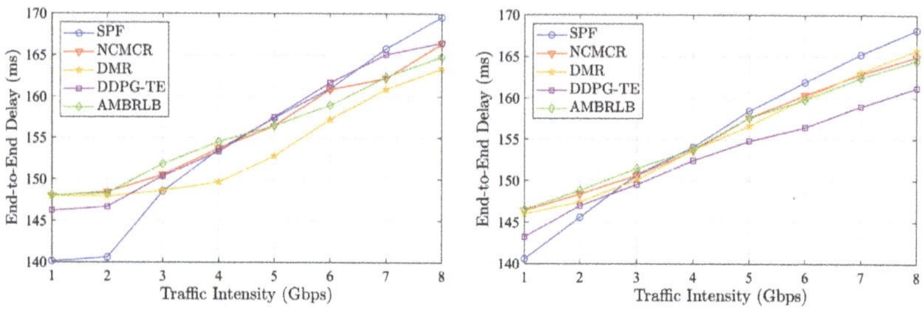

(a) (b)

Figure 8. Average end-to-end delay for different constellations with various traffic intensities. (**a**) Average end-to-end delay for Iridium constellation. (**b**) Average end-to-end delay for OneWeb constellation.

6. Conclusions

This paper proposed a deep reinforcement learning (DRL)-based multipath routing (DMR) solution for satellite networks. The objective was to address the multipath routing transmission problem in LEO megaconstellation networks. The proposed DMR approach split multipath routing into two subproblems: multipath route selection and multipath traffic scheduling. In the context of the multipath path selection subproblem, this paper proposed a centralized routing scheme, MHMRD, which generates a set of available paths for end-to-end nodes. This is achieved by collecting node state information based on the current end-to-end minimum hop count and link utilization. In the multipath traffic scheduling subproblem, GNN-based multipath traffic scheduling (GMTS) is proposed. Each routing node is controlled by a DQN agent, and the GNN is used to address the issue of topologically varying satellite network topologies. The optimal multipath traffic scheduling model was obtained through iterative training. The proposed DMR scheme was finally validated on two different sizes of megaconstellations in the NS3 simulation environment. The simulation results demonstrate that the proposed DMR scheme exhibited superior performance in terms of the average network throughput and average flow completion ratio compared with the baseline scheme. Furthermore, the DMR scheme effectively addressed satellite network routing under load balancing while maintaining minimal increases in delay costs. Furthermore, the DMR scheme devised in this paper exhibited favorable scalability and adaptability in the context of evolving constellation topologies. It is also capable of being extended to satellite networks with changing topologies without the necessity for additional training.

It should be noted that as the size of the constellation continues to expand, the computational requirements of DQN-based online routing also rise. In the future, we intend to explore the potential of adopting solutions such as multi-controller deployment to enhance the responsiveness of the network and further validate its efficacy in real-world scenarios.

Author Contributions: Conceptualization, C.H.; methodology, C.H.; software, R.Y.; validation, W.X.; formal analysis, C.H.; investigation, W.X.; resources, W.X.; data curation, C.H.; writing—original draft preparation, C.H.; writing—review and editing, C.H.; visualization, C.H.; supervision, W.X.; project administration, W.X. and R.Y.; funding acquisition, W.X. and R.Y. All authors have read and agreed to the published version of the manuscript.

Funding: This research was funded by the National Key Laboratory Fund Project under grant number 614201003022207 and the Aerospace Discipline Education New Engineering Project under grant number 145AXL250004000X.

Data Availability Statement: The relevant data and source code of this paper can be accessed by contacting hanchi@hgd.edu.cn.

Conflicts of Interest: The authors declare no conflicts of interest.

Abbreviations

The following abbreviations are used in this manuscript:

DRL	Deep reinforcement learning
DMR	DRL-based multipath routing
MHMRD	Minimum hop count-based multipath routing discovery algorithm
GNN	Graph neural network
GMTS	GNN-based multipath traffic scheduling
MCN	Megaconstellation network
SAGIN	Space-air-ground integrated network
ISLs	Inter-satellite links
SDN	Software-defined network
RAAN	Right ascension of ascending node
NCC	Network control center
LEO	Low Earth orbit
DQN	Deep Q network

References

1. Jiang, X.; Huang, Y.; Li, J.; He, H.; Chen, S.; Yang, F.; Yang, J. Spatio-Temporal Routing, Redundant Coding and Multipath Scheduling for Deterministic Satellite Network Transmission. *IEEE Trans. Commun.* **2023**, *71*, 2860–2875. [CrossRef]
2. Kodheli, O.; Lagunas, E.; Maturo, N.; Sharma, S.K.; Shankar, B.; Montoya, J.F.M.; Duncan, J.C.M.; Spano, D.; Chatzinotas, S.; Kisseleff, S.; et al. Satellite Communications in the New Space Era: A Survey and Future Challenges. *IEEE Commun. Surv. Tutor.* **2021**, *23*, 70–109. [CrossRef]
3. Rakhmanov, A.; Wiseman, Y. Compression of GNSS Data with the Aim of Speeding up Communication to Autonomous Vehicles. *Remote Sens.* **2023**, *15*, 2165. [CrossRef]
4. Correia, S.D.; Perez, R.; Matos-Carvalho, J.; Leithardt, V.R.Q. μJSON, a Lightweight Compression Scheme for Embedded GNSS Data Transmission on IoT Nodes. In Proceedings of the 2022 5th Conference on Cloud and Internet of Things (CIoT), Marrakech, Morocco, 28–30 March 2022; pp. 232–238.
5. Xue, C.; Li, W.; Yu, L.; Shang, J.; Chen, X.; Lu, S. SERO: A Model-Driven Seamless Roaming Framework for Wireless Mesh Network With Multipath TCP. *IEEE Trans. Commun.* **2019**, *67*, 1284–1296. [CrossRef]
6. Liu, X.; Ma, T.; Qin, X.; Zhou, H.; Zhao, L. A DRL Empowered Multipath Cooperative Routing for Ultra-Dense LEO Satellite Networks. In Proceedings of the GLOBECOM 2023, Kuala Lumpur, Malaysia, 4–8 December 2023; pp. 5961–5966.
7. Li, R.; Zhang, J.; Zheng, S.; Wang, K.; Wang, P.; Zhang, X. LEO Mega-Constellations Routing Algorithm Based on Area Segmentation. In Proceedings of the 2023 IEEE Wireless Communications and Networking Conference (WCNC), Glasgow, UK, 26–29 March 2023; pp. 1–6.
8. Li, Y.; Zhang, Q.; Yao, H.; Gao, R.; Xin, X.; Yu, F.R. Stigmergy and Hierarchical Learning for Routing Optimization in Multi-Domain Collaborative Satellite Networks. *IEEE J. Select. Areas Commun.* **2024**, *42*, 1188–1203. [CrossRef]
9. Perkins, C.E.; Royer, E.M. Ad-Hoc on-Demand Distance Vector Routing. In Proceedings of the WMCSA'99. Second IEEE Workshop on Mobile Computing Systems and Applications, New Orleans, LA, USA, 25–26 February 1999; pp. 90–100.
10. Jiang, F.; Zhang, Q.; Yang, Z.; Yuan, P. A Space–Time Graph Based Multipath Routing in Disruption-Tolerant Earth-Observing Satellite Networks. *IEEE Trans. Aerosp. Electron. Syst.* **2019**, *55*, 2592–2603. [CrossRef]
11. Huang, Y.; Yang, D.; Feng, B.; Tian, A.; Dong, P.; Yu, S.; Zhang, H. A GNN-Enabled Multipath Routing Algorithm for Spatial-Temporal Varying LEO Satellite Networks. *IEEE Trans. Veh. Technol.* **2024**, *73*, 5454–5468. [CrossRef]
12. Huang, Y.; Jiang, X.; Chen, S.; Yang, F.; Yang, J. Pheromone Incentivized Intelligent Multipath Traffic Scheduling Approach for LEO Satellite Networks. *IEEE Trans. Wirel. Commun.* **2022**, *21*, 5889–5902. [CrossRef]
13. Tian, A.; Feng, B.; Zhou, H.; Huang, Y.; Sood, K.; Yu, S.; Zhang, H. Efficient Federated DRL-Based Cooperative Caching for Mobile Edge Networks. *IEEE Trans. Netw. Serv. Manag.* **2023**, *20*, 246–260. [CrossRef]
14. Huang, X.; Yuan, T.; Qiao, G.; Ren, Y. Deep Reinforcement Learning for Multimedia Traffic Control in Software Defined Networking. *IEEE Netw.* **2018**, *32*, 35–41. [CrossRef]
15. Lei, K.; Guo, P.; Wang, Y.; Wu, X.; Zhao, W. Solve Routing Problems with a Residual Edge-Graph Attention Neural Network. *Neurocomputing* **2022**, *508*, 79–98. [CrossRef]
16. Zheng, X.; Huang, W.; Li, H.; Li, G. Research on Generalized Intelligent Routing Technology Based on Graph Neural Network. *Electronics* **2022**, *11*, 2952. [CrossRef]
17. Wei, H.; Zhao, Y.; Xu, K. G-Routing: Graph Neural Networks-Based Flexible Online Routing. *IEEE Netw.* **2023**, *37*, 90–96. [CrossRef]
18. Rusek, K.; Suarez-Varela, J.; Almasan, P.; Barlet-Ros, P.; Cabellos-Aparicio, A. RouteNet: Leveraging Graph Neural Networks for Network Modeling and Optimization in SDN. *IEEE J. Select. Areas Commun.* **2020**, *38*, 2260–2270. [CrossRef]
19. Bi, Y.; Han, G.; Xu, S.; Wang, X.; Lin, C.; Yu, Z.; Sun, P. Software Defined Space-Terrestrial Integrated Networks: Architecture, Challenges, and Solutions. *IEEE Netw.* **2019**, *33*, 22–28. [CrossRef]

20. Hu, M.; Xiao, M.; Hu, Y.; Cai, C.; Deng, T.; Peng, K. Software Defined Multicast Using Segment Routing in LEO Satellite Networks. *IEEE Trans. Mob. Comput.* **2024**, *23*, 835–849. [CrossRef]
21. Han, Z.; Xu, C.; Zhao, G.; Wang, S.; Cheng, K.; Yu, S. Time-Varying Topology Model for Dynamic Routing in LEO Satellite Constellation Networks. *IEEE Trans. Veh. Technol.* **2023**, *72*, 3440–3454. [CrossRef]
22. Giambene, G.; Luong, D.K.; De Cola, T.; Le, V.A.; Muhammad, M. Analysis of a Packet-Level Block Coding Approach for Terrestrial-Satellite Mobile Systems. *IEEE Trans. Veh. Technol.* **2019**, *68*, 8117–8132. [CrossRef]
23. Chen, Z.; Zhou, W.; Wu, S.; Cheng, L. An Adaptive On-Demand Multipath Routing Protocol With QoS Support for High-Speed MANET. *IEEE Access* **2020**, *8*, 44760–44773. [CrossRef]
24. Tang, F.; Zhang, H.; Yang, L.T. Multipath Cooperative Routing with Efficient Acknowledgement for LEO Satellite Networks. *IEEE Trans. Mob. Comput.* **2019**, *18*, 179–192. [CrossRef]
25. Li, G.; Zhou, H.; Feng, B.; Zhang, Y.; Yu, S. Efficient Provision of Service Function Chains in Overlay Networks Using Reinforcement Learning. *IEEE Trans. Cloud Comput.* **2022**, *10*, 383–395. [CrossRef]
26. Kato, N.; Fadlullah, Z.M.; Mao, B.; Tang, F.; Akashi, O.; Inoue, T.; Mizutani, K. The Deep Learning Vision for Heterogeneous Network Traffic Control: Proposal, Challenges, and Future Perspective. *IEEE Wirel. Commun.* **2017**, *24*, 146–153. [CrossRef]
27. Mao, B.; Fadlullah, Z.M.; Tang, F.; Kato, N.; Akashi, O.; Inoue, T.; Mizutani, K. Routing or Computing? The Paradigm Shift Towards Intelligent Computer Network Packet Transmission Based on Deep Learning. *IEEE Trans. Comput.* **2017**, *66*, 1946–1960. [CrossRef]
28. Geyer, F.; Carle, G. Learning and Generating Distributed Routing Protocols Using Graph-Based Deep Learning. In Proceedings of the 2018 Workshop on Big Data Analytics and Machine Learning for Data Communication Networks, Budapest, Hungary, 20 August 2018; pp. 40–45.
29. Liu, C.; Xu, M.; Yang, Y.; Geng, N. DRL-OR: Deep Reinforcement Learning-Based Online Routing for Multi-Type Service Requirements. In Proceedings of the IEEE INFOCOM 2021—IEEE Conference on Computer Communications, Vancouver, BC, Canada, 10–13 May 2021; pp. 1–10.
30. Almasan, P.; Suárez-Varela, J.; Rusek, K.; Barlet-Ros, P.; Cabellos-Aparicio, A. Deep Reinforcement Learning Meets Graph Neural Networks: Exploring a Routing Optimization Use Case. *Comput. Commun.* **2022**, *196*, 184–194. [CrossRef]
31. Sun, P.; Guo, Z.; Li, J.; Xu, Y.; Lan, J.; Hu, Y. Enabling Scalable Routing in Software-Defined Networks With Deep Reinforcement Learning on Critical Nodes. *IEEE/ACM Trans. Netw.* **2022**, *30*, 629–640. [CrossRef]
32. Kim, G.; Kim, Y.; Lim, H. Deep Reinforcement Learning-Based Routing on Software-Defined Networks. *IEEE Access* **2022**, *10*, 18121–18133. [CrossRef]
33. Jiang, Z.; Wu, Q.; Li, H.; Wu, J. scMPTCP: SDN Cooperated Multipath Transfer for Satellite Network With Load Awareness. *IEEE Access* **2018**, *6*, 19823–19832. [CrossRef]
34. De Santis, E.; Giuseppi, A.; Pietrabissa, A.; Capponi, M.; Delli Priscoli, F. Satellite Integration into 5G: Deep Reinforcement Learning for Network Selection. *Mach. Intell. Res.* **2022**, *19*, 127–137. [CrossRef]
35. Iridium Constellation. Available online: https://celestrak.org/NORAD/elements/table.php?GROUP=iridium (accessed on 30 July 2024).
36. Oneweb Constellation. Available online: https://celestrak.org/NORAD/elements/table.php?GROUP=oneweb (accessed on 30 July 2024).
37. MAWI Working Group Traffic Archive. Available online: https://mawi.wide.ad.jp/mawi/ (accessed on 30 July 2024).
38. Francis Antony Selvi, P.; Manikandan, M.S.K. Ant Based Multipath Backbone Routing for Load Balancing in MANET. *IET Commun.* **2017**, *11*, 136–141. [CrossRef]
39. Chen, B.; Sun, P.; Zhang, P.; Lan, J.; Bu, Y.; Shen, J. Traffic Engineering Based on Deep Reinforcement Learning in Hybrid IP/SR Network. *China Commun.* **2021**, *18*, 204–213. [CrossRef]

electronics

MDPI

Article

Performance Evaluation of Routing Algorithm in Satellite Self-Organizing Network on OMNeT++ Platform

Guoquan Wang, Jiaxin Zhang *, Yilong Zhang, Chang Liu and Zhaoyang Chang

Key Laboratory of Universal Wireless Communication, School of Information and Communication Engineering, Beijing University of Posts and Telecommunications, Ministry of Education, Beijing 100876, China; gqwang@bupt.edu.cn (G.W.); ylzhang@bupt.edu.cn (Y.Z.); 2019210087@bupt.edu.cn (C.L.); changczy@bupt.edu.cn (Z.C.)

* Correspondence: jxz@bupt.edu.cn

Abstract: Self-organizing networks of small satellites have gradually gained attention in recent years. However, self-organizing networks of small satellites have high topological change frequency, large transmission delay, and complex communication environments, which require appropriate networking and routing methods. Therefore, this paper, considering the characteristics of satellite networks, proposes the shortest queue length-cluster-based routing protocol (SQL-CBRP) and has built a satellite self-organizing network simulation platform based on OMNeT++. In this platform, functions such as the initial establishment of satellite self-organizing networks and cluster maintenance have been implemented. The platform was used to verify the latency and packet loss rate of SQL-CBRP and to compare it with Dijkstra and Greedy Perimeter Stateless Routing (GPSR). The results show that under high load conditions, the delay of SQL-CBRP is reduced by up to 4.1%, and the packet loss rate is reduced by up to 7.1% compared to GPSR. When the communication load is imbalanced among clusters, the delay of SQL-CBRP is reduced by up to 12.7%, and the packet loss rate is reduced by up to 16.7% compared to GPSR. Therefore, SQL-CBRP performs better in networks with high loads and imbalance loads.

Keywords: satellite self-organizing; routing algorithm; OMNeT++

1. Introduction

With the popularization and development of the Internet, the Internet of Things, and mobile terminals, people's demand for accessing the network anytime and anywhere is becoming increasingly urgent. Satellite networks, due to their wide coverage, large communication capacity, and reliable and secure characteristics, play an increasingly important role in communication and have attracted extensive attention from academia and industry [1]. Traditional "all-in-one" satellites are expensive to build, and their launch technology is complex, making them unsuitable for mass production and launch. Compared to "all-in-one" satellites, small satellites have the advantages of low cost, simple launch and use, and high flexibility [2]. However, one small satellite has a weaker performance in power and computing capabilities. To address this issue, the concept of wireless self-organizing networking can be combined with satellite communication to form a satellite self-organizing network by combining multiple small satellites. In this model, when a task arrives, satellites within a certain area can construct a network and complete the task together like a large "virtual satellite". In this way, multiple small satellites can integrate resources, improving overall communication and computing capabilities to complete tasks successfully [3].

Currently, satellite self-organizing networks can be categorized into three types based on the spatial distribution of satellites: self-organized single constellation, self-organized satellite formation, and self-organized hybrid mega-constellation [4,5].

Self-organized single constellations are formed by temporarily integrating satellites within one constellation according to the scale of the mission and the coverage area required, to jointly complete tasks through self-organized networking. In this scenario, satellites are isomorphic, with the same functionality and payload.

Self-organized satellite formation integrates satellites within a formation to create a network. There are two types of satellites in a satellite formation based on the function and payload: the leader satellite and the follower satellite. The follower satellites move around the leader satellite and have inter-satellite links with the leader satellite [6]. The leader satellite has stronger computing and communication capability and a more complete protocol stack, hence it often serves as the control node and network node for the formation. Data generated by the follower satellites are all forwarded through the leader satellite.

Self-organized hybrid mega-constellations are constructed by satellites with different functions distributed at different orbital altitudes. Similar to a single constellation self-organizing network, this type of network selects multiple satellites to form a self-organizing network to jointly complete tasks based on mission requirements. Due to the dense distribution of satellites in a heterogeneous mega-constellation and their placement at various altitudes, a single satellite may establish more than just four inter-satellite links. For example, Yan et al. proposed the Grid+ type inter-satellite link connection strategy, which allows a satellite to establish more than four inter-satellite links [7].

Despite the advantage small satellites have, the frequent topological changes, high transmission delays, and complex communication environment in networks of small satellites pose challenges to satellite self-organizing networks, making the routing schemes applied in traditional terrestrial networks no longer suitable for satellite self-organizing networks. In detail, The high-speed movement of satellites leads to frequent link switching and topology updates. If the routing strategies of terrestrial networks are applied to satellite self-organizing network directly, it would result in significant signaling overhead when re-building a routing table, occupying limited communication resources and affecting the performance of the network, especially in large-scale networks [5]. The frequent topology changes also lead to a high packet loss rate because of the outdated routing table. Additionally, the complex and harsh space environment, including multipath fading, the ionosphere, atmosphere, and solar activity, also severely impact inter-satellite links, potentially leading to the destruction of the links and even the nodes themselves [8]. Additionally, owing to the long-distance transmission between satellites, the delay of transmission between satellites is high [9].

Therefore, this paper proposes a shortest queue length-cluster-based routing protocol (SQL-CBRP) for the satellite self-organizing network scenario. This algorithm is based on a cluster-based network structure, dividing routing into inter-cluster and intra-cluster routing. Intra-cluster routing is forwarded by the cluster head, while inter-cluster routing selects the node with the shortest queue length among neighboring cluster heads as the next hop. This effectively avoids packet loss due to queue overflow (i.e., a new packet arrives when the queue is full) and reduces queuing delay. Furthermore, to prevent network collapse caused by the destruction of the cluster head, a sub-head is established. When the cluster head is destroyed, the sub-head can take over the original cluster head's role in data forwarding and cluster management. Additionally, this paper has constructed a satellite self-organizing network simulation platform based on OMNeT++ and has validated the performances of SQL-CBRP on this platform, comparing it with Dijkstra's algorithm and the GPSR (Greedy Perimeter Stateless Routing) algorithm.

The main works of this paper are as follows:

1. A novel routing strategy for satellite self-organizing network, shortest queue length-cluster-based routing protocol is proposed, which considers nodes' queue length when making routing decisions;
2. A satellite self-organizing network simulation platform is built based on OMNeT++. In this platform, some functions are designed and implemented, including initial network forming, cluster maintenance, data generation and routing, etc.;
3. The delay and packet loss performance of the proposed algorithm are tested and compared with GPSR and Dijkstra.

2. Related Works

The routing technologies currently applied to wireless self-organizing networks can generally be divided into three types: topology-based routing algorithms, location-based routing algorithms, and cluster-based routing algorithms. These three types of algorithms and their typical algorithms are shown in Table 1.

2.1. Topology-Based Routing Algorithms

Topology-based routing algorithms make decisions based on the network's topological structure (such as the number of hops, connectivity). This type of routing strategy is divided into proactive and reactive approaches. Proactive routing algorithms require nodes to maintain a global routing table locally and generate path selection schemes based on this routing table, such as optimized link state routing (OLSR) and its improved algorithms [10]. Reactive routing algorithms, on the other hand, calculate routes on demand based on real-time node communication requests, such as ad hoc on-demand distance vector routing (AODV), and its improved algorithms [11]. Yang tested the performances of AODV and OLSR in wireless self-organizing networks. The results showed that the maximum latency of AODV is about 4 s, and the packet delivery ratio is approximate 85%; the maximum latency of OLSR is nearly 2 s, and the packet delivery ratio fluctuates between 20% and 80% [12].

Table 1. Three types of self-organizing routing algorithms.

Self-Organizing Routing Algorithm	Characteristics	Delay	Packet Delivery Rate
Topology-based routing algorithm	Route calculation is based on the network's topological structure and is divided into active and passive methods.	OLSR: up to 2 s [12] AODV: up to 4 s [12]	OLSR: 20–80% [12] AODV: about 85% [12]
Location-based routing algorithm	Routing is calculated based on the geographical location of the nodes.	GPSR and its improved algorithm: up to 0.45 s [13]	GPSR and its improved algorithm: up to 95% [13]
Cluster-based routing algorithm	Divide the network into layers and categorize routing into inter-cluster and intra-cluster.	CBRP: up to 2.2 s [14]	CRBP: higher than 90% [14]

2.2. Location-Based Routing Algorithm

The basic idea of location-based routing algorithms is to make routing decisions based on the geographical location information of nodes. Greedy Perimeter Stateless Routing (GPSR) is a typical location-based routing algorithm. In networks using GPSR, nodes do not need to maintain the topological information of the network. Instead, they only need to maintain the location information of neighboring nodes and the destination node to select the neighbor closest to the destination node as the next hop. Dong improved GPSR based on Double Deep Q-Network (D-DQN) and tested the performance of the new algorithm and GPSR in mobile self-organizing networks. The results show that the delay of GPSR and its improved routing algorithm is up to 0.45 s, and the packet delivery rate can reach 95% at maximum [13].

2.3. Cluster-Based Routing Algorithm

The cluster-based routing algorithm divides routing into intra-cluster routing and inter-cluster routing. Members within a cluster only maintain routing information within their cluster, while inter-cluster routing is decided by the cluster head or gateway through on-demand routing. In large-scale networks, this type of algorithm can effectively reduce the number of nodes involved in routing calculations and shrink the size of the routing table, thereby reducing storage overhead. Therefore, in large-scale networks, it has better performance than other algorithms [14]. For example, Cluster-Based Routing Protocol

(CBRP) confines flooding to cluster heads and gateway nodes, thereby reducing the overhead compared to flooding across all nodes in the entire network [15]. Zhang proposed the Cluster-Based On-demand Multi-path Routing Protocol (COMRP) which searches for multiple paths that meet QoS requirements through the cluster structure. The results showed that the delay of COMRP is lower than 1.5 s, while the delay of CBRP is up to 2.2 s [14]. Xu et al. proposed Efficient Clustering V2V Routing Based on PSO (Particle Swarm Optimization) in VANETs (Vehicular Ad Hoc Networks), whose packet delivery rate is higher than 90% in VANETs [16].

Although the above algorithms have solved the end-to-end routing problem in self-organizing networks through various methods, they do not consider the overhead of the nodes, which may lead to high queuing delay and high loss rate of packet. Therefore, this paper proposes a new routing algorithm, the shortest queue length-cluster-based routing protocol (SQL-CBRP), which calculates routes based on the queue length of neighbor nodes.

3. Shortest Queue Length-Cluster-Based Routing Protocol

In the traditional satellite scenario, one satellite can establish up to four inter-satellite links. But some researchers have proposed encountering inter-satellite links (eISLs), allowing satellites to communicate without predetermined inter-satellite links when they are visible to each other [17]. In this way, one satellite can establish more than four ISLs. Based on the concept of eISLs, the scenario considered in this article is as shown in Figure 1. The satellite nodes are divided into multiple cluster structures, with members within a cluster directly connected to the cluster head. The cluster heads are interconnected to form a satellite self-organizing network. Users connect to the satellite and communicate via the satellite network.

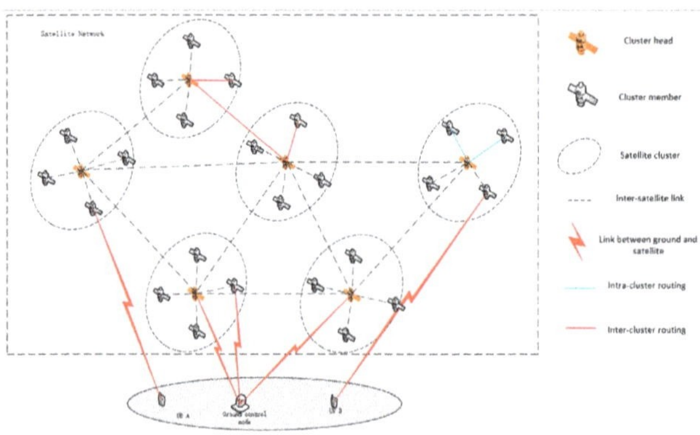

Figure 1. Satellite self-organizing network scenario.

Based on this scenario, a new routing algorithm called Shortest Queue Length-Cluster-Based Routing Protocol (SQL-CBRP) is proposed. In this algorithm, different nodes perform different functions. The specific functions are as follows:

- Cluster head: The cluster heads are generally located at the center of the cluster and are responsible for managing the cluster structure, maintaining network topology information, routing computation, and forwarding data packets;
- Cluster member: The cluster members orbit around the cluster head, directly connected to the cluster head. When a node has data to forward, it sends the data to the cluster head;
- Ground control node: The ground control node can communicate with all satellites in the network. When receiving a request to join the network from a satellite node, the ground control node assigns it to the nearest cluster head or creates a new cluster head, and allocates time slots for it;

- User: Users on the ground send data to a certain satellite. The data can be forwarded to the destination through the satellite network, and finally arrive at another user.

3.1. Initial Cluster Creation Process

This paper designs a satellite self-organizing network clustering process based on the high-rate wireless self-organizing network clustering method from reference [18]. A ground control node is introduced, and satellites send network access requests and their information (mainly position information) to the ground node. The ground node completes the division of cluster structures and the assignment of cluster heads and then sends the cluster affiliation information of each node back respectively.

The details of the clustering process executed by the ground control node are as follows:

1. The ground control node receives a network access request from a node;
2. The ground node determines whether there is a cluster head that is close enough to the node. If there is, the node is assigned to the nearest cluster head; if not, the node becomes a new cluster head;
3. After the assignment is completed, the ground node updates the local cluster structure information and ends the node's cluster allocation process.

3.2. Topology Information Maintenance Mechanism

Despite the routing decision process mainly happening in cluster heads, the source node also needs to know the cluster where the destination node is located to establish a path from the source node's cluster to the destination node's cluster. Therefore, each cluster head needs to maintain the global cluster structure information of the network. However, due to the mobility of the satellite self-organizing network, the cluster structure may change over time. Therefore, it is necessary for the cluster heads to periodically flood their connection table which contains the topology information of the cluster such as members and neighbors. The period is consistent with the cluster maintenance cycle.

The details of the topology information maintenance mechanism are as follows:

1. After the periodic cluster maintenance is completed, if the local cluster structure has been changed, the cluster head updates the connection table and sends it to neighboring cluster heads;
2. After receiving the broadcast connection table, the neighboring cluster head first checks if it is a new connection table. If it is, create a new connection table locally and update the local connection table list. If not, proceed to the next step;
3. Determine whether the connection table is the latest table of the corresponding cluster. If it is, update the local connection table list. Otherwise, discard this table;
4. Repeat the above three steps in a cycle until the connection table floods over the entire network.

3.3. Specific Routing Process of SQL-CBRP

As mentioned before, satellite self-organizing network routing faces issues such as high dynamics, high latency, and high signaling overhead. Therefore, SQL-CBRP adopts the concept of cluster-based routing, dividing the network into multiple cluster structures and separating routing into inter-cluster and intra-cluster routing. This approach can reduce the signaling overhead for network management and to some extent mitigate the impact of topology changes on routing, thereby reducing packet loss. Additionally, unlike traditional cluster-based routing protocols, SQL-CBRP does not use on-demand routing to establish paths passively. This is because in satellite networks with frequent topology changes, on-demand routing could lead to the frequent transmission of route discovery packets, which not only result in significant signaling overhead but also introduce latency in the process of searching for the routes. Accordingly, SQL-CBRP adopts a greedy algorithm to select the neighbor node with the shortest queue as the next hop, which not only eliminates the need for frequent route discovery flooding but also reduces queuing delay and packet loss due to queue overflow.

The specific routing decision process of SQL-CBRP is as follows: Every time a cluster head node u forwards a data packet, the next hop cluster head node v sends a message back to node u, informing it of its current queue length. When node u needs to forward data the next time, it compares the queue lengths sent back by all neighbor nodes last time according to the locally maintained neighbor queue length table and selects the neighbor with the shortest queue length to forward, as shown in Algorithm 1. The process of this algorithm is as follows:

1. When a cluster head node CH has a data packet to forward, it determines whether the destination node is within its cluster. If it is, the packet is sent directly to the destination node. If not, it consults the connection table;
2. If the local connection table indicates that there is a neighbor node closer to the destination node, then proceed to the next step. Otherwise, it is determined that the routing has failed;
3. Among the neighbor nodes closer to the destination node, compare their queue lengths and select the node with the shortest queue length as the next hop.

Algorithm 1 Routing Algorithm Based on Shortest Queue Length

Input: destination node *Dest*, neighbor cluster heads'set *N*

Output: *Nexthop;*

Calculate the distance between this node and the destination node *DistanceLocal*
if *Dest* is in this cluster **then**
    Nexthop <- *Dest*
else
    for each neighbor $n \in N$ **do**
        Calculate distance between neighbor *n* and destination node
    end for
    if there are neighbors closer to destination than this node **then**
        Nexthop <- id of the neighbor with shortest queue length
    else
        Nexthop <- −1
    end if
end if
return *Nexthop*

After the next hop node receives the data packet, it sends back its current queue length to node CH, and based on this, CH updates the queue length table of its neighbors. The updates of local queue length information are frequent as the cluster heads forward packets continuously, so a cluster head can determine next hop based on the queue length information updated recently, which can lead to a relevant reliable routing decision.

Although the information of nodes' queue lengths adds overhead, it does not need to be flooded over the entire network. Its transmission is limited between cluster heads, and it only needs to be sent to the last hop cluster head instead of being broadcasted. Therefore, the increased overhead is not severe. Compared to routing protocols that require global flooding of signaling, this algorithm actually reduces the overall network overhead.

4. Satellite Self-Organizing Network Simulation Platform Based on OMNeT++

In this part, a satellite self-organizing network simulation platform based on OMNeT++ is built to evaluate the performance of the satellite routing algorithm. OMNeT++ is an open-source, component-based discrete event simulation tool widely used for network simulation. It has the following features:

1. Component-based simulation models: OMNeT++ models are based on components, which means that models can be constructed from multiple modular components, facilitating the building and expansion of models.
2. Flexible simulation process design: OMNeT++ allows users to freely design the simulation process, including the implementation of network protocols and the transmission of packets. Therefore, users can design the network structure and node behavior according to the specific conditions of their research network scenarios.
3. Graphical User Interface (GUI): OMNeT++ provides a GUI that enables users to visually observe the simulation process. By observing, users can confirm whether the designed and implemented functions operate correctly, thus verifying and modifying the design process.

Given these advantages of OMNeT++, this paper utilizes it to build a satellite self-organizing network simulation platform. In this platform, initial network formation clustering, periodic cluster maintenance, data generation and routing, and other functions are designed and implemented for the satellite self-organizing network. In the following sections, the simulation platform is introduced from three domains: network domain, module domain, and function domain.

4.1. Network Domain

In OMNeT++, the distribution of the network is set in the *.ned and *.ini files, where the number, mobility, and distribution of nodes can be set. In this paper, the distribution of satellite nodes in the network simulates a leader–follower satellite formation flying scenario, with one leader node and four follower nodes in a formation. The follower nodes follow the leader node in orbiting flight, as shown in Figure 2. A total of 20 formations are set up, with a total of 80 satellites. The distance between the master and slave nodes is 100 km, and the distance between the master nodes is between 600 km and 700 km [19]. During the simulation process, the topology between the master satellites remains relatively stable.

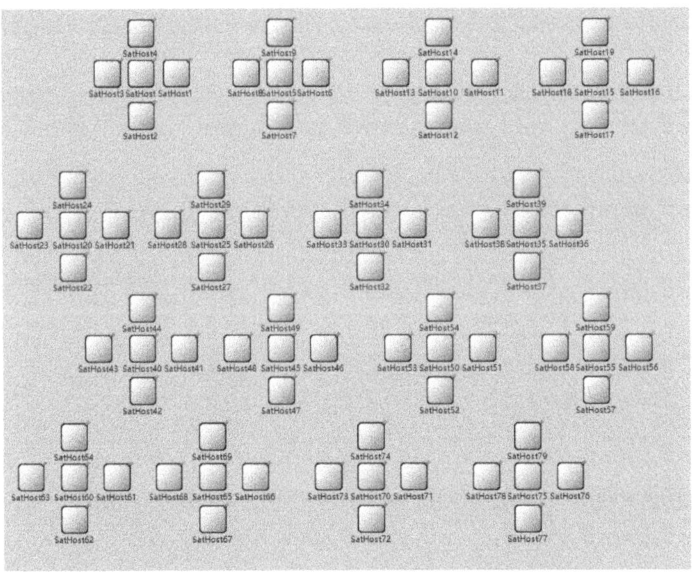

Figure 2. Node distribution in simulation.

4.2. Module Domain

In OMNeT++, a module is a functional entity that can be programmed to perform certain functions. Modules can communicate through artificially defined interfaces. Additionally, modules can be nested to form a hierarchical structure. In the satellite self-organizing network system implemented in OMNeT++, the module structure of the ground nodes and satellite nodes is consistent, both inheriting from the NodeBase module of the INET Framework. Based on the NodeBase module, network layer modules and Mac layer modules are added, as shown in Figure 3. The downward interface of the network layer module and the upward interface of the Mac layer module are bidirectionally connected, and the downward interface of the Mac is bidirectionally connected with the external interface of the node module. Although the structures of the two types of nodes are consistent, the functional design of their network layers and Mac layers is not the same, as shown in Figure 4.

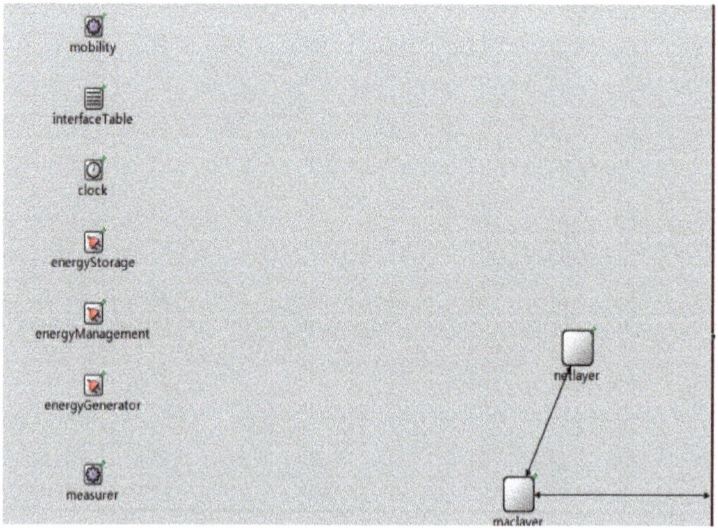

Figure 3. Structure of satellite and ground node.

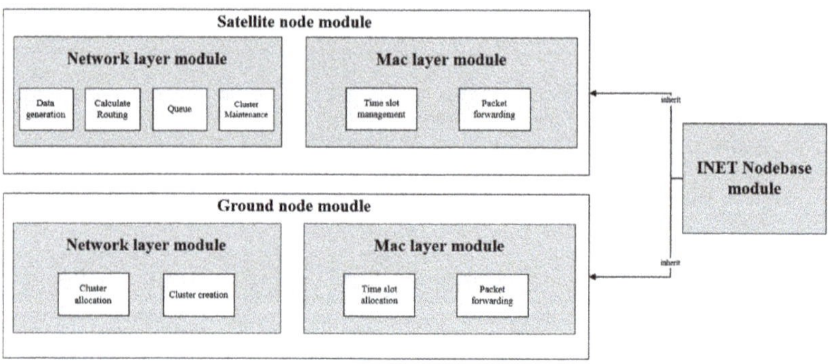

Figure 4. Functions of network layer and Mac layer of satellite and ground node.

The functions of satellite nodes' network layer module include cluster maintenance, data generation, and routing. The Mac layer module of the satellite nodes is responsible for sending messages passed down from the network layer to other nodes, which includes end-to-end transmission and broadcasting functions.

The network layer module of the ground nodes is responsible for cluster allocation to each node when the network is initially established, and the Mac layer module is responsible for allocating time slots for each node.

The specific function is illustrated in the following part.

4.3. *Function Domain*

The function domain mainly has four parts: initial network formation, cluster maintenance, time slots allocation, and data generation and forwarding.

4.3.1. Initial Network Formation

Based on the high-speed wireless self-organizing network clustering method from reference [18], this paper implements a satellite self-organizing network clustering process, which introduces a ground control node for clustering. Satellites send network access requests and their information (mainly location information) to the ground node, which completes the division of the cluster structure and the allocation of cluster heads and then sends back the cluster allocation information for each node separately. The initial network formation process is deployed at the network layer module of the ground node.

The specific network formation process is as follows:

1. The joining satellite node sends a network access request message to the ground node;
2. Based on the location information of the joining node and current network cluster structure, the ground node performs an initial cluster allocation for the new joining node. After the allocation is completed, the ground node sends a message to the joining node carrying the assigned node ID, cluster information, whether it is a cluster head, time slots, and other information;
3. The joining node takes corresponding actions based on its assigned status. If it is assigned as a member node, it sends a message to the cluster head node, informing its own ID information and time slot. If it is assigned as a cluster head, it broadcasts a "hello" message to its neighbors to announce its presence, while also listening for messages from neighboring cluster heads and new joining member nodes in its cluster to establish a local connection table.

4.3.2. Cluster Maintenance

After the initial cluster structure is established, due to the movement of the satellites, the relative positions of the nodes may change. These changes can affect the structure of the clusters, which may lead to network instability. Therefore, it is necessary to establish a cluster maintenance mechanism to maintain the relative stability of the clusters.

The content of cluster maintenance mainly involves monitoring and responding to the behavior of nodes, such as leaving clusters, joining clusters, and damage. The satellite nodes in the network are divided into three types: cluster head nodes, member nodes, and isolated nodes. Different nodes perform different cluster maintenance methods to work together to complete the cluster maintenance. This process is deployed at network layer modules of satellite nodes.

The specific cluster maintenance operations performed by nodes of different types are as follows:

1. Cluster head node. The cluster head node maintenance process is shown in Figure 5a. The cluster maintenance of the cluster head is periodically triggered. By using the "ScheduleAt()" method in OMNeT++, the network layer module of the cluster head can send a message object to itself after a certain period. This method can be used to set a timer to periodically trigger cluster maintenance. After the cluster head triggers the cluster maintenance operation, it first broadcasts an "isConnected" message to the member nodes to confirm whether the member nodes are still within their connection range. If any node has left the cluster, the cluster head updates the local connection table and floods this new connection table. Meanwhile, the member with the highest connection degree (i.e., it has the most neighbors in the cluster) is elected as the

sub-head, and a message is sent to this member to notify it of becoming the sub-head. When the cluster head is damaged due to certain circumstances, the sub-head can become the new cluster head to manage other member nodes. This can effectively avoid the direct demise of the cluster after the cluster head is damaged, thereby reducing the frequency of cluster structure changes and enhancing the stability of the network;

2. Member nodes. The cluster maintenance process of member nodes is shown in Figure 5b. A timer is also set within the member nodes. If a member node receives an "isConnected" message from the cluster head before the timer is up, it resets the timer and sends its own position information and node ID in a "Reply" message to the cluster head for updating the local connection table and for the next sub-head election. If the timer is up and the member node still has not received a message from the cluster head, it is considered that the node has left the communication range of the cluster head due to its movement. Therefore, the node changes its status to "isolate" and switches to performing the cluster maintenance methods of an isolated node;
3. Isolate nodes. The process for isolated nodes to join a cluster is shown in Figure 5c. An isolated node listens for broadcast messages from cluster heads. If it receives a broadcast message from a cluster head, it sends a network access request to the cluster head. Then, a timer starts counting. If the node receives a confirmation message from the cluster head before the timer is up, the node changes its state to "member". If the timer is up and the node still has not received a confirmation message from the cluster head, it continues to listen for cluster head broadcasts.

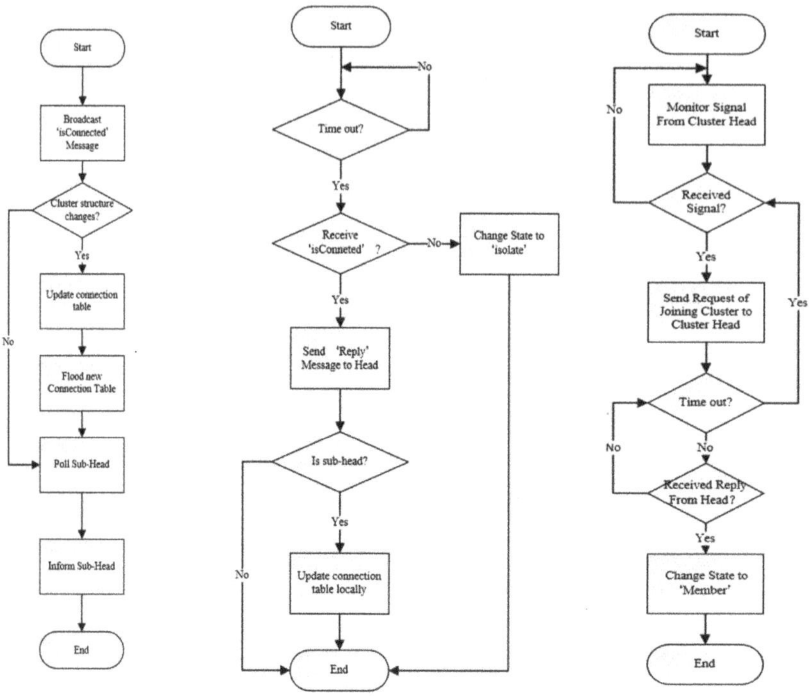

(a) (b) (c)

Figure 5. Cluster maintenance process of different nodes. (**a**) Cluster maintenance process of cluster head. (**b**) Cluster maintenance process of member. (**c**) Cluster maintenance process of isolate node.

4.3.3. Time Slot Allocation

In order to avoid congestion and packet loss caused by multiple nodes sending messages simultaneously during message transmission, it is necessary to allocate a time slot to each node. Nodes send messages within their allocated time slots to prevent sending messages simultaneously. Since the network has a layered structure and different clusters are relatively independent, time slots can be independently allocated within each cluster to achieve the reuse of time slot resources and improve resource utilization. In the scenario considered in this paper, the number of nodes in a cluster is limited to under 9, so 1 s can be divided into ten time slots, each 100 ms, and they can be allocated to the member nodes within the cluster. Time slot allocation can be divided into time slot allocation for nodes initially joining the network and time slot allocation for isolated nodes joining the network.

The specific process of time slot allocation for initial network access is as follows:

1. When the ground node receives the satellite node's network access request, after assigning the cluster affiliation in the network layer, the Mac layer allocates an idle time slot to the node and sends the time slot information to the node through a message;
2. After the node receives the message containing the time slot information, it informs the cluster head of its time slot information through a message;
3. The cluster head updates the utilization of time slot resources in the cluster.

The time slot allocation process for isolated nodes joining the network is as follows:

1. The isolated node sends a network access request to the cluster head;
2. After receiving the access request, the cluster head checks if there are any remaining idle time slots in the cluster. If there are idle time slots, proceed to the next step;
3. The cluster head allocates an idle time slot to the new joining node and notifies the node through a message.

The time slot allocation process is deployed at the Mac layer modules of satellite nodes and the ground node. Additionally, the Mac layer modules of the satellite nodes also have the function of managing time slots, where the cluster heads manage the time slots of the members within the cluster, and the member nodes send data based on their time slots.

4.3.4. Data Generation and Forwarding

Each member node has a certain probability of generating a data packet every second. The node sends this data packet to the cluster head within its time slot. If the cluster head's queue is full when it receives the data packet, the packet is discarded. If the data queue is not full when receiving a data packet, the data are added to the queue for processing. Once all the data packets ahead in the queue have been forwarded, the cluster head parses the destination node of the data packet, calls the routing algorithm function to calculate the next hop, and forwards it. In the previous text, the specific process of the SQL-CBRP algorithm has been introduced. To implement this algorithm on the OMNeT++ platform, nodes need to be able to obtain the location information of other nodes. Since the nodes inherit from the INET Framework's NodeBase module, which includes a mobility module, the location information of each node can be obtained from it. In addition, the algorithm requires maintaining a neighbor table locally at each node. Therefore, in this platform, a neighbor class that includes the neighbor's ID and its queue length is created, and a neighbor list composed of neighbor objects is created in the network layer module of each satellite node. Each time a neighbor sends back the queue length information, the queue length of the corresponding neighbor object is updated for routing decisions. When a cluster head receives a data packet forwarded from another cluster head, it executes the same queuing and forwarding process, until the packet arrives at its destination node.

The packet queue and routing calculation functions are deployed at network layer modules, and forwarding is implemented by Mac layer modules.

5. Simulation Analysis

The simulation of this paper is based on the satellite self-organizing network platform introduced before, mainly focusing on the performance of packet delay and packet loss rate when making routing decisions through SQL-CBRP, and comparing it with the Dijkstra algorithm based on the minimum number of hops and the GPSR algorithm based on the shortest distance.

5.1. Simulation Parameters

The simulation parameters are shown in Table 2. The bandwidth setting for satellite nodes is adapted from the reference [20]. This paper uses 15 Mbps as the message service transmission rate. According to ZTE's white paper '5G Video Uplink Transmission Applications and Challenges' published in 2021, the transmission rate for high-definition video is generally between 8 and 15 Mbps. This paper uses 8 Mbps as the video service transmission rate. In [21], it is mentioned that the satellite data uplink rate in emergency scenarios is 13.8 Mbps. In this paper, 15 Mbps is adopted as the transmission rate for message services. Additionally, since the standard rate used for telephony voice communication in G.711 is 64 kbps [22], this paper also uses 64 kbps for the voice service transmission rate. Since all data must be forwarded by the cluster heads, the cluster heads are prone to be overwhelmed. To avoid a queue overflow at the cluster head, the number of member nodes in a cluster should not be too large. Therefore, without loss of generality, in this paper the maximum number of cluster member nodes is set to 4.

Table 2. Simulation parameters.

Parameter Name	Parameter Value
Number of satellite nodes	80
Distance between leaders	600–700 km
Distance between leader and follower	100 km
Nodes' speed	7.6 km/s
Nodes' longest queue length	9
Rate of packet generation	1 packet/s
Bandwidth of cluster head	30 Mbps [20]
Simulation duration	1500 s
Message service's data rate	15 Mbps [21]
Voice service's data rate	64 kbps [22]
Video service's data rate	8 Mbps

5.2. Simulation Results

5.2.1. Performance of Different Types of Packets

The packet loss rates for the three types of services are shown in Figure 6a. It can be observed that the packet loss rate of the Dijkstra algorithm is significantly higher than the other two algorithms. For different services, the algorithm proposed in this paper performs better than the traditional GPSR in terms of packet loss rate. Specifically, the average loss rate of SQL-CBRP is about 21.5% lower than GPSR relatively. This is because the SQL-CBRP algorithm selects the neighbor node with the shortest queue as the next hop, which can effectively prevent packets from being forwarded to a node whose queue is already full, thereby reducing the probability of packet loss.

The average delay results for the three types of services are shown in Figure 6b. It can be seen from the figure that for the three different service scenarios, the routing algorithm proposed in this study has shown some optimization in terms of delay compared to the GPSR algorithm. Specifically, the reduction in average delay is about 90 ms, which is about 11.0% lower than GPSR relatively. The emergence of this result is because SQL-CBRP takes into account both the queue length and the geographical location, which can reduce the queuing delay and propagation delay simultaneously.

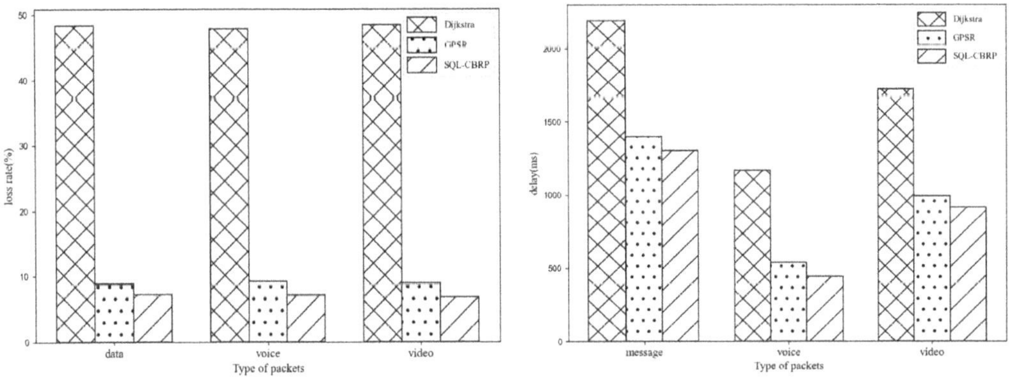

(a) (b)

Figure 6. Packet loss rate and delay of voice, message, and video service in Dijkstra, GPSR, and SQL-CBRP. (**a**) Packet loss rate of different types of packets; (**b**) Delay of different types of packets.

5.2.2. Performance of Different Packet-Sending Probability

The SQL-CBRP proposed in this paper makes routing decisions based on the queue length of neighboring nodes. Therefore, the number of packets in the network is likely to affect the performance of the algorithm. To test the impact of the number of network packets, this section sets different packet-sending probabilities and tests the performances of the Dijkstra algorithm, GPSR algorithm, and SQL-CBRP algorithm under different sending probabilities.

According to Figure 7, when the packet-sending probability is 60%, the delay and packet loss rate of SQL-CBRP are slightly higher than those of GPSR, but the difference is negligible. When the packet-sending probability is 80% and 100%, the delay and packet loss rate of SQL-CBRP are reduced to some extent compared to GPSR. Specifically, the delay is reduced by up to 18.0% relative to GPSR, and the packet loss rate is reduced by up to 22.1% relative to GPSR.

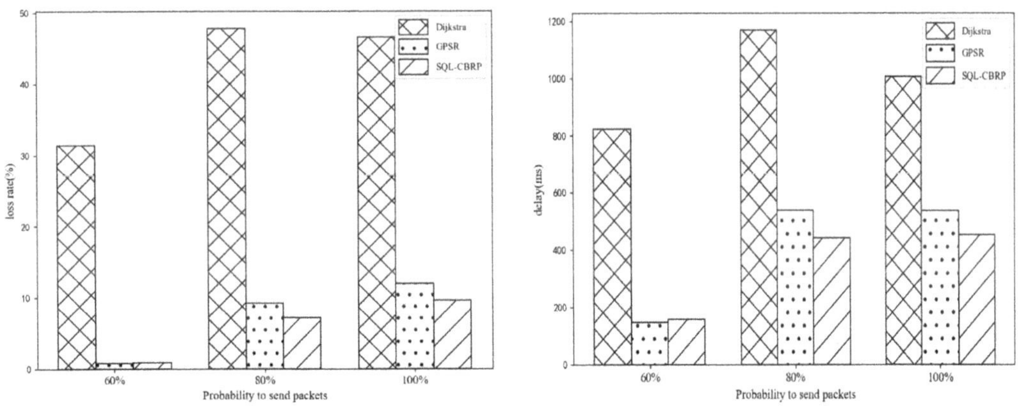

(a) (b)

Figure 7. Packet loss rate and delay of Dijkstra, GPSR, and SQL-CBRP when the probability of sending packet is 60%, 80%, and 100%. (**a**) Packet loss rate of different packet-sending probability; (**b**) Delay of different packet-sending probability.

5.2.3. Performance under Load Imbalance Conditions

To test the performance of SQL-CBRP under load imbalance conditions, the packet-sending probability of nodes within several clusters is set to 90%, while the packet-sending probability of the remaining nodes within clusters is set to 60%. The latency and packet loss rate of the three algorithms are tested under these conditions.

Figure 8 illustrates the comparison of delay and packet loss rate for the three algorithms when the number of high-load clusters is 4, 8, and 12. It is evident that under the condition of load imbalance, the SQL-CBRP algorithm has a lower delay and packet loss rate compared to Dijkstra and GPSR. Moreover, compared to GPSR, the delay is reduced by up to 24.5%, and the packet loss rate is reduced by up to 32.7% relatively. This is because SQL-CBRP selects nodes with shorter queuing queues (i.e., nodes with lower load) as the next hop. Therefore, in a network with load imbalance, SQL-CBRP can play a role in load balancing to a certain extent, thereby directing packets to areas with lower load, which in turn reduces the network's latency and packet loss rate.

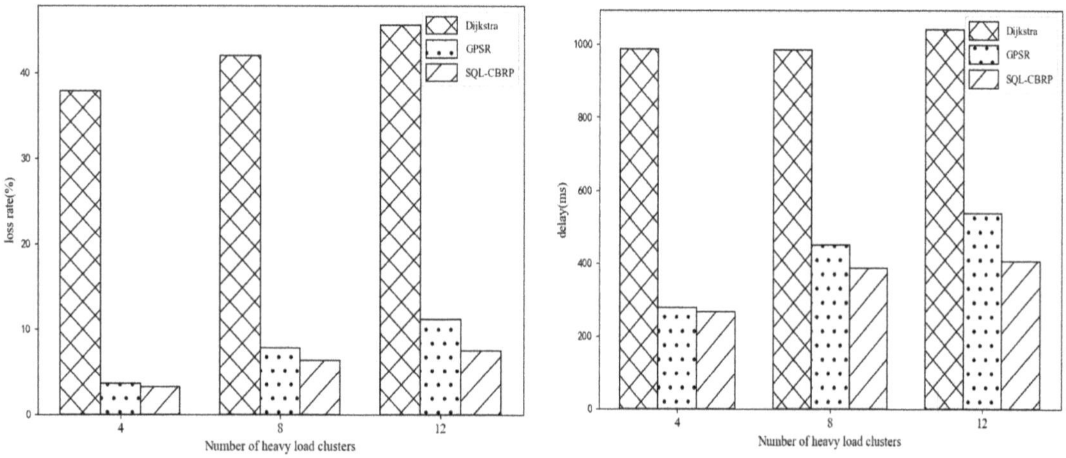

(a) (b)

Figure 8. Packet loss rate and delay of Dijkstra, GPSR, and SQL-CBRP when the number of heavy load clusters is 4, 8, and 12. (**a**) Packet loss rate of different numbers of heavy load clusters; (**b**) Delay of different numbers of heavy load clusters.

According to the analysis above, it is evident that since the SQL-CBRP algorithm selects the neighbor with the shortest queue as the next hop, it reduces queuing delay and effectively avoids packet loss due to queue overflow. Therefore, compared to the Dijkstra algorithm and the GPSR algorithm, the SQL-CBRP algorithm has lower packet loss rates and delays for services with different data rates. Specifically, when the network has high loads or the load distribution is imbalanced, the advantages of SQL-CBRP become more pronounced.

6. Conclusions

This paper proposes the SQL-CBRP, a new routing strategy for satellite self-organizing networks, and tests the performance of this routing strategy through a satellite self-organizing network simulation platform based on OMNeT++. SQL-CBRP combines the cluster-based routing and greedy algorithm, selecting the next hop by the queue lengths of neighbor nodes, which can reduce delay, packet loss rate, and the signaling overhead in the whole network. The results show that for the three types of services with specified transmission rates, the packet loss rate and delay performance of SQL-CBRP are generally better than Dijkstra and GPSR. Furthermore, this paper also tests the performance of the

SQL-CBRP under different packet-sending probabilities. The results indicate that under conditions of high packet-sending probability and imbalance load, the performance of the SQL-CBRP algorithm is superior to the other two algorithms, confirming the superiority of SQL-CBRP in terms of performance under higher and imbalanced network loads.

Author Contributions: Conceptualization, G.W. and J.Z.; methodology, G.W. and J.Z.; software, G.W.; validation, G.W. and C.L.; formal analysis, G.W.; investigation, G.W. and J.Z.; resources, J.Z.; data curation, G.W. and Y.Z.; writing—original draft preparation, G.W.; writing—review and editing, J.Z. and C.L.; visualization, Z.C. and Y.Z.; supervision, J.Z.; project administration, J.Z. All authors have read and agreed to the published version of the manuscript.

Funding: This research was funded by "the Fundamental Research Funds for the Central Universities", grant number 2023ZCJH09.

Data Availability Statement: The data presented in this study are available upon request from the corresponding author.

Conflicts of Interest: The authors declare no conflicts of interest.

References

1. Zhang, J.; Chang, C.; Zhang, Y.; Li, R.; Wang, K.; Zhang, X. Survey on Routing Technology of Mega Constellation. *Space-Integr.-Giround Inf. Netw.* **2024**, *5*, 2–13.
2. Liu, J.; Zhang, X.; Zhang, R.; Huang, T.; Yu, F.R. Reliable and low-overhead clustering in LEO small satellite networks. *IEEE Internet Things J.* **2021**, *9*, 14844–14856. [CrossRef]
3. Radhakrishnan, R.; Edmonson, W.W.; Afghah, F.; Rodriguez-Osorio, R.M.; Pinto, F.; Burleigh, S.C. Survey of inter-satellite communication for small satellite systems: Physical layer to network layer view. *IEEE Commun. Surv. Tutor.* **2016**, *18*, 2442–2473. [CrossRef]
4. Hui, L.; Xue, Z.; Ma, W. A TDMA-based Combined MAC Protocol forthe Small Satellite Formation System. *Space Electron. Technol.* **2016**, *13*, 23–28.
5. Zhang, J.; Wang, K.; Li, R.; Chang, Z.; Zhang, X.; Wang, W. MaCro: Mega-Constellations Routing Systems with Multi-Edge Cross-Domain Features. *IEEE Wirel. Commun.* **2023**, *30*, 69–76. [CrossRef]
6. Liu, G.-P.; Zhang, S. A Survey on Formation Control of Small Satellites. *Proc. IEEE* **2018**, *106*, 440–457. [CrossRef]
7. Yan, F.; Nie, H.; Xia, W.; Shen, L. Inter-satellite link connection strategy for large-scale low earth orbit satellite networks. *J. Commun.* **2024**, *45*, 100–109.
8. Wei, W.; Fu, L.; Wang, K.; Lu, X.; Zhou, Z. Survey of routing technologies for the satellite Internet. *J. Xidian Univ.* **2024**, accepted.
9. Cao, X.; Li, Y.; Xiong, X.; Wang, J. Dynamic Routings in Satellite Networks: An Overview. *Sensors* **2022**, *22*, 4552. [CrossRef] [PubMed]
10. Singh, J.; Singh, G.; Gupta, D.; Muhammad, G.; Nauman, A. OCI-OLSR: An Optimized Control Interval-Optimized Link State Routing-Based Efficient Routing Mechanism for Ad-Hoc Networks. *Processes* **2023**, *11*, 1419. [CrossRef]
11. Yang, B. Research on the Improvement of AODV Routing Protocol in ad Hoc Networks of UAVs. Master's Thesis, Xian University of Technology, Xian, China, 2023.
12. Yang, M. Research and Optimization of Routing Link Selection Method in Wireless ad Hoc Network. Master's Thesis, Harbin University of Science and Technology, Harbin, China, 2023.
13. Dong, L. Research on Mobile ad Hoc Network Routing Technology. Master's Thesis, Beijing University of Post and Telecommunication, Beijing, China, 2023.
14. Zhang, Y. Research on the Networking Technology of the Large-Scale Small Satellites Cluster. Master's Thesis, University of Chinese Academy of Sciences, Beijing, China, 2021.
15. Xu, J. Research on Centralized Adaptive Hybrid Routing Mechanism for ad Hoc Network. Master's Thesis, Beijing University of Post and Telecommunication, Beijing, China, 2015.
16. Bao, X.; Li, H.; Zhao, G.; Chang, L.; Zhou, J.; Li, Y. Efficient clustering V2V routing based on PSO in VANETs. *Measurement* **2020**, *152*, 107306. [CrossRef]
17. Wang, K. Research on Highly Reliable Routing Control Systems and Methods for Satellite Network. Master's Thesis, Beijing University of Post and Telecommunication, Beijing, China, 2024.
18. Guo, G. Research and Implementation on a Network Layer Protocol of High-Rate Wireless ad Hoc Network. Master's Thesis, Beijing University of Post and Telecommunication, Beijing, China, 2019.
19. Cheng, C.; Ren, X.; Xu, S. Starlink and its operational analysis. *Command Control Simul.* **2024**, *46*, 154–160.
20. Wang, C.; XU, P.; Zhang, S.-b.; Wang, L.-q.; Wang, W.-d. Research and implementation of dynamic routing protocol for LEO satellites based on Linux system. *J. Beijing Univ. Posts Telecommun.* **2020**, *43*, 94.

21. Xiao, K.; Liu, X.; Wei, H.; Fei, Z. Application of the on the move phased array satellite communication antenna in emergency communication. *Chang. Inf. Commun.* **2023**, *36*, 187–190.
22. Feng, B. Research on Preprocessing Algorithm of G.722 Wideband Speech Signal. Master's Thesis, Dalian University of Technology, Dalian, China, 2009.

electronics

MDPI

Article

A Power Control and Intervention Algorithm for Co-Existing IMT Base Stations and Satellite Services

Min Jia *,†, Shiyao Meng †, Hui Wang, Zhouhao Tang and Ziliang Jin

Electronics and Communications Engineering, Harbin Institute of Technology, Harbin 150080, China; 23b905019@stu.hit.edu.cn (S.M.); 23s005015@stu.hit.edu.cn (H.W.); 24b905008@stu.hit.edu.cn (Z.T.); 24s105219@stu.hit.edu.cn (Z.J.)

* Correspondence: jiamin@hit.edu.cn

† These authors contributed equally to this work.

Abstract: IMT-2020 (International Mobile Telecommunications-2020) is the prevailing mobile communication technology at the moment, significantly affecting societal progress. Nevertheless, the roll-out of the IMT-2020 system has introduced numerous interferences to existing services. The coexistence with fixed satellite services has become a topical issue currently under consideration. This paper discusses the compatibility and interference issues between IMT-2020 and the 14 GHz FSS (fixed-satellite service) uplink, as well as the spectrum access issue solved by artificial intelligence methods. The study shows that the interference from IMT-2020 macro-base stations to FSS space stations exceeds the ITU standard by approximately 10 dB. To control the interference, a partition-based power control algorithm is proposed, which divides ground base stations into multiple areas and virtualizes each area's base stations into a single large base station then applies power control to maximize the total transmission power of the base stations within the area. Furthermore, three intra-partition power control algorithms are introduced: average power allocation, power allocation based on channel gain, andna power allocation method based on the maximum intra-partition sum rate. Additionally, under the assumption that dynamic satellite nodes are available in the system for ground user access, a spectrum access algorithm utilizing deep reinforcement learning is designed. Simulation results confirm the effectiveness of the proposed scheme, which can reduce the interference from the IMT-2020 system to the FSS service below the threshold, ensuring harmonious coexistence of the two services.

Keywords: IMT-2020; fixed-satellite service (FSS); interference management; power control

1. Introduction

Over the past few decades, the development of mobile communications has significantly influenced societal progress. However, the increasing data traffic demands further exploration of spectrum resources [1]. These resources must support extensive and continuous channel bandwidth.

Spectrum sharing between networks has emerged as a key area of research [2]. However, deploying IMT-2020 systems across various frequency bands may interfere with other existing services [3]. In 2019, the World Radiocommunication Conference (WRC23) addressed this issue. It coordinated the 12.75–13.25-GHz frequency band, used by earth stations on aircraft and ships, for communication with geostationary space stations in the Fixed Satellite Service (FSS) [4–6]. This coordination ensures coexistence between existing services and IMT-2020 in these bands.

Current research focuses extensively on the coexistence possibilities of IMT-2020 with other existing radio services. Several studies have explored co-channel sharing between IMT-2020 and Fixed Service (FS) systems [7]. Jo et al. proposed analytical expressions for frequency-dependent suppression using CP-OFDM, windowed orthogonal frequency-division multiplexing, and filtered orthogonal frequency-division multiplexing. These methods aim to assess the interference of IMT-2020 on existing radar systems. Extensive analyses have been conducted on the interference between IMT-2020 and satellite services.

However, most of these studies focus on analyzing the interference between IMT-2020 and satellite earth stations, neglecting the interference received by satellite systems [8–16]. This is because satellite systems are located at high altitudes, where the interference from IMT-2020 is minimal. However, for interference-sensitive satellite systems, the aggregated interference from numerous IMT-2020 base stations distributed over a large area may still have significant impacts [17–19]. Therefore, to achieve complete coexistence of both services, it is necessary to estimate the interference received by existing satellite systems. Generally, there are two main methods for calculating aggregated interference: The first method involves calculating the interference power received by the satellite system from a single IMT-2020 system and then multiplying it by the number of IMT-2020 stations distributed within the satellite's main beam coverage [20]. The second method models a single cell of the IMT-2020 network using a hexagonal grid, calculates the interference received by the FSS satellite from this network, and finds the maximum number of networks that satisfy the protection standards [21]. The advantage of these methods is that after considering the interference from IMT-2020 distributed within a single or very small area, they can predict the interference power received from multiple IMT-2020 systems in a short period by multiplying by the coverage area. Both methods can serve as interference evaluation metrics for IMT-2020 and satellite services, demonstrating the aggregated interference received by satellites from IMT-2020.

Although there have been studies on IMT-2020 interference received by satellite systems, research on interference avoidance remains insufficient. Suppressing IMT-2020 interference with satellite FSS services while minimizing or eliminating impacts on IMT-2020 systems, thus achieving complete coexistence of both services, remains an open issue [22]. Additionally, in GEO satellite FSS services, due to the high orbital altitude and large coverage of GEO satellites, there is a significant number of IMT-2020 base stations within their coverage area. Conventional power control or optimization methods face immense computational complexity and large-scale signaling interactions when dealing with numerous IMT-2020 base stations, which is nearly difficult to achieve in practical deployments [23]. Therefore, achieving low-complexity interference avoidance solutions to promote coexistence between services is also a key issue to be addressed.

To address these needs, this paper first presents an interference analysis method to calculate the interference received by satellite systems from IMT-2020. It models and analyzes the actual interference received by satellites from terrestrial IMT-2020, considering factors such as IMT-2020 antennas, satellite antennas, and satellite-ground channels. The calculated interference-to-noise ratio is compared with existing protection standards to suggest the possibility of coexistence between IMT-2020 and FSS services. Based on the results of aggregated interference calculations, an innovative optimization algorithm based on a zoning strategy is proposed. This algorithm divides the vast number of ground base stations into zones and virtualizes the base stations within each zone into a virtual macro base station, allowing for power control of this macro-base station. Furthermore, three intra-zone power control methods are considered: average power control, power control based on channel gain, and power control for maximum sum rate, enabling fine-grained optimization of power control within each zone. This approach effectively reduces the complexity of IMT-2020 base station power control while avoiding extensive signaling interactions.

The structure of the remaining sections is as follows. In Section 2 of this article, the interference scenarios are introduced, and the methods for calculating the number of IMT-2020 base stations within the FSS coverage area, the antenna models, and the calculation of aggregate interference are clarified. In Section 3, three specific algorithms are presented. Section 4 analyzes and discusses the simulation results, comparing the performance of severalschemes. Section 5 provides the conclusion of this article.

2. System Model

2.1. Satellite-Ground Interference Scenario

When the satellite–ground service of FSS operates normally in the 6-GHz frequency band, its space stations can receive signals as well as interference. As a new service in this channel, IMT-2020 may cause such harmful interference. The interference scenario of the IMT-2020 system on the FSS space stations is depicted in the Figure 1. First, we have calculated the coverage area of the FSS satellite on Earth. Based on the interference geometry, the coverage area, S, can be given by the following formula:

$$S = 2\pi R^2(1 - \cos\alpha), \quad (1)$$

where R is the radius of the Earth, and α is the central angle, which is The angle between the line connecting the Earth to the edge of the coverage area and the line connecting Earth to the satellite. It can be calculated as:

$$\alpha = \cos^{-1}(R/(R+H)), \quad (2)$$

Then, the deployment density value of IMT-2020 macro-base stations in sharing and compatibility research can be calculated according to the following formula:

$$Dl = Ds \times Ra \times Rb, \quad (3)$$

Among them, D_s is the density value of the coverage area, which is the density of the number of base stations per square kilometer. R_a represents the ratio of the coverage area to the city/built-up area/region in the study, and R_b represents the ratio of the built-up area to the total area of the region. IMT-2020 base stations located within the satellite coverage area are represented as:

$$N_{IMT} = S \times Dl \times f \times f_{\text{TDD}}, \quad (4)$$

Among them, f is the factor of network load and f_{TDD} is the time division duplex activity factor.

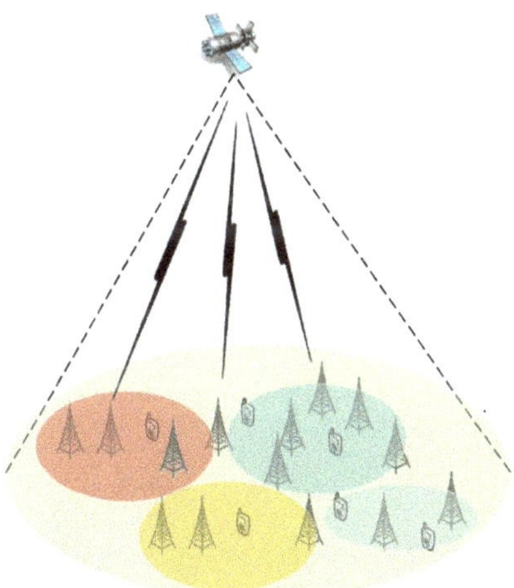

Figure 1. The interference scenario.

2.2. IMT-2020 Antenna Model

Active Antenna System (AAS) technology is crucial for achieving high transmitter gain in the IMT-2020 network. ITU-R recommendation M.2101 typically provides an IMT-AAS antenna model for IMT-2020 spectrum sharing research. The antenna array consists N_h of and N_v elements placed along the y-axis and z-axis respectively, as shown in Figure 2.

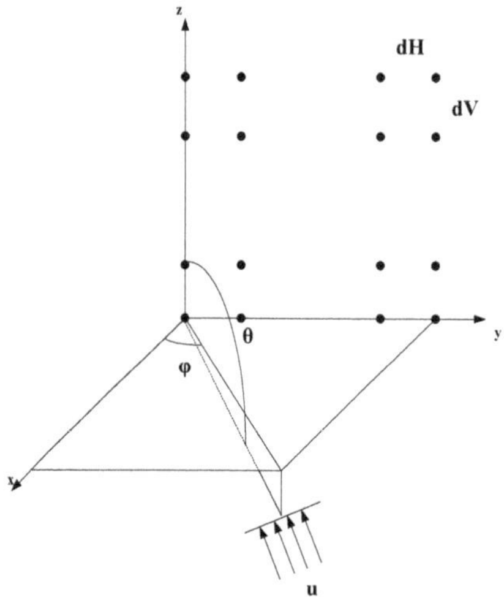

Figure 2. AAS antenna unit.

The composite gain of IMT-2020 AAS antenna is calculated by adding the component gain and array gain, and its mathematical formula is:

$$A(\theta,\varphi) = A_E(\theta,\varphi) + 10\log_{10}\left(\left|\sum_{m=1}^{Nh}\sum_{n=1}^{Nv} w_{i,n,m}\cdot v_{n,m}\right|^2\right), \quad (5)$$

$A_E(\theta,\varphi)$ represents the gain of a single element in an antenna array; more specifically, it depends on various factors such as the azimuth and elevation of the signal, the maximum gain of the element, and the horizontal and vertical aspect ratios of the antenna. The characteristics of IMT-2020 AAS are shown in Table 1.

Table 1. IMT-2020 AAS Characteristics.

Parameter	City	Suburban
Antenna pattern (m)	based on Recommendation ITU-R M.2101 Attachment 1 Section 5	based on Recommendation ITU-R M.2101 Attachment 1 Section 5
Component gain (dBi)	6.4	5.5
Horizontal/vertical	90° H	90° H
3 dB beam width of a single element (in degrees)	65° V	90° V
Horizontal/vertical front to back ratio	30	30
Antenna polarization	linear ± 45°	linear ± 45°
Antenna array configuration (row × column)	16 × 8	16 × 8
Array Ohmic Loss (dB)	2	2
Maximum coverage angle on the horizontal plane of the base station (degrees)	±60	±60
BS vertical coverage range	90–120	90–120
Mechanical Downtilt (Degrees)	6	10

2.3. FSS Satellite Antenna Model

According to IMT-2020, for circular or elliptical beam spacecraft antennas with a single feed source in satellite fixed services, the following radiation patterns should be used as design indicators outside the coverage area.

$$G(\psi) = \begin{cases} G_m - 3(\psi/\psi_0)^{\alpha}, & \psi \leq \psi \leq a\psi_0 \\ G_m + L_S + 20\log_{10} z, & a\psi_0 \leq \psi \leq 0.5b\psi_0 \\ G_m + L_S, & 0.5b\psi_0 \leq \psi \leq b\psi_0, \\ X - 2.5\log_{10}\psi, & b\psi_0 \leq \psi \leq Y \\ 0, & Y \leq \psi \end{cases} \tag{6}$$

In the above equation, $Y = b\psi_0 10^{0.04(G_m+L_S)}$, $G(\psi)$ represents the gain (dBi) at an angle deviating from the main beam direction, G_m represents the peak gain of the main lobe (dBi), ψ_0 represents half of the 3 dB beam width, L_S represents the paraxial sidelobe level (dB) relative to the peak gain required by the system design, and z represents the (principal/secondary axis) of the radiation beam. The radiation pattern of a single feed circular beam antenna can be represented as follows:

$$G(\psi) = \begin{cases} G_m - 3(\psi/\psi_0)^2, & \psi_0 \leq \psi \leq a\psi_0 \\ G_m + L_s, & a\psi_0 \leq \psi \leq b\psi_0 \\ G_m + L_s + 20 - 25\log_2(\psi/\psi_0), & b\psi_0 \leq \psi \leq \psi_1 \\ 0, & \psi_1 < \psi \end{cases}, \tag{7}$$

In the formula, ψ_1 represents the value ψ of time $G(\psi) = 0$ dBi. Table 2 shows the values of the sidelobe levels L_S for −20 dB, −25 dB, and −30 dB.

Table 2. Valuable of a , b and α.

L_s (dB)	a	b	α
−20	2.58	6.32	2
−25	2.88	6.32	2
−30	3.16	6.32	2

2.4. Satellite Ground Channel Model

Basic path loss PL_b includes signal free space path loss ($FSPL$), clutter loss (CL), and shadow fading (SF). The unit of free space path loss is dB, which satisfies:

$$FSPL(d, f_c) = 32.45 + 20\log_{10}(f_c) + 20\log_{10}(d), \tag{8}$$

where, the tilt distance between the satellite and the ground terminal is d, f_c is the carrier frequency (in GHz) and d is related to the satellite's orbital altitude, μ, geocentric angle, and α,communication elevation. The specific relationship is as follows:

$$d = (R+h)\sin(\mu)/\cos(\alpha), \tag{9}$$

By solving the equation simultaneously, it can be seen that the relationship between d and α is as follows

$$d = \sqrt{(R\sin\alpha)^2 + h^2 + 2Rh} - R\sin(\alpha), \tag{10}$$

Clutter loss $CL(\alpha, f_c)$ simulates signal power attenuation caused by surrounding buildings and ground objects. This depends on the elevation angle, carrier frequency, and specific environment. When the UE is in LOS state, the clutter loss can be ignored and set to 0 dB in the basic path loss model. Shadow fading SF is a zero mean normal distribution, measured in dB, with a standard deviation of σ_{SF}^2. The following table provides the sum values of partial elevation angles in different scenarios. The remaining elevation angles can refer to the values corresponding to the nearest reference angles in Tables 3 and 4.

Table 3. Shadow fading and clutter loss in urban environments.

Elevation Angle	σ_{SF}^2 in the S-Band LOS Scenario	σ_{SF}^2 in the S-Band NLOS Scenario	CL in the S-Band NLOS Scenario	σ_{SF}^2 in the Ka-Band LOS Scenario	σ_{SF}^2 in the Ka-Band NLOS Scenario	CL in the Ka-Band NLOS Scenario
10	4	6	34.3	4	6	44.3
20	4	6	30.9	4	6	39.9
30	4	6	29.0	4	6	37.5
40	4	6	27.7	4	6	35.8
50	4	6	26.8	4	6	34.6
60	4	6	26.2	4	6	33.8
70	4	6	25.8	4	6	33.3
80	4	6	25.5	4	6	33.0
90	4	6	25.5	4	6	32.9

Table 4. Shadow fading and clutter loss in suburban environments.

Elevation Angle	σ_{SF}^2 in the S-Band LOS Scenario	σ_{SF}^2 in the S-Band NLOS Scenario	CL in the S-Band NLOS Scenario	σ_{SF}^2 in the Ka-Band LOS Scenario	σ_{SF}^2 in the Ka-Band NLOS Scenario	CL in the Ka-Band NLOS Scenario
10	1.79	8.93	19.52	1.9	10.7	29.5
20	1.14	9.08	18.17	1.6	10.0	24.6
30	1.14	8.78	18.42	1.9	11.2	21.9
40	0.92	10.25	18.28	1.3	11.6	20.0
50	1.42	10.56	18.63	2.7	11.8	18.7
60	1.56	10.74	17.68	2.1	10.8	17.8
70	0.85	10.17	16.50	3.0	10.8	17.2
80	0.72	11.52	16.30	3.6	10.8	16.9
90	0.72	11.52	16.30	0.4	10.8	16.8

2.5. Aggregate Interference Calculations

In order to evaluate the aggregated interference caused by multiple IMT macro-base stations to FSS spatial receivers, the interference from each BS sector is first calculated as follows:

$$I_i^{DL} = P_t + G_t(\theta_t) + G_r(\theta_r) - PL - OL, \tag{11}$$

where I_i denotes the single incoming interference power from the base station i in dBm/MHz. P_t is the signal power density of one base station in dBm/MHz. G_t (θ_t) is the base station antenna gain in the direction of the FSS spatial receiver after viewing the beamforming antenna direction map in dBi. G_r (θ_r) is the gain of the FSS space receiver antenna in the direction of the base station after considering the off-axis angle, in dBi. PL is the propagation loss, including polarisation discrimination, clutter loss, free-space path loss and atmospheric gas attenuation, in dB. OL is the transmitter station array ohmic loss in dB. The overall interference level caused by all IMT-2020 base stations after considering TDD is:

$$I_{\text{total}} = 10\log\left(0.75 * \sum_{i=1}^{i=N_{DL}} 10^{I_i^{DL}/10}\right), \tag{12}$$

The thermal noise power density of the FSS spatial receiver can be calculated as

$$N_0 = 10 * \log(k * T * 1000 * B) = -110.61 \quad (\text{dBm / MHz}), \tag{13}$$

Finally, I/N can be expressed as $I_total - N_0$.

3. Optimisation Algorithm

The flowchart of the power allocation algorithm based on the partitioning strategy is given in Figure 3. The input is the coverage area and the number of base stations. The output is the allocated power, aggregate interference, and throughput.

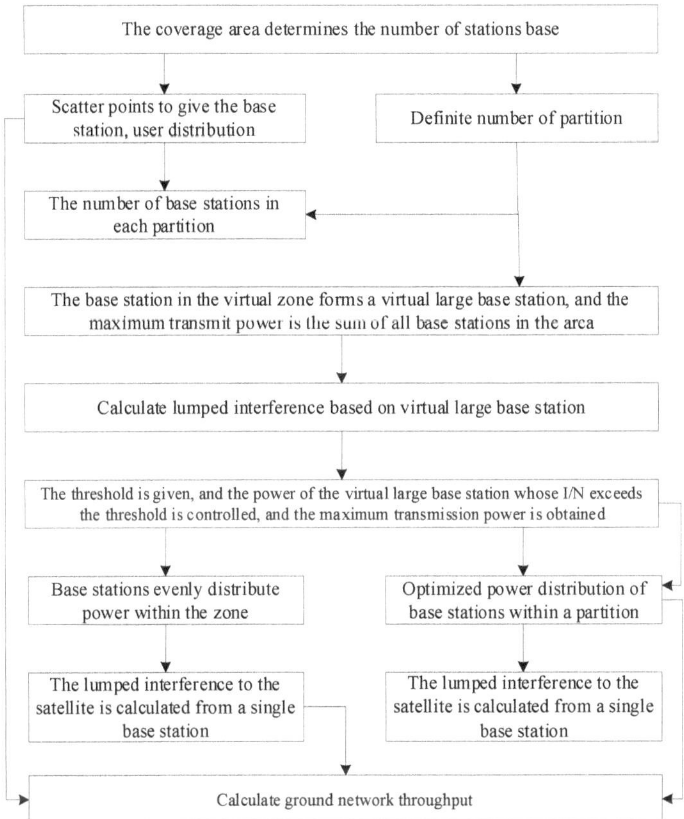

Figure 3. Flowchart of power allocation algorithm based on partitioning strategy.

3.1. Average Power Allocation Algorithm Based on Partitioning Strategy

When the maximum transmit power of the virtual large IMT-2020 base station in the sub-area is determined, the average power distribution scheme distributes the maximum transmit power of this virtual large base station equally to all IMT-2020 base stations in the area, thus determining the transmit power of the IMT-2020 base station. The following table describes the Algorithm 1.

Algorithm 1 Average Power Allocation Algorithm Based on Partitioning Strategy.

1: Determine the total number of base stations based on the coverage area
2: Obtain the number of base stations within the partition
3: Calculate the aggregate interference of the virtual large base station based on each partition
4: Solve for the maximum transmission power of the base station based on the threshold value
5: Perform average power allocation

3.2. Power Allocation Algorithm Based on Channel Gain

The core idea of the method is to allocate power according to the channel relationship between the user and the base station, i.e., the better the channel conditions, the higher the transmit power.

3.3. Maximum and Rate-Based Power Allocation

The maximum sum rate based power allocation algorithm can be mathematically represented as:

$$\begin{aligned} \max \quad & \sum_{j=1}^{N_u} R_j \\ \text{s.t.} \quad & \sum_{j=1}^{J_N} P_j \leq P_{\max}^n \quad , \\ & \frac{I_{total}}{N} \leq I_{threshold} \end{aligned} \tag{14}$$

In the above equation, R_j denotes the rate of the user j , which can be expressed as:

$$R_j = \log_2(1 + \frac{P_j G_t G_r |\mathbf{h}_j|^2}{\sigma^2}), \tag{15}$$

In addition, P_j denotes the transmit power of the IMT-2020 base station j, P_{max}^n denotes the maximum transmit power of the virtual IMT-2020 large base station in the sub-area n, I_total denotes the aggregate interference to the FSS satellites from all IMT-2020 base stations in the sub-area, and $I_threshold$ denotes the interference threshold for the aggregate interference to the FSS in the sub-area. Since the objective function in the optimisation problem is non-convex, in order to solve the problem, the non-convex objective function is transformed into a convex function as follows: Introducing the slack variable x_j satisfies: $x_i \leq P_j G_t G_r |h_j|^2$

$$e^{x_i} \leq P_j G_t G_r |\mathbf{h}_j|^2, \tag{16}$$

Then the optimisation objective function of the original problem can be written as

$$\log_2 e(\sum_{j=1}^{N_u} x_j - \ln(\sigma^2)), \tag{17}$$

At this point the problem is convex and an optimised solution can be obtained by the solver.

4. Simulation Verification

4.1. Simulation Parameters

The simulation parameters are given in this section in Table 5 below.

Table 5. Simulation parameters.

Parameters	Value
Number of base stations	439,860
Satellite maximum antenna gain	22 dBi
Network load factor	20%
TDD activity factor	75%
Noise temperature	630 K
Satellite antenna half-beam angle	15°
Antenna bypass level	−25 dB
Satellite altitude	35,786 km

Table 5. *Cont.*

Parameters	Value
Maximum transmit power of the base station	46 dBm/100 MHz
Bandwidths	100 MHz
Antenna front-to-back ratio	30 dB
Bypass Level Limit	30 dB
Antenna array element gain	5.5 dBi
3 dB pitch angle	65°
3 dB direction angle	65°
Mechanical downward angle	10°
Antenna Horizontal Spacing	0.5 wavelength
Antenna vertical spacing	0.5 wavelength
Pitch/Direction	Input parameters
Base station coverage diameter	300 m
Array of antennas	8H16V

4.2. *Performance of Average Power Allocation Algorithm Based on Partitioning Strategy*

In Figure 4, it can be seen that the aggregate interference I/N value from the terrestrial network to the satellite without interference control is concentrated in the range of 5.2–6.8 dB, which exceeds the interference threshold by about 13 dB. Therefore, an appropriate interference control scheme is needed to satisfy the I/N threshold requirement needed for network coexistence.

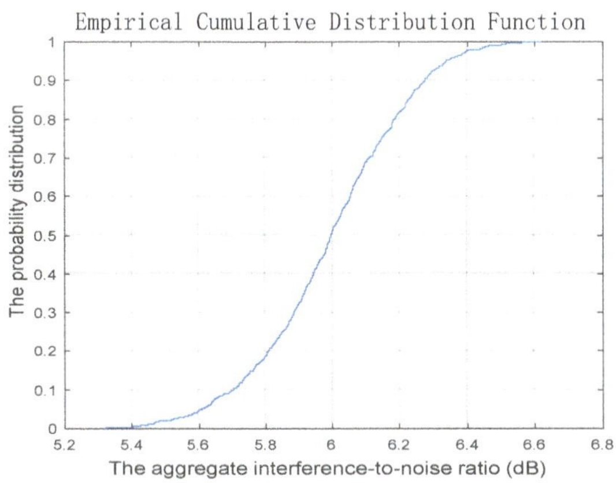

Figure 4. The empirical cumulative distribution function (CDF) of the aggregate interference-to-noise ratio (I/N) caused by ground networks on satellites.

In Figure 5, the simulation results of the proposed partition-based interference avoidance method are presented. The ground base stations are divided into 1000 partitions. Considering the long distance between the satellite and the ground, the deployment positions of the ground base stations are relatively small compared to the satellite-ground distance, thus the elevation angles of the base stations within the same partition to the satellite can be approximately the same. The base stations within a partition are equivalent to a virtual large base station with a transmit power equal to the sum of all base stations within that partition, and the interference caused by this equivalent large base station to the satellite is measured. The figure shows the interference values from 1000 virtual large base stations to the satellite, and based on experience, an interference threshold of −34 dB is set for each. The blue line in the figure represents the partitions that meet the interference threshold, meaning no power control is required, while the red line represents the partitions that require interference control.

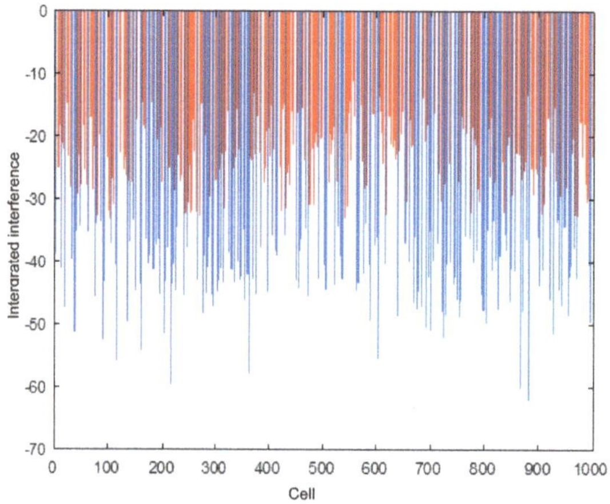

Figure 5. I/N (dB) of each region to the satellite when the number of partitions is $K = 1000$, where the blue line indicates the partitions that do not require power control. (Threshold = −34 dB, empirical value).

Figure 6 shows the transmission power values of base stations in each partition after power control when the number of partitions is K = 1000. In the figure above, an average power scheme is adopted for the base stations within each partition, so the transmission power of the base stations within the same partition is the same. The figure only shows the transmission power of base stations in different partitions. As can be seen, the transmission power of most base stations is greater than 25 dBm/100 MHz. Meanwhile, due to the average distribution of power among base stations, there are no sleeping base stations. Correspondingly, Figure 7 presents the CDF curve of the base station transmission power. From the figure, we can conclude that 60% of the base stations have a transmission power less than 46 dBm, which means 40% of the base stations in the network can transmit signals at their maximum transmission power.

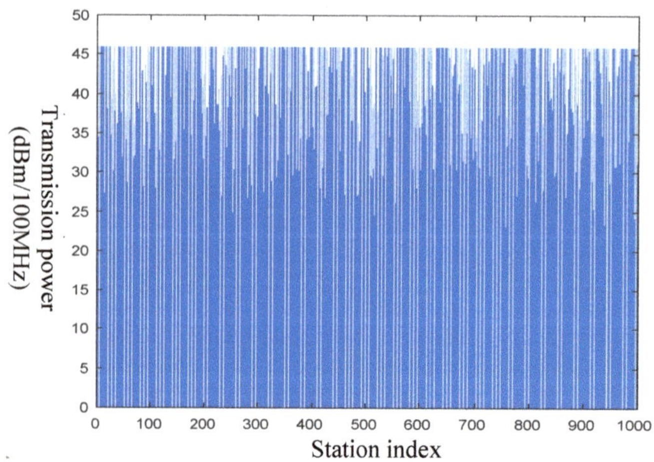

Figure 6. Transmission Power Values of Base Stations in Each Region after Power Control When the Number of Partitions is K = 1000.

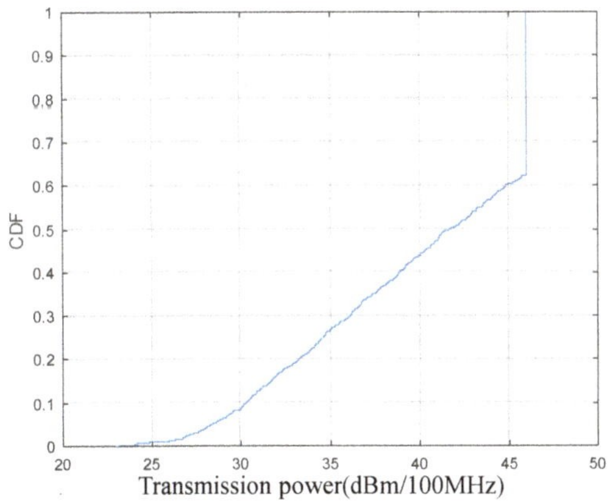

Figure 7. CDF Curve of the Transmission Power of Base Stations in Each Region when the Number of Partitions is K = 1000.

4.3. Performance of the Channel Gain-Based Power Allocation Algorithm Based on Partitioning Strategy

As observed in Figure 8, under this power allocation scheme, a small portion of base stations receive the majority of power due to their favorable channel conditions, resulting in significant rates for their served users. However, the remaining base stations transmit information with extremely low power, leading to lower user rates. According to statistics, under a 100-MHz bandwidth, the coverage rate for 50 Mbps is only 34.95%.

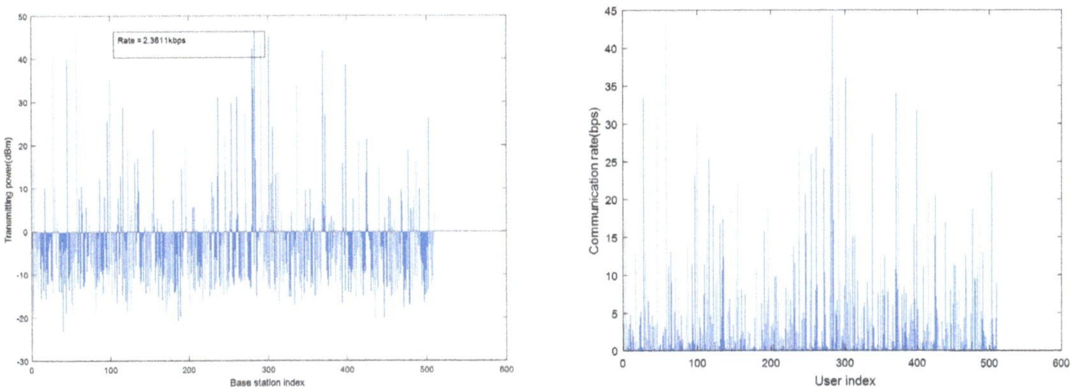

(a) (b)

Figure 8. Performance of the Channel Gain-Based Power Allocation Algorithm with Partitioning Strategy. (**a**) Transmission Power of Base Stations within a Single Partition. (**b**) User Rate within a Single Partition.

For ease of observation, Figure 9 takes a single partition as an example and shows the CDF curves of user rates within a partition and for all users. Table 6 compares the performance of three different power control methods. The results in the table show that without power control, the maximum terrestrial network sum rate can be achieved, approximately 9.2447 kbps/Hz. However, at this point, the aggregate interference-to-noise ratio from terrestrial IMT-2020 base stations to the satellite is approximately 6 dB, far exceeding the threshold requirement of −6 dB. Among the three power control algorithms, the power control algorithm based on maximizing the sum rate can achieve the highest sum rate performance, with the smallest performance loss compared to no power control. However, this scheme requires solving an optimization problem, resulting in higher computational complexity. The average power allocation scheme can achieve a sum rate performance close to the maximum sum rate-based method, while requiring minimal computational complexity.

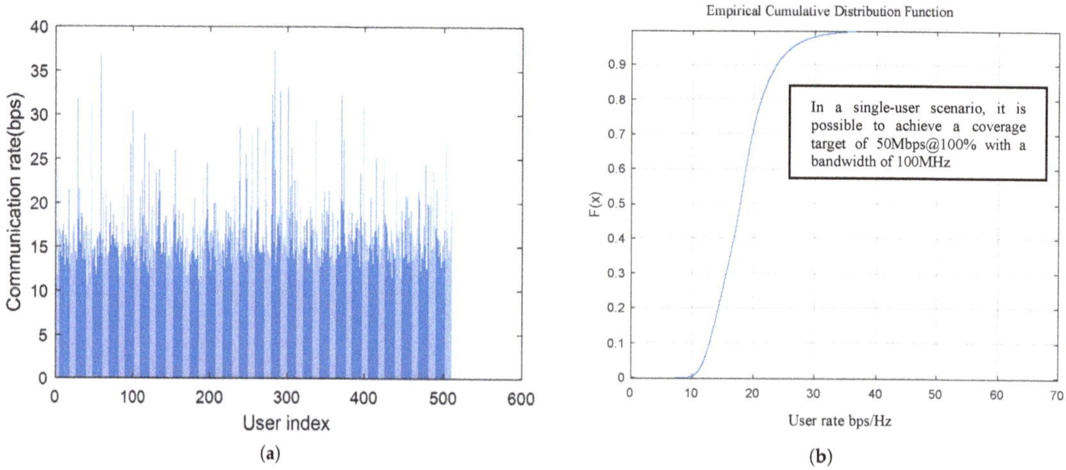

Figure 9. Performance of the Maximum Sum Rate-Based Power Allocation Algorithm with Partitioning Strategy. (**a**) User Rate within a Single Partition. (**b**) User Rate CDF Curve.

Table 6. Simulation parameters.

Scheme	Sum Rate	Performance Loss	100 MHz Coverage	Performance
Channel Gain-Based Power Allocation Algorithm with Partitioning Strategy	3.0370 kbps/Hz	67.14%	50 Mbps@50.19%	−6.8977
Average Power Allocation Algorithm with Partitioning Strategy	8.0024 kbps/Hz	13.43%	50 Mbps@100%	−6.8052 dB
Maximum Sum Rate-Based Power Allocation Algorithm with Partitioning Strategy	8.2096 kbps/Hz	11.19%	50 Mbps@100%	−6.8443 dB
No Power Control	9.2447 Mbps/Hz	0%	50 Mbps@100%	6.1725 dB

4.4. Research on Self-Organizing Multi-Hop NGSO Access Based on Reinforcement Learning

Based on the scenario and algorithm described earlier, it is assumed that there are a huge number of LEO satellites connected to ground base stations in space to form a communication network, and each satellite node has communication and forwarding functions. Different satellite nodes complete power allocation and ground base station access according to the algorithm described earlier. It is very important to determine the spectrum of the network and the connection order of relay satellites, and to complete the overall specific point-to-point communication network design based on the spectrum access information of each node. This problem can be optimized as a sequential decision-making problem. For each node, the model is unknown and needs to be explored through multi-step learning to make it very suitable for reinforcement learning, as shown in Figure 10 shows.

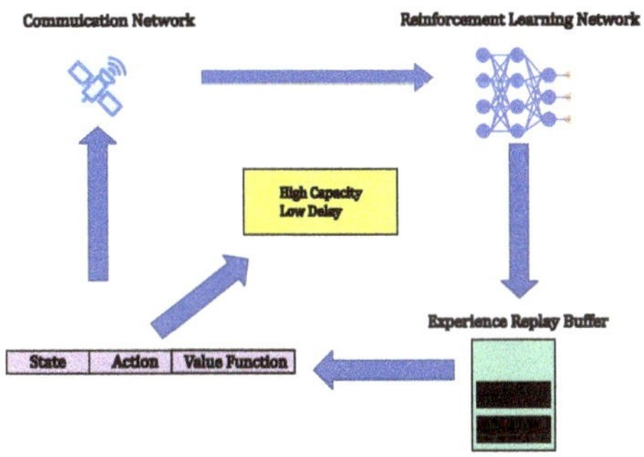

Figure 10. Spectrum access strategy update map based on Dueling DQN and empirical playback.

When considering the design of the spectrum access scheme for the satellite wireless autonomous network, the physical layer environment of the satellite network should be taken into account. Attention should be paid to the self-organizing spectrum and node access design scheme for point-to-point communication of network nodes within the satellite communication network [24]. For the reinforcement learning process of path and spectrum selection access design, assuming that adaptive coding and modulation are carried out in each link, the maximum total throughput that can be supported in the spectrum path is the minimum rate of the link connected by all points, so it is a function of the lowest signal-to-noise ratio as the bottleneck in each link of the data flow. Such bottleneck goals are not easily adapted to the Markov decision process framework, so reinforcement learning agents need to be trained to maximize the bottleneck goals [25].

Define the action space and state space of a reinforcement learning agent for a frequency band. The action of the agent is to select the next hop node for transmission. Since this paper uses a physical layer network model where the connections are not predetermined, we consider a fixed number of c available neighbors with the strongest channel environment as candidates. In addition, in order to ensure sufficient detection capability, we also add a re-detection action, which means that if the agent determines that no current strongest neighbor is a suitable next hop, it will continue to detect c strongest neighbors until it finds a suitable next hop. Thus, on a single band, the agent's action space consists of c+1 actions: (1) c independent action: Select each of the c strongest neighbors. (2) reprobate: Repeated detection of the new c strongest neighbors. If the agent reexplores, the agent stays at the current boundary node at the current time. In the next time step, the agent explores the next c nearest neighbors and collects the status information of these new neighbor sets.

We consider a wireless AD hoc network in a 1000 km $\times$ 1000 km region with F = 3 data streams. The data stream will be accessed for spectrum selection using B = 8 available frequency bands, each with 5 MHz bandwidth. The node density distribution is specified on nine equally divided subregions, each of which is randomly distributed with 6, 8, 7, 6, 5, 10, 8, 9, and 6 nodes.

We randomly generated 290,000 satellite wireless AD hoc network layouts to train our agents. Of these, 20,000 layouts are used for spectrum and communication node selection based on random exploration to generate the initial experience of the agent, 250,000 layouts are used to train the agent based on spectrum selection access based on the -greedy strategy, and the remaining 20,000 layouts are used for extended training, where 0 is set. To test, we randomly generate 500 new satellite wireless AD hoc network layouts based on the same distribution, in which case the number of satellite nodes accessing the network is about 65.

This paper uses the bottleneck SINR in the path as the future cumulative reward assigned Q to each state-action pair. For actual training, SINR is represented on the dB scale to get a more appropriate range of values. In addition, we added a constant bias to ensure that rewards are almost always positive. This extends the reward definition to situations where you can later add a delay-related discount factor. Specifically, the Q function is defined in terms of SINR in actual training as follows:

$$\tilde{Q}\left(S_t, a_t\right) = \min 10log_{10}\left(SINR_{n_i^{(f)}, n_{i+1}^{(f)}, b_i^{(f)}}\right) + bias, \tag{18}$$

Next, we present the total rate and minimum rate test results of the deep reinforcement learning agent, that is, the method using Dueling DQN, and compare them with a complete list of benchmarks. These benchmarks are greedy in nature because there is currently no effective global optimal algorithm for selecting access to the spectrum of AD hoc networks based on physical layer properties, so compare several currently common algorithms: (1) Strongest neighbor: Select the strongest neighbor of the wireless channel from the boundary node. (2) Nearest neighbor: Select the neighbor closest to the target node. If all neighbors are further from the destination than the agent, the agent will reject. (3) Minimum interference between neighbors: Select the neighbor with the least interference. (4) Maximum data rate between neighbors: Select the neighbor with the highest link capacity.

All methods follow multi-round sequential spectrum selection access. For Dueling Deep Q Network (DDQN) agents, the spectrum access solution is described above. For all benchmarks, the best frequency band is selected based on state characteristics. Figure 11 shows the cumulative probability distribution curve of the total speed. As shown in the figure, the proposed method has better performance than other algorithms in terms of total rate performance and minimum rate performance. We emphasize that this performance is achieved by reusing a single agent for all data streams, all frequency bands, and all test network layouts.

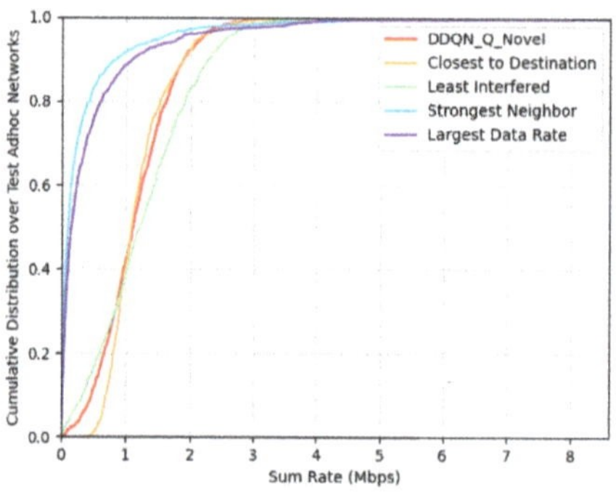

Figure 11. Performance comparison diagram of different algorithms.

5. Conclusions

This paper investigates the compatibility and sharing analysis between IMT-2020 in the 12.75–13.25-GHz frequency band and the uplink of Fixed Satellite Service (FSS). Based on IMT-2020 antenna and channel models, the analysis of the aggregate interference from IMT-2020 to FSS services reveals that, without any interference control measures, the interference from IMT-2020 base stations to FSS services exceeds the ITU standard threshold by approximately 10 dB. To achieve coexistence between the two services, this paper proposes a power control scheme based on partitioning, dividing ground base stations into several regions and virtualizing the base stations within each region into a virtual macro base station. Through power control of these virtual macro-base stations within the partitions, the maximum transmission power sum of all base stations within the region is obtained. Meanwhile, three intra-partition power control algorithms are proposed, including average power allocation, channel gain-based power allocation, and power allocation based on maximizing the sum rate within the partition. Simulation comparisons of the sum rate performance of different power allocation schemes show that average power allocation and power allocation based on maximizing the sum rate within the partition result in relatively small performance losses for the IMT-2020 system, achieving 100%

coverage of 50 Mbps rates under a 100-MHz bandwidth. In summary, this study provides an effective solution and theoretical support for the coordinated development of IMT-2020 and FSS services.

Author Contributions: Methodology, M.J.; validation, S.M.; investigation, H.W.; writing—original draft preparation, Z.T.; writing—review and editing, Z.J. All authors have read and agreed to the published version of the manuscript.

Funding: This work was supported by National Natural Science Foundation of China No. 62231012.

Data Availability Statement: Data is contained within the article.

Acknowledgments: The authors would like to thank all reviewers for their helpful comments and suggestions regarding this paper.

Conflicts of Interest: The authors declare no conflicts of interest.

References

1. Asmad, R.; Abdul, B.; Baiocchi, A. Analysis of Status Update in Wireless Networks with Successive Interference Cancellation. *arXiv* **2024**, arXiv:2409.00277.
2. Yuan, Y.; Zhu, L. Application Scenarios and Enabling Technologies of 5G. *China Commun.* **2014**, *11*, 69–79.
3. Zhang, X.; Chen, Z.; Zhang, Y.; Liu, Y.; Jin, M.; Qiu, T. Deep reinforcement learning-based distributed dynamic spectrum access in multi-user multi-channel cognitive radio internet of things networks. *IEEE Internet Things J.* **2024**, *11*, 17495–17509. [CrossRef]
4. Aderonmu, A.I.; Olanrewaju, O.A. Artificial intelligence-based spectrum allocation strategies for dynamic spectrum access in 5G and IMS networks. *ATBU J. Sci. Technol. Educ.* **2024**, *12*, 482–493.
5. Doke, K.; Abedi, A.; Hollingsworth, M.; Zheleva, M.; Sahai, A.; Grunwald, D.; Gremban, K. Towards Data-Driven Policies in Spectrum Management. In Proceedings of the IEEE International Symposium on Dynamic Spectrum Access Networks, Washington, DC, USA, 13–16 May 2024; pp. 163–168.
6. Pachler, N.; Crawley, E.F.; Cameron, B.G. Avoiding Self-Interference in Megaconstellations through Cooperative Satellite Routing and Frequency Assignment. *IEEE J. Sel. Areas Commun.* 2024, *Early Access*. [CrossRef]
7. Aliaga, S.; Petrov, V.; Jornet, J.M. Modeling Interference from Millimeter Wave and Terahertz Bands Cross-links in Low Earth Orbit Satellite Networks for 6G and Beyond. *IEEE J. Sel. Areas Commun.* **2024**, *42*, 1371–1386.
8. Ding, X.; Lei, Y.; Zou, Y.; Zhang, G.; Hanzo, L. Interference Management by Harnessing Multi-Domain Resources in Spectrum-Sharing Aided Satellite-Ground Integrated Networks. *IEEE Trans. Veh. Technol.* **2024**, *73*, 8306–8321. [CrossRef]
9. ITU-R. *Characteristics of Terrestrial Component of IMT for Sharing and Compatibility Studies in Preparation for WRC-23*; Document 5D/TEMP/422; ITU-APT SECRETARIAT ITU-APT Foundation of India 504: Delhi, India, 2021.
10. Hassan, W.A.; Jo, H.-S.; Tharek, A.R. The feasibility of coexistence between 5G and existing services in the IMT-2020 candidate bands in Malaysia. *China Commun.* **2017**, *5*, 14867–14888. [CrossRef]
11. Hattab, G.; Visotsky, E.; Cudak, M.; Ghosh, A. Coexistence of 5G mmWave users with incumbent fixed stations over 70 and 80 GHz. In Proceedings of the IEEE Globecom Workshops (GC Wkshps), Singapore, 4–8 December 2017; pp. 1–5.
12. Hassan, W.A.; Jo, H.-S.; Ikki, S. Spectrum-sharing method for co-existence between 5G OFDM-based system and fixed service. *IEEE Access* **2019**, *7*, 77460–77475. [CrossRef]
13. Kim, J.; Xian, L.; Sadri, A.S. Numerical simulation study for frequency sharing between micro-cellular systems and fixed service systems in millimeter-wave bands. *IEEE Access* **2016**, *4*, 9847–9859. [CrossRef]
14. Park, J.; Lee, E.; Park, S.-H. Modeling and analysis on radio interference of OFDM waveforms for coexistence study. *IEEE Access* **2019**, *7*, 35132–35147. [CrossRef]
15. Lim, J.W.; Jo, H.S.; Yoon, H.G.; Yook, J.G. Interference mitigation technique for the sharing between IMT-advanced and fixed satellite service. *J. Commun. Netw.* **2007**, *9*, 159–166. [CrossRef]
16. Guidolin, F.; Nekovee, M. Investigating spectrum sharing between 5G millimeter wave networks and fixed satellite systems. In Proceedings of the IEEE Globecom Workshops (GC Wkshps), San Diego, CA, USA, 6–10 December 2015; pp. 1–7.
17. Chen, J.; Qian, Z.; Wang, T.; Li, X. Analysis on the protection distance for spectrum sharing between IMT-2020(5G) and EESS systems in 25.5–27 GHz band. In Proceedings of the IEEE 2nd Information Technology, Networking, Electronic and Automation Control Conference (ITNEC), Chengdu, China, 15–17 December 2017; pp. 970–975.
18. Tan, H.; Liu, Y.; Feng, Z. Coexistence analysis between 5G system and fixed-satellite service in 3400–3600 MHz. *China Commun.* **2018**, *15*, 25–32. [CrossRef]
19. Guidolin, F.; Nekovee, M.; Badia, L.; Zorzi, M. Study on the coexistence of fixed satellite service and cellular networks in a mmWave scenario. In Proceedings of the IEEE International Conference on Communications (ICC), London, UK, 8–12 June 2015; pp. 2444–2449.
20. Kim, S.; Visotsky, E.; Moorut, P.; Bechta, K.; Ghosh, A.; Dietrich, C. Coexistence of 5G with the incumbents in the 28 and 70 GHz bands. *IEEE J. Sel. Areas Commun.* **2017**, *35*, 1254–1268. [CrossRef]

21. Cho, Y.; Kim, H.; Tettey, D.K.; Lee, K.J.; Jo, H.S. Modeling method for interference analysis between IMT-2020 and satellite in the mmWave band. In Proceedings of the IEEE Globecom Workshops (GC Wkshps), Waikoloa, HI, USA, 9–13 December 2019; pp. 1–6.
22. Liu, R.; Wu, Q.; Di Renzo, M.; Yuan, Y. A Path to Smart Radio Environments: An Industrial Viewpoint on Reconfigurable Intelligent Surfaces. *IEEE Wirel. Commun.* **2022**, *29*, 202–208. [CrossRef]
23. Kliks, A.; Kulacz, L.; Kryszkiewicz, P.; Bogucka, H.; Dryjanski, M.; Isaksson, M.; Koudouridis, G.P.; Tengkvist, P. Beyond 5G: Big Data Processing for Better Spectrum Utilization. *IEEE Veh. Technol. Mag.* **2020**, *15*, 40–50. [CrossRef]
24. ITU-R. *WP 3K Chairman's Report. Working Document Towards a Draft Revision of Recommendation ITU-R P.2108-1-Prediction of Clutter Loss*; Document 3K/178; ITU-APT SECRETARIAT ITU-APT Foundation of India 504: Delhi, India, 2022.
25. *3GPP TR 37.840*; Technical Specification Group Radio Access Network; Study of Radio Frequency (RF) and Electromagnetic Compatibility (EMC) Requirements for Active Antenna Array System (AAS) Base Station (Release 12). 3GPP: Sophia Antipolis, France, 2014.

electronics

MDPI

Article

Structure Design and Reliable Acquisition of Burst Spread Spectrum Signals Without Physical Layer Synchronization Overhead

Shenfu Pan [1], Leyu Yin [1,*], Yunhua Tan [2] and Yan Wang [1]

1 The 54th Research Institute of China Electronics Technology Group Corporation, Shijiazhuang 050081, China; 15651879079@163.com (S.P.); 287888326@163.com (Y.W.)
2 School of Electronics, Peking University, Beijing 100871, China; tanggeric@pku.edu.cn
* Correspondence: yinly0531@163.com

Abstract: In order to improve the concealment and security of a point-to-point transparent forwarding satellite communication system, a signal structure based on aperiodic long code spread spectrum is designed in this paper. This structure can achieve reliable signal acquisition without special physical layer synchronization overhead, which can effectively shorten signal transmission time and improve the concealment of communication. In addition, the performance of burst spread spectrum signal acquisition is analyzed in detail by establishing a mathematical model, and the influencing factors and design criteria of the matching filter length for aperiodic long code acquisition are determined. On this basis, a matched filter acquisition method based on high-power clock multiplexing and an adaptive decision threshold design method based on an auxiliary channel are proposed. The above methods effectively reduce hardware complexity and resource consumption caused by long code acquisition, and realize reliable acquisition under the condition of low SNR. The simulation results show that under the condition of Eb/N0 = 3 dB, the transmission efficiency for a 128-symbol burst frame can be increased by 50%, thereby significantly reducing the burst communication time. Furthermore, the acquisition success probability can reach 99.99%.

Keywords: burst spread spectrum; no synchronization overhead; acquisition; adaptive threshold

1. Introduction

Satellite communication has many advantages. It has a large communication coverage area, long communication distance, and the cost of building a station is almost independent of the communication distance. Besides this, it also has flexible networking, large communication capacity, good communication quality, strong reliability, etc. This paper focuses on the satellite communication system of UAVs and submarines, which uses a geosynchronous high orbit (GEO) satellite. It is well known that three satellites can achieve global coverage, which means only one satellite can achieve long-distance communication within 2000 km. The scenario in this paper adopts transparent forwarding and uses a custom private satellite communication protocol. We take the satellite-drone communication model as an example, as shown in Figure 1 below. The space and weight of platforms such as drones are severely limited, so a single input and a single output are usually used. Besides this, the caliber of the airborne antenna is usually less than 0.3 m. The signal frequency is in the ka/ku band. The SNR received by the airborne terminal is about 5–8 dB. Due to the influence of rain fade in the channel, the attenuation of the SNR should be considered. Therefore, in practical scenarios, reliable communication in the SNR of 3 dB should be satisfied at least.

For submarine and unmanned aerial vehicle satellite communication systems, as well as submarines and unmanned aerial vehicles used as stealth combat platforms, the communication system is required to have strong concealment, anti-interference and reliability. This is especially true for satellite communication based on transparent forwarding, because the channel is open, the signal in the communication process is vulnerable to eavesdropping and interception, and there are serious security threats such as the positioning of the combat platform and the acquisition of business information by the other party. The longer the exposure time, the higher the security risk. Burst spread spectrum combines the advantages of the low spectral density of spread spectrum signals and the short duration of burst communication, which is an effective anti-interception communication technique. Therefore, most UAV and submarine satellite communication systems adopt the burst spread spectrum communication mode at present.

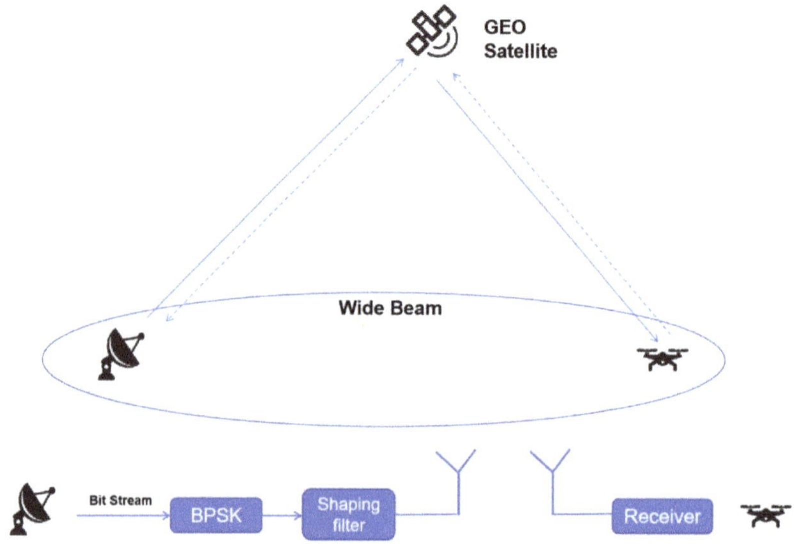

Figure 1. Satellite communication scenario.

Due to the randomness and temporality of burst spread spectrum communication, synchronization overhead is usually added to the physical layer to realize synchronization. The frame structure diagram of the physical layer of a typical burst spread spectrum signal is shown in Figure 2. A burst physical frame is usually composed of synchronous overhead and effective data, in which the effective data refer to modulation symbols after channel coding and symbol mapping. Synchronous overhead usually adopts a specific modulation symbol for a spread spectrum, which is generally composed of three parts, as follows: synchronization field 1 is used for spread spectrum code synchronization; synchronization field 2 is used for carrier frequency offset estimation and carrier initial phase estimation; synchronization field 3 is used to locate the start time of valid data. L is the total length of overhead 1, 2 and 3. $L1$ must be at least 64 bytes, and $L = L1 + L2 + L3$ must be at least 128 bytes [1,2].

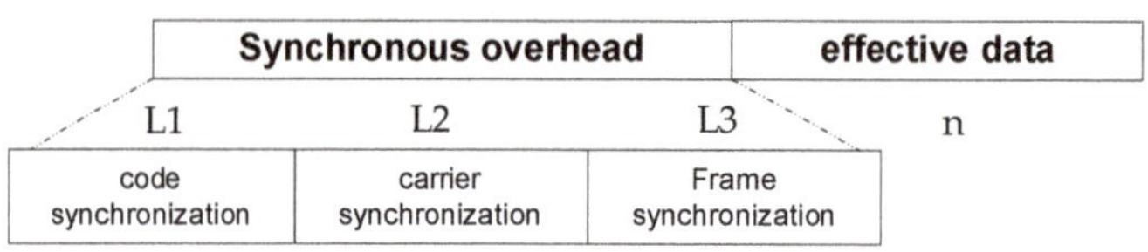

Figure 2. Physical layer frame structure of a typical burst spread spectrum signal.

The burst spread spectrum based on synchronous overhead usually adopts periodic code, and the burst spread spectrum signal is a fixed-length frame signal structure. The baseband signal processing process at the sending end (before he shaping filter) is shown in Figure 3. Here, N is the spread spectrum multiple, n is the length of effective data, and Rc is the code rate.

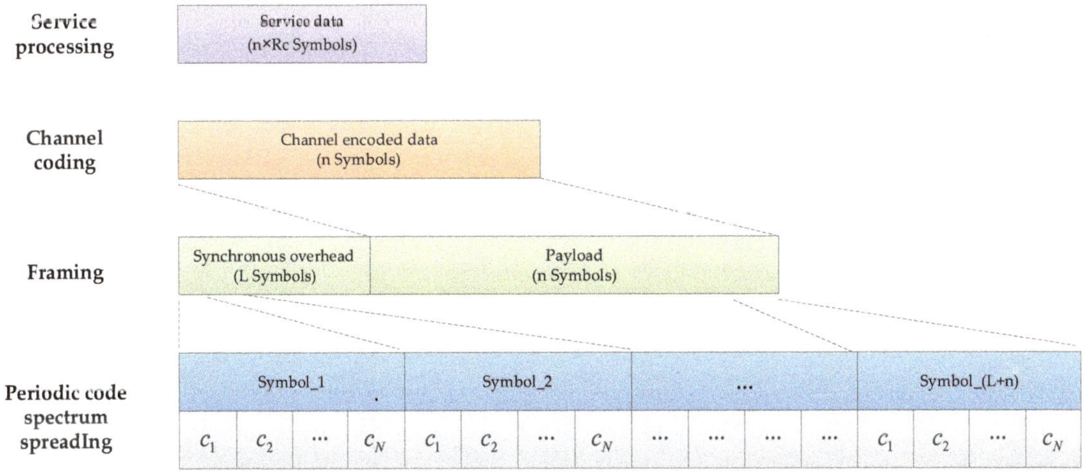

Figure 3. The processing flow of burst spread spectrum signal based on synchronous overhead.

The method based on synchronization overhead can make the process of signal synchronization simpler, but it will reduce the efficiency of information transmission. In particular, submarine and UAV satellite communication systems have the special conditions of low SNR and large carrier frequency offset application [3,4]. In order to achieve the correct synchronization of the carrier, usually a long synchronization header is needed, that is, long auxiliary data for capturing and synchronizing the carrier must be included in a burst frame. This will increase the synchronization overhead of the burst communication system, affect the communication efficiency, increase the signal space exposure time, and increase the probability of being intercepted. Removing the overhead of the synchronous head can greatly improve the transmission efficiency, shorten the airborne exposure time of the signal, and further improve the anti-interception performance of the communication.

At present, most burst spread spectrum systems realize synchronization based on synchronization overhead. Paper [5–11] uses the synchronization overhead of the leading code, frame header, special synchronization sequence and pilot frequency to realize system synchronization. However, no synchronization overhead is only realized in multiple-input multiple-output (MIMO) systems, which can achieve no synchronization overhead through the correlation of multiple signals [12]. The removal of the synchronization cost in single-input single-output spread spectrum systems only focuses on the carrier synchronization part, such as by using non-data-assisted methods [13], or it just focuses on removing frame synchronization overhead, as in paper [14]. The proposed method in this paper can improve the frame transmission efficiency by removing the synchronization overhead of the physical layer, and better complement the gaps in related fields.

The length and complexity of synchronization overhead are related to the acquisition mode, frequency offset, dynamics, SNR and so on. Burst spread spectrum communication has the characteristics of low power spectrum density, short burst time and random transmission time. Therefore, the system response time is required to be short enough, that is, the receiver should be able to quickly acquire the signal, demodulate the original signal, and wait for the arrival of the next signal. This puts forward a high requirement for the real-time operation and probability of spread spectrum code acquisition.

To remove the physical layer synchronization overhead and ensure synchronization, an aperiodic long code spread spectrum is used. However, the acquisition of aperiodic long codes has the problem of high hardware resource usage, so it is necessary to ensure the capture success rate and reduce the hardware resource usage.

Typical acquisition methods include sliding correlation, a matching filter, FFT, PMF-FFT, and so on. However, the sliding correlation method is slow and the search time is long, which can meet the requirement of fast acquisition set in this paper. The hardware complexity of the matching filter method is high, and the time domain correlation value is large, which is proportional to the square of the code period length, so it is not suitable for the long code acquisition in this paper. The FFT algorithm mainly deals with the acquisition of spread spectrum signals under the condition of high dynamic and high SNR [15], and is not applicable to the situation of low SNR in this paper. The PMF-FFT method can shorten the length of a single matching filter by increasing the number of matching filters, but it cannot effectively reduce the hardware resource consumption [16,17].

Aiming at long code acquisition, paper [18–20] studies the indirect acquisition method of short-code-assisted long code acquisition. However, when indirect acquisition is adopted, the short code with poor anti-interference ability is easily disturbed. Long codes cannot be captured once interfered with, which means they cannot meet the anti-interference requirements of the transparent forwarding satellite communication system in this paper. In addition, refs. [21–24] adopt the direct acquisition method of XFAST and use overlapping codes to perform correlation operations on the received signals. However, the folding process of local pseudo-codes will increase the mutual interference between pseudo-codes and reduce the sensitivity. Therefore, this method is more suitable for acquisition under the condition of high SNR, but not suitable for the condition of low SNR in this paper.

None of the existing acquisition methods can meet the requirements of acquisition success rate and hardware resource occupation at low SNR. Therefore, a matching filter acquisition method based on high-power clock multiplexing is proposed, and the adaptive decision threshold design method based on auxiliary channels is used to achieve the reliable acquisition of a spread spectrum code under the condition of low SNR.

Based on the above ideas, the main contributions of this paper are as follows:

- The synchronization head overhead is removed to improve transmission efficiency—a burst spread spectrum signal design method using an aperiodic long code spread spectrum is proposed, and a signal model without physical layer synchronization overhead is constructed;
- Reliable acquisition is ensured while eliminating synchronization overhead—the acquisition strategy of the aperiodic long code spread spectrum signal is given, and the factors affecting the acquisition performance are analyzed in detail;
- The hardware resource complexity problem caused by aperiodic long code acquisition is effectively solved—the matching filter length design method is analyzed, and a matching filter implementation method based on high-power clock multiplexing is proposed. This method can meet the requirements of resource-limited engineering applications and high security, and has wide application value;
- The reliable acquisition of spread spectrum code under the condition of low SNR is realized—an adaptive decision threshold design method based on auxiliary channels is proposed.

The rest of this paper is organized as follows. Section 2 introduces the design of the structure of a burst spread spectrum signal based on an aperiodic long code. In Section 3, the acquisition performance is analyzed, and the acquisition method of the aperiodic length code without physical layer synchronization overhead is designed. In Section 4, the design and implementation of matching filter length based on high-power clock multiplexing are introduced. In Section 5, an adaptive decision threshold design method based on auxiliary channels is presented. In Section 6, simulation and experimental tests are carried out, and the results are given and discussed. A summarization of this paper is given in Section 7.

2. Burst Spread Spectrum Signal Design Based on Aperiodic Long Code

In order to improve the efficiency of frame transmission, this paper presents a signal design method to remove the synchronous overhead. To remove the synchronization overhead, the signal structure needs to be redesigned. The burst spread spectrum signal without physical layer synchronization overhead adopts an aperiodic long code spread spectrum. The baseband processing process at the signal sender (before shaping filter) is shown in Figure 4.

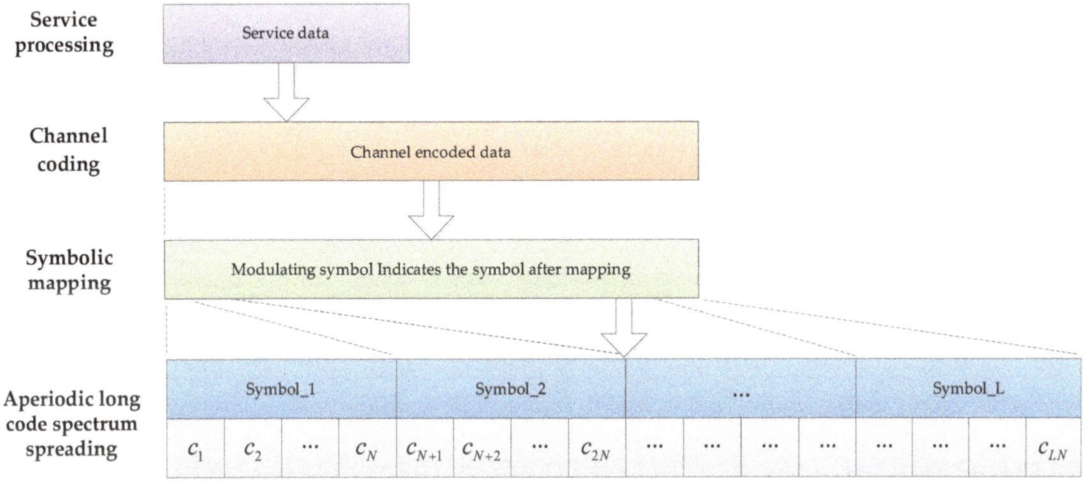

Figure 4. Processing flow of aperiodic long code signal sender without synchronization overhead.

Compared with the burst spread spectrum signal based on synchronous head, there are two main differences. First, there is no special synchronous head. Second, the spread spectrum code adopts an aperiodic length code, which does not repeat in a burst frame. Therefore, as long as the receiver realizes the synchronization of the spread spectrum code and locates the spread spectrum code, the beginning time of the burst signal and the position of the signal in the burst frame are determined. This enables bit synchronization and frame synchronization, and restores the signal structure.

The receiver processing block diagram of the aperiodic long code spread spectrum burst signal without physical layer synchronization overhead is shown in Figure 5. The specific work flow is as follows:

1. Cache the sampled data;
2. Spread spectrum code acquisition—Configure the matching filter parameters to start the spread spectrum code acquisition. The beginning time of the burst signal can be located after the spread spectrum code is captured;
3. Spread spectrum code tracking—After the spread spectrum code synchronization, the data of certain symbol length are taken from the cache for processing to complete the spread spectrum code tracking;
4. Carrier frequency offset correction—The best sampling point data corresponding to a certain symbol length are taken from the cache, and the local spread spectrum code is de-expanded. After the de-expanding, the carrier frequency offset is estimated based on the algorithm assisted by no data, and the estimated results are used to correct the frequency offset of the data in the cache;
5. Carrier phase recovery—After carrier frequency offset correction, the data are read from the cache for frequency conversion, de-expansion and carrier recovery. Because there is no data assistance, the phase ambiguity problem exists in the data after carrier recovery;

6. Extract valid data for channel decoding—Phase ambiguity needs to be identified and corrected during decoding.

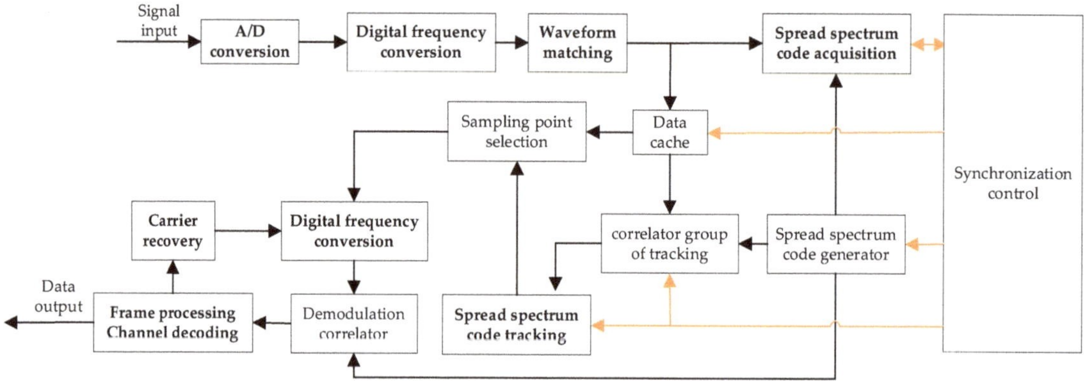

Figure 5. Block diagram of the receiver of burst spread spectrum signal.

The fast and reliable acquisition of an aperiodic long spread spectrum code is the key and most difficult point of receiving and processing burst spread spectrum signals without synchronization overhead of the physical layer. According to the application background of burst spread spectrum acquisition, the acquisition method based on matching filter is adopted. Compared with the burst spread spectrum receiving based on a synchronous head, a higher-order matched filter is needed and the hardware implementation complexity is higher.

The basic principle of using matching filter to achieve spread spectrum code acquisition is as follows. The correlation between the input signal and the local spread spectrum code is carried out in a modulation symbol period, and then the test statistics are obtained by modularization and multiple incoherent accumulations. Then, the test statistics and the threshold are judged to determine whether the input signal and the local spread spectrum code are synchronized, and the acquisition performance is determined by the incoherent accumulation and the decision threshold. Due to the short duration of the burst spread spectrum signal, it is necessary to complete the acquisition quickly and ensure the correctness of the acquisition. Therefore, an incoherent accumulation length and synchronous decision threshold should be reasonably designed for the given SNR constraints and acquisition success probability requirements.

3. Strategy and Performance Analysis of Spread Spectrum Code Acquisition

There is delay and rain attenuation in signal propagation, and satellite communication is exposed to various electromagnetic interferences in space. The satellite channel model can be designed as an ideal Gaussian channel; the SNR reflects the influence of rain decline, the anti-interference ability is enhanced by spread spectrum technology, and the signal synchronization is carried out by acquisition technology.

3.1. Signal Model

The spread spectrum signal waveform can be expressed as $s(t)$, and the chip period is T_c. The receiving end uses the root raised cosine filter to match the $s(t)$ signal, after which the channel response has the raised cosine characteristic [25]. The signal after passing through the matched filter can be expressed as $r(t)$, assuming that the local spread spectrum code is not strictly aligned with the received signal. Then, the normalized spread spectrum code synchronization error is τT_c, where τ is the normalized signal transmission delay.

$r(t)$ is sampled at the chip rate at time $t = kT_c$. Considering the complex baseband model under an ideal Gaussian channel, the sampled baseband signal can be expressed as

$$r(k) = As(k)e^{j(2\pi fkT_c+\varphi)} + n(k) \tag{1}$$

where A is the signal amplitude, f is the carrier frequency and φ is the carrier phase after matched filtering. $n(k)$ is the noise after sampling, where $n \sim N(0,\sigma_n^2)$. For detailed expressions, please refer to Appendix A.

3.2. Aperiodic Long Code Acquisition Process

The block diagram of burst spread spectrum code acquisition based on a matched filter is shown in Figure 6.

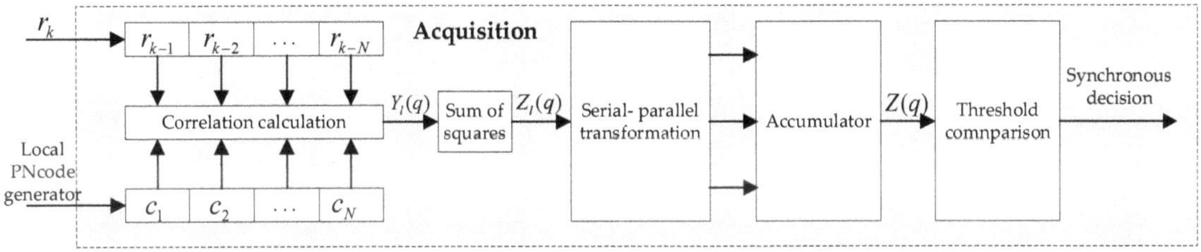

Figure 6. Block diagram of spread spectrum code acquisition based on matched filter.

Assuming that the spread spectrum multiple is N, the following strategies are generally adopted for synchronous acquisition in engineering:

The first step—The local spread spectrum code sequence is correlated with the received signal of different delays, and the sampling of the 2× bit rate clock is performed. There can be $2N$ correlation values in 1 symbol period. Since the spread spectrum code is an aperiodic long code, there are $2NL$ possible correlation values. The expression after correlation is

$$\begin{aligned} Y_l(q) &= \frac{1}{N}\sum_{k=lN}^{(l+1)N-1} c^*(k)r(k-q) \approx Ad_l h(-\tau T_c)\frac{\sin(\pi fT)}{\pi fT}e^{j\varphi_l} + \frac{1}{N}\sum_{k=lN}^{(l+1)N-1} c^*(k)n(k) + n_{2,l}(q) \\ &= Be^{j\varphi_l} + n_{1,l}(q) + n_{2,l}(q) \end{aligned} \tag{2}$$

where $B = Ah(-\tau T_c)\sin c(fT)$. $n_{1,l} \sim N(0,\sigma_n^2/N)$, where $n_{2,l}$ is the noise caused by inter-code crosstalk, which can be ignored. The local spread spectrum code is $c(k)$. q is the number of shift chips, $q = lN, lN+1/2, \ldots, (l+1)N-1$. $h(t)$ is the raised cosine function, and $h(-\tau T_c)$ is the amplitude change caused by the code synchronization error. $\sin c(fT)$ is the amplitude change caused by frequency offset. φ_l is code phase. For the relevant calculations, please refer to Appendix B.

Binary hypothesis testing is adopted, where

$$\begin{cases} Y_l(q) = n_{1,l}(q), & H_0 \\ Y_l(q) = Be^{j\varphi_l} + n_{1,l}(q), & H_1 \end{cases} \tag{3}$$

Here, H_0 is set under the condition that the sender does not send a spread spectrum signal or the local code is not aligned with the received signal, and H_1 is set under the condition of the local code being basically aligned with the received signal.

Y_l is a complex signal, where the signal in I channel is Y_I and thesignal in Q channel is Y_Q, and $Y_l = Y_I + jY_Q$. The conjugate multiplication of Y_l yields

$$Z_l(q) = Y_l(q)Y_l^*(q) = \sum Y_I^2(q) + Y_Q^2(q) \tag{4}$$

In the second step, the correlation values, Z_l, obtained in the first step are incoherently accumulated L times, and $2q$ test statistics are obtained,

$$Z(q) = \sum_{l=0}^{L-1} Z_l(q) = \sum_{l=0}^{L-1} Y_I^2(q) + Y_Q^2(q) \tag{5}$$

In the third step, the test statistic $Z(q)$ obtained in the second step is compared with the threshold V_{th}. If the value is greater than the threshold, it is considered as synchronization. If the threshold is not exceeded, the next statistics and judgment are carried out. It is hoped that only 1 of the $2LN$ possible correlation values that are fully synchronized will exceed the threshold.

Let P_n present the noise power σ_n^2/N, and so the symbol signal-to-noise ratio is $\mu = B^2/P_n$. p_0 is the probability density function under the condition of H_0 and p_1 is the probability density function under the condition of H_1. By integrating p_0 and p_1, the unit false alarm probability P_f and unit detection probability P_d of each decision can be obtained. Let the normalized threshold be $\lambda = V_{th}/\sigma_0^2$, according to [5,26]

$$P_f = \int_{V_{th}}^{\infty} p_0(Z)dZ = \int_{\lambda}^{\infty} p_0(z)dz = \int_{\lambda}^{\infty} \frac{x^{L-1}e^{-z}}{(L-1)!}dz = e^{-\lambda}\sum_{k=0}^{L-1}\frac{\lambda^k}{k!} \tag{6}$$

$$P_d = \int_{V_{th}}^{\infty} p_1(Z)dZ = \int_{\lambda}^{\infty} \left(\frac{x}{L\mu}\right)^{(L-1)/2} e^{-(x+L\mu)}\Phi_{L-1}(2\sqrt{L\mu x})dx = Q_L(\sqrt{2L\mu}, \sqrt{2\lambda}) \tag{7}$$

where $Q_L(a,b)$ is an L-order Marcum Q function, which can be solved by calling function $marcumq(a,b,L)$ in matlab. The specific form of this function is [27]

$$Q_L(a,b) = a^{-(L-1)} \int_b^{\infty} x^L e^{-\frac{x^2+a^2}{2}} \Phi_{L-1}(ax)dx \tag{8}$$

The synchronous acquisition of spread spectrum signals based on the above strategies has the following characteristics: a synchronous decision is made for every L symbols, and $2NL$ decisions can traverse all possible spread spectrum code phases. The probability of false acquisition can be reduced under the condition of low SNR.

3.3. Acquisition Success Probability and Incoherent Accumulation Length Analysis

The system false alarm probability is the same as the unit false alarm probability, that is,

$$P_F = P_f \tag{9}$$

As can be seen from Figure 7, if a false alarm occurs in the frame burst period before the successfully captured decision position, the receiver will directly transfer to the tracking state and cancel the acquisition. This will miss the successful acquisition decision position of the burst signal, resulting in a missed signal. If the number of decisions in a period is $M = 2NL$, there will be $M-1$ false alarms of the decision position in the previous frame period of the signal, which may lead to missing alarms. The probability of no missing alarms due to false alarms in the previous frame period is $(1-P_f)^{M-1}$. Then, the system missing probability, that is, the acquisition failure probability, is the missing probability of the correct decision position plus the failure probability caused by false alarms,

$$P_M = (1-P_f)^{M-1}P_m + 1 - (1-P_f)^{M-1} \tag{10}$$

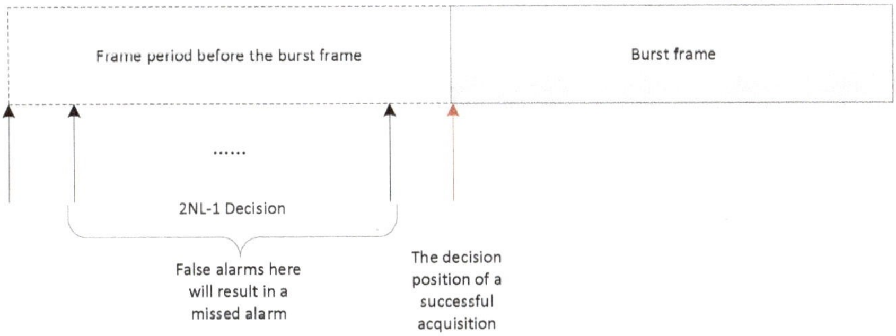

Figure 7. Signal acquisition decision diagram.

Therefore, the acquisition success probability is the system detection probability, which is $M-1$ (no false alarm) and 1 (no missing alarm), and its calculation formula is as follows:

$$\begin{aligned} P_{acq} &= P_D = 1 - P_M = (1-P_f)^{M-1}P_d \\ &= \left(1 - e^{-\lambda}\sum_{k=0}^{L-1}\frac{\lambda^k}{k!}\right)^{2NL-1} Q_L(\sqrt{2L\mu}, \sqrt{2\lambda}) \end{aligned} \tag{11}$$

In addition, it can be seen from Equation (11) that the acquisition success probability is jointly determined by the incoherent accumulation length L and the normalized decision threshold λ. The larger the number of incoherent accumulations, the higher the maximum value of P_{acq}, but an increase in L will increase the hardware implementation complexity. Based on the goal of improving the success probability of the acquisition of burst spread spectrum signal, the length L determination method is presented in Section 4.1; a matched filter acquisition implementation method based on high-power clock multiplexing is proposed in Section 4.2 to effectively reduce hardware complexity and resource consumption; and an adaptive decision threshold design method based on the auxiliary channel is proposed in Section 5. Based on the above designs, the success probability of burst spread spectrum acquisition can be optimized.

4. Design and Implementation of Matched Filter Based on High-Power Clock Multiplexing

4.1. Matched Filter Length Design

According to Formula (11), when the length of incoherent accumulations L is fixed, the values of P_{acq} under different normalization thresholds are different, and P_{acq} can reach the maximum at a certain threshold. The maximum value of P_{acq} is different under different incoherent accumulation lengths, and is denoted as $P_{max}(L)$. Increasing the length of incoherent accumulations can improve the success probability of acquisition at suitable thresholds. As can be seen from Figure 8, the larger the length of incoherent accumulations, the larger the $P_{max}(L)$ value, and it gradually tends to 1. According to the requirement of acquisition success probability, when the maximum acquisition success probability under a certain incoherent accumulation length is large enough, the appropriate length of incoherent accumulation can be obtained. A detailed simulation is shown in Section 5.1.

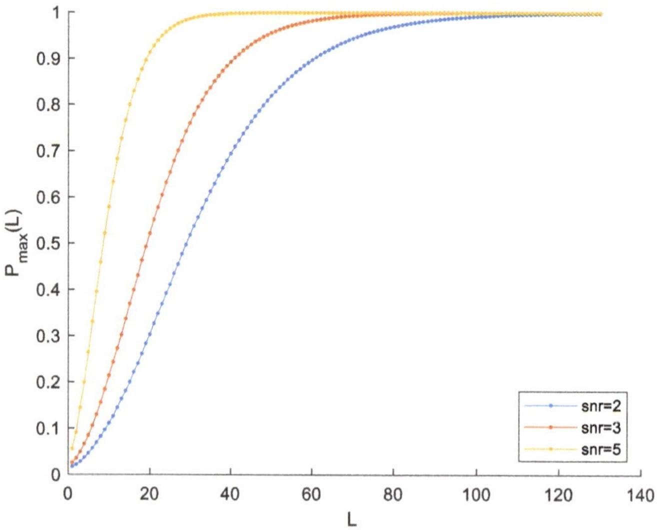

Figure 8. The relationship between the maximum value of P_{acq} and L.

4.2. Implementation of Matched Filter Based on High-Power Clock Multiplexing

Because the increase in length will increase the complexity of hardware implementation, this paper presents a solution of high-power clock multiplexing, which can effectively reduce the resource consumption. Taking $8\times$ clock multiplexing as an example, the structural difference between multiplexing and non-multiplexing is shown in Figure 9.

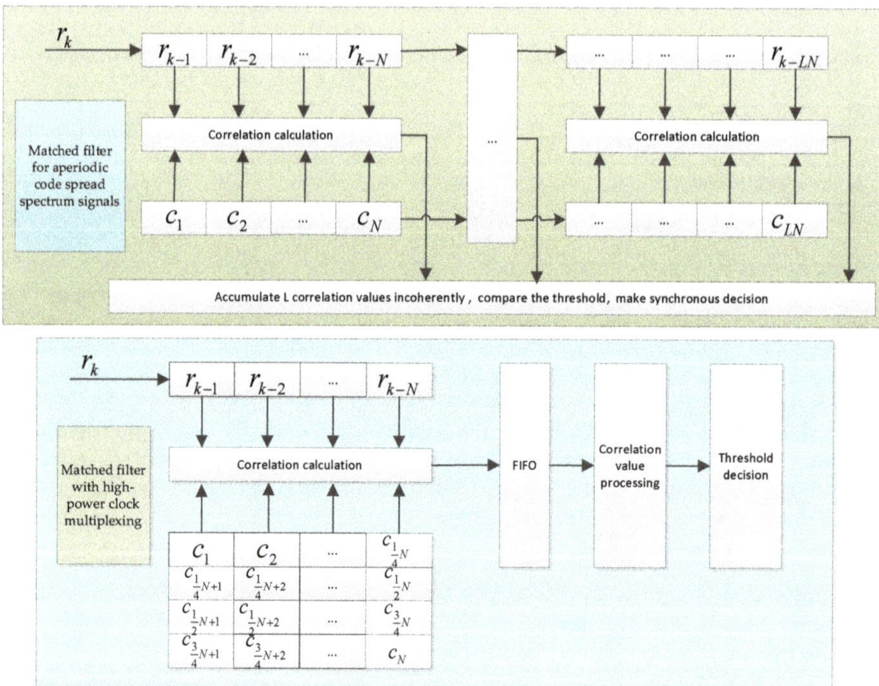

Figure 9. Comparison of non-multiplexed and multiplexed long code matched filters.

We assume that the spread spectrum ratio is N, the number of incoherent accumulations is L, and the system is driven by an $8\times$ code clock (denoted as $sclk$). The minimum code phase step during code acquisition is half the chip width, corresponding to the sampling of the $2\times$ code clock, that is, two sample points per chip. Since the system is driven by $8\times$ code clock (denoted as $sclk$), the duration of each sample point is four $sclk$ cycles. Therefore, the length of the shift register of the matched filter input data is designed to be $N/4$. The input signal $r(k)$ is shifted once every four $sclk$ cycles, and the register enters a sample point. The length of a symbol after the spread spectrum is N chips, with $2N$ sample points. Therefore, stepping through $2N$ times yields the complete signal after spreading for one symbol.

Since X_l is only updated every four $sclk$ cycles, the data in the register can be completely updated every $2Nsclk$ cycles. When the correlation operation is performed, one sample point is taken for each code slice. Using $sclk$ drive, four partially relevant values can be obtained during this period,

$$Y_{I,l}(q) = X_l(q)C_I = \sum_{k=IlN/4}^{I(l+1)N/4-1} r(k-q)c_I{}^*(k),\ I = 0,1,2,3 \tag{12}$$

where $q = IlN/4, IlN/4 + \frac{1}{2}, \ldots, I(l+1)N/4 - 1$.

After the data in the register are fully updated four times, that is, after $8N$ $sclk$ cycles, the fully correlated value of a symbol can be obtained, which can be expressed as

$$Y_l(q) = \sum_{k=lN}^{(l+1)N-1} r(k-q)c^*(k),\ \ where\ q = lN, lN + \frac{1}{2}, \ldots, (l+1)N - 1 \tag{13}$$

In order to save storage resources, FIFO with a depth of $2N$ is designed, and $sclk$ is used as the working clock. If the partial correlation value $Y_{I,l}(IlN/4)$ is directly taken as the current input of the FIFO, then the current output of the FIFO is $Y_{I-1,l}((I-1)lN/4)$. Assuming that the input of FIFO is a and the output is b, in order to obtain the fully correlated value of (13), the input design of FIFO is as follows:

$$a = \begin{cases} Y_{I,l}(q) + b, & I \neq 3 \\ 0, & I = 3 \end{cases} \tag{14}$$

Thus, every eight $sclk$ cycles (corresponding to one sampling cycle), we can obtain a fully correlated value corresponding to a certain code phase,

$$Y_l(q) = Y_{I,l}(q) + b\,|_{I=3} \tag{15}$$

According to Formula (5), in order to improve the reliability of the synchronous decision, the correlation values of L symbols are incoherently accumulated before the decision. An FIFO with a depth of $2N$ is designed and driven by the sampling clock (2x code clock). If $Z_l(lN)$ is directly taken as the current input to the FIFO, then the current output of the FIFO is $Z_{l-1}((l-1)N)$. Assuming that the input of FIFO is a and the output is b, in order to obtain the decision statistics after cumulation, the input design of FIFO is as follows:

$$a = \begin{cases} Z_l(q) + b, & count < 2N(L-1) \\ 0, & count \geq 2N(L-1) \end{cases} \tag{16}$$

In this way, the $2N$ cumulative decision statistics corresponding to each code phase are obtained in each L symbol period,

$$Z(q) = Z_l(q) + b\,\Big|_{count \geq 2N(L-1)} \tag{17}$$

The result of (17) can be compared with the set threshold to make a synchronous decision. Therefore, in the following section, we design the input threshold.

5. Adaptive Decision Threshold Design Method Based on Auxiliary Channel

In the actual burst spread spectrum communication system, the noise intensity of the channel is constantly changing with time. Therefore, in signal detection, if the fixed threshold is used, it will be difficult to detect the signal correctly and accurately. Therefore, it is necessary to adaptively change the decision threshold according to the strength of the current background noise to maintain a high detection probability [28,29]. At this time, the PN code acquisition of the wireless mobile communication system can be completed only by real-time estimation of the decision threshold according to the statistical characteristics of the test statistics Z [30,31]. Based on the statistical characteristics Z of the test statistic, and the relationship between unit false probability, unit detection probability, acquisition success probability and decision threshold in (6), (7) and (11), an adaptive design method of a decision threshold based on the auxiliary channel is proposed.

The block diagram of the acquisition algorithm based on adaptive threshold decisions is shown in Figure 10. The acquisition threshold Vth is related to the normalized threshold λ and noise power P_n.

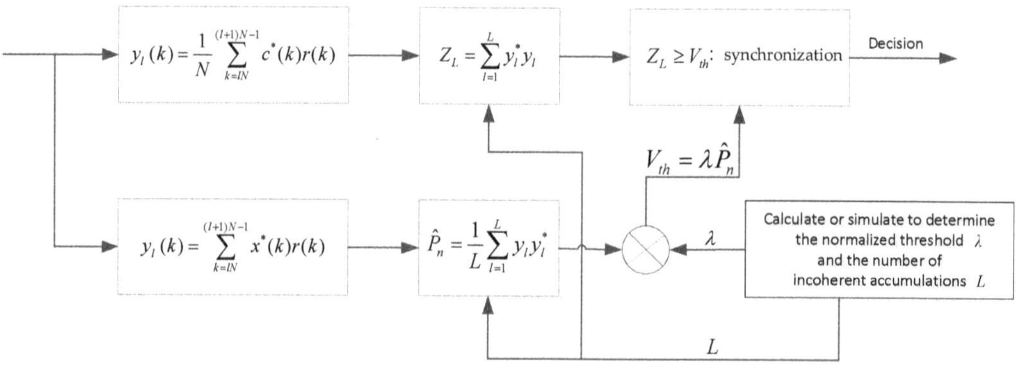

Figure 10. Acquisition algorithm based on adaptive threshold decision.

5.1. Adaptive Threshold Design

The normalization threshold λ and the acquisition threshold V_{th} are defined by (6) and (7) in Section 3. The normalization threshold does not vary with the SNR, while the acquisition threshold is the decision threshold of the test statistic Z. The relationship between λ and V_{th} is $\lambda = V_{th}/P_n$.

The expression of P_{acq} in relation to λ can be obtained from (11), and the relationship between λ and P_{acq} when other variables are fixed is shown in Figure 11. If $P_{acq}(\lambda) > \gamma$, the target requirement is met. Therefore, the minimum and maximum λ are obtained by $P_{acq}(\lambda) = \gamma$. The first solution of the equation is $\lambda_{\min}$, and the second solution is λ_{max}. $\lambda_{\min}$ and λ_{max} are related to the incoherent accumulation length L, the number of decisions M, the symbol signal-to-noise ratio μ and the target acquisition probability γ, that is, $\lambda_{min} = f(L, M, \mu, \gamma)$, $\lambda_{max} = g(L, M, \mu, \gamma)$. We can use γ to constrain the threshold, obtain the maximum and minimum λ by solving the equation, and select a suitable intermediate value as the λ. When the P_{acq} is maximum, λ is λ_{opt}, which is only related to L, M and μ.

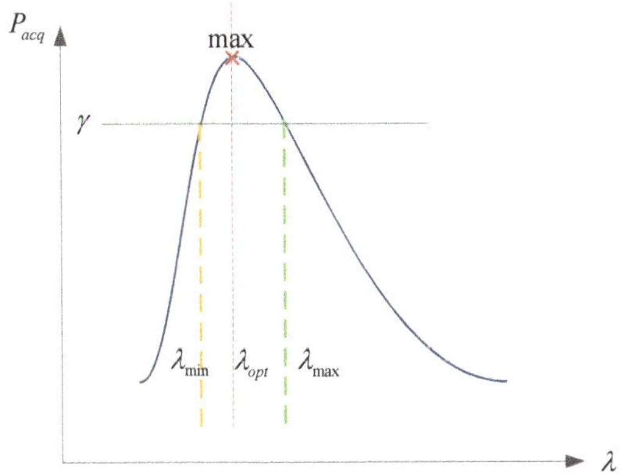

Figure 11. The relationship between λ and P_{acq}.

If we choose λ_{opt} at the maximum P_{acq}, we need to ensure that the selected V_{th} is still within the acceptable range after being affected by the noise estimation error. Because $V_{th} = \lambda P_n$, the impact of noise estimation error $\Delta\hat{P}_n$ on V_{th} at this time (that is, the acquisition threshold estimation error) is $\Delta V = \Delta\hat{P}_n \times \lambda_{opt}$. When the error is within the required range of P_{acq}, its relationship with $V_{\min}$ and $V_{\max}$ is as shown in Figure 12.

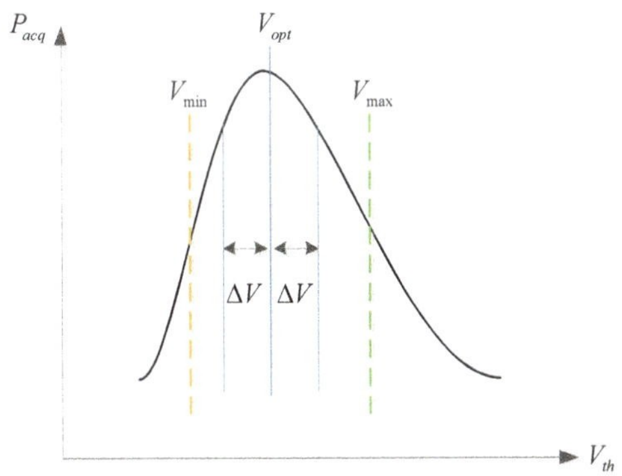

Figure 12. Acquisition threshold estimation error ΔV diagram.

V_{th} is related to L, M, μ and P_n, so we still need to analyze μ and P_n.

5.2. Symbolic SNR μ and Noise Power P_n Estimation

According to Section 5.1, λ is related to μ. The frequency error and code synchronization error will lead to the deterioration of μ, which will affect the selection of V_{th}. Therefore, the influence of frequency error and code synchronization error on SNR is calculated through formula derivation. See Appendix B for the derivation process.

SNR after despreading is

$$SNR = \left[\frac{\sin(\pi f T)}{\pi f T}\right]^2 \frac{NA^2h^2(-\tau T_c)}{\sigma_n^2} = \left[\frac{\sin(\pi f T)}{\pi f T}h(-\tau T_c)\right]^2 \left(\frac{E_s}{N_0}\right) \quad (18)$$

According to (18), μ is affected by the synchronization error of the spread spectrum code and carrier frequency difference. It is assumed that the threshold SNR required by the system is $(E_s/N_0)_{th}$ dB. In the signal acquisition phase, it can be assumed that $|fT| < 1/4$, $|\tau T_c| < 1/4$. Then, through calculation, when $0 < \alpha \leq 1$, SNR deteriorates 1.82 dB $\leq d \leq 2.34$ dB. The signal-to-noise ratio of the front receiving channel of the receiver deteriorates to d' and is considered at a maximum of 0.5 dB. Then, the actual working SNR based on logarithmic representation is

$$\mu \geq \left(\frac{E_s}{N_0}\right)_{th} - d - d' \text{ dB} \tag{19}$$

where $\frac{E_s}{N_0} = \frac{E_b}{N_0} + 10\lg m + 10\lg N(\text{dB})$, $m = \begin{cases} 1, & when\ BPSK \\ 2, & when\ QPSK \end{cases}$.

5.3. Noise Estimation Based on Auxiliary Channels

Since $V_{th} = \lambda P_n$, when λ is constant, V_{th} is only related to P_n. The estimator of P_n is $\hat{P}_n$. The real-time estimation of noise variance can be realized based on auxiliary channels. We multiply the received sample sequence with unrelated pseudo-noise sequences (values of ± 1) and add them up in a symbolic period to obtain

$$y_l = \frac{1}{N} \sum_{k=(l-1)N+1}^{lN} x^*(k) r(k) \tag{20}$$

Assuming that the observation length is L_{obs} and the number of auxiliary channels is 1, the estimation result of the noise variance of the auxiliary channel is as follows:

$$\hat{P}_n = M_1 = \frac{1}{L_{obs}} \sum_{l=1}^{L_{obs}} y_l {y_l}^* \tag{21}$$

If based on p auxiliary channels, the estimated noise M_p of p auxiliary channels can be calculated using p uncorrelated pseudo-noise sequence $x(k)$, and its mean is the estimated value of P_n

$$\hat{P}_n = \frac{1}{p} \sum_{i=1}^{p} M_p \tag{22}$$

When the SNR changes at high speed, the number of auxiliary channels can be increased, and the length of observation can be shortened, so the noise estimation can become more accurate.

6. Results and Discussion

The burst spread spectrum based on synchronous overhead has some problems, such as low transmission efficiency, long space exposure time and decline in anti-interception performance. So, the optimal design of the burst spread spectrum is carried out. With P_{acq} better than 0.9999 and SNR of 3 dB before despreading as the design constraint, the design method of burst spread spectrum without synchronization overhead is studied using the BPSK modulation method. According to the relevant background project situation, the data length of 128 symbols can meet the needs of submarine and UAV burst transmission. Therefore, this paper assumes that the design requirements of burst spread spectrum are shown in Table 1:

Table 1. Design requirements.

Design Requirements	Parameters
E_b/N_0	3 dB
data length	16 bytes, 128 symbols
spread spectrum ratio	128
encoding mode	LDPC (1/2)
symbol rate	32 ksps
modulation mode	BPSK
demodulation mode	coherent demodulation
signal duration	4 ms

In addition, $\alpha = 0.3$, and demodulation deterioration is not considered temporarily, while only the carrier frequency difference and code synchronization error are considered, which are both 1/4, then $d = 1.87$ dB.

6.1. Incoherent Accumulation Length Design

A P_{acq} of at least 0.9999 is required for successful communication. When SNR before despreading is 2, 3, and 5 dB, P_{acq} is simulated according to formula $P_{acq} = (1 - P_f)^{2NL-1} P_d$ (11). The image of P_{acq} changing with λ can be obtained as shown in Figure 13, and the simulation data under various SNRs are given in Table 2.

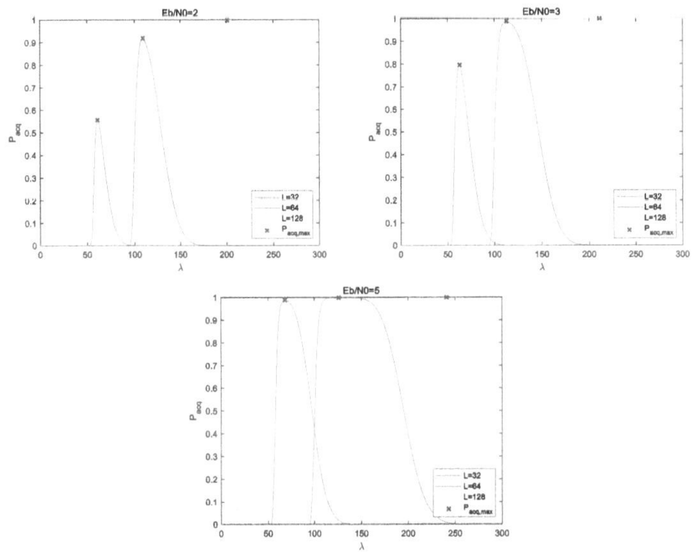

Figure 13. P_{acq} of different L and different SNR varying with λ.

Table 2. Data of the P_{acq} maximum points when SNR before despreading is 2, 3 and 5 dB.

Eb/N0 (dB)	L (Symbols)	λ_{opt}	P_{acq}	P_d	P_f
2	32	61.3	0.5586	0.63	1.4693×10^{-5}
	64	108.7	0.9202	0.9413	1.3865×10^{-6}
	128	201.3	0.9988	0.9992	1.2863×10^{-8}
3	32	62.8	0.796	0.8416	6.7883×10^{-6}
	64	112.7	0.9881	0.9920	2.3684×10^{-7}
	128	**211.4**	**1**	1	2.4605×10^{-10}
5	32	68.2	0.991	0.994	3.6968×10^{-7}
	64	**126.2**	**1**	1	3.5931×10^{-10}
	128	**241.5**	**1**	1	3.8858×10^{-16}

According to the simulation results, the incoherent accumulation length needs to reach 128 symbols under the 3 dB condition. Under the condition of 5 dB, the incoherent accumulation length of 64 symbols can meet $P_{acq} \geq 0.9999$. (Results that meat P_{acq} requirement are shown in bold in the table.) Monte Carlo simulation is carried out for the acquisition with an incoherent accumulation length of 128 at 3 dB. It can be observed in Figure 14 that the Monte Carlo simulation results are basically consistent with the theoretical curve, and the maximum acquisition success probability meets $P_{max} \geq 0.9999$.

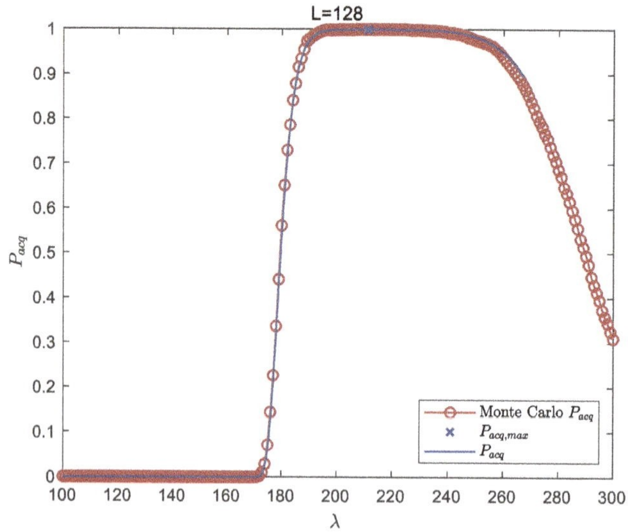

Figure 14. Comparison of theoretical curve and Monte Carlo simulation curve.

However, the adaptive constant false alarm threshold [1,2] did not consider the false alarm caused by a false alarm, and only selected the threshold according to P_f rather than the acquisition success probability. According to this method, the threshold obtained when P_f = 0.0001 is 166. It can be observed from the above figure that the acquisition success probability is close to 0 at this time (the theoretical value is 2.3122×10^{-14}), and the acquisition success probability of the Monde Carlo simulation results is 0. This threshold cannot achieve successful acquisition, and the false alarm value must continue to be reduced. Based on the acquisition success probability, the threshold setting method in this paper is equivalent to giving the exact index basis of false alarm setting on the basis of a constant false alarm threshold, so as to ensure the acquisition success probability of 99.99%.

6.2. Adaptive Threshold Under the Probability of Target Acquisition Success Probability

As can be seen from Table 2, when the SNR before despreading is 3 dB, and L = 128, $P_{acq} \geq 0.9999$. In Figure 15, λ_{min} and λ_{max} are obtained at P_{acq} = 0.9999. In this case, 100,000 Monte Carlo simulations are performed, and the simulation results are shown in Figure 16. Due to the fluctuation of noise variance, the acquisition thresholds also change. Similar to Figure 15, Table 3 can be obtained through simulation.

Table 3. Normalized threshold λ range when Eb/N0 = 3 dB, L = 128.

γ	λ_{opt}	λ_{min}	λ_{max}
0.9999		205.1	220.9
0.999	211.4	199	231.2
0.99		192.6	246.9

According to Section 5.1, because $V_{th} = \lambda P_n$, the acquisition threshold estimation error is $\Delta V = \Delta \hat{P}_n \times \lambda_{opt}$. Taking the median noise $P_n = 0.2535$ of 100,000 simulations as an example, Table 4 can be calculated based on the data in Table 3. The noise power estimation error requirements $\Delta \hat{P}_n$ under different acquisition success probabilities can be obtained by referring to Tables 4 and 5. For example, if we want $\gamma = 0.9999$, then the noise power estimation errors need to be $0.01 \leq \Delta \hat{P}_n \leq 0.005$. This method can effectively offset the negative effect of noise estimation error on V_{th}.

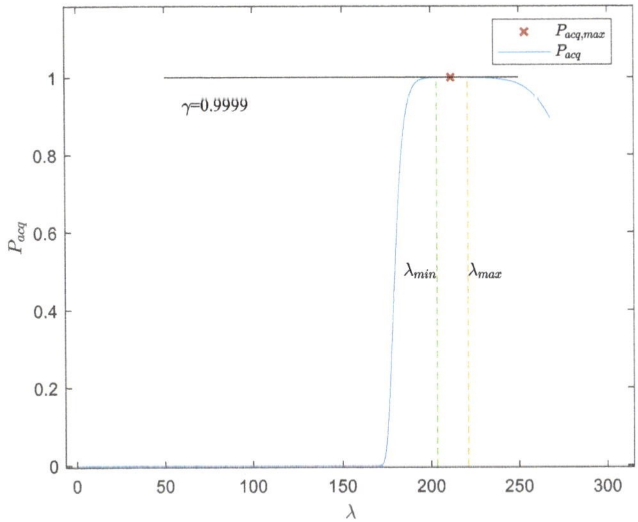

Figure 15. The maximum and minimum normalized thresholds when the target acquisition success probability γ is 0.9999.

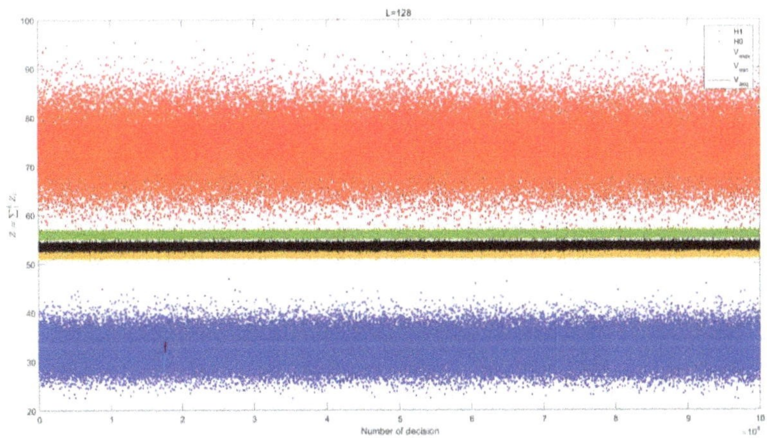

Figure 16. Acquisition simulation diagram when Eb/N0 = 3 dB, L = 128.

Table 4. The acquisition threshold range under different acquisition success probability requirements γ.

P_n	V_{opt}	γ	$V_{\min}$	V_{max}
0.2535	53.5899	0.9999	51.9928	55.9982
		0.999	50.4465	58.6092
		0.99	48.8241	62.5892

Table 5. The estimation acquisition threshold range under different noise estimation errors $\Delta\hat{P}_n$.

$\Delta\hat{P}_n$	ΔV	$V_{opt}-\Delta V$	$V_{opt}+\Delta V$
0.01	2.114	51.4735	55.7015
0.005	1.0570	52.5329	54.6469
0.001	0.2114	53.3785	53.8013

Tables 6 and 7 show the simulation results of noise power under different observation times and numbers of auxiliary channels. Based on the experimental data in Tables 6 and 7, the noise observation error meets the requirement of $P_{acq} \geq 0.9999$. With the increase in observation time and the number of auxiliary channels, the estimated noise error decreases gradually.

Table 6. The estimation of noise power under different observed lengths.

Observed Length L_{obs}	err	bias	RMSE
1	0.0011	2.3631×10^{-4}	0.0333
64	1.6873×10^{-5}	4.8980×10^{-5}	0.0041
128	8.4187×10^{-6}	3.3503×10^{-6}	0.0029

Table 7. The estimation of noise power under different numbers of auxiliary channels.

Numbers of Auxiliary Channels	err	bias	RMSE
1	3.4355×10^{-5}	-3.5409×10^{-5}	0.0059
4	8.7349×10^{-6}	2.2638×10^{-5}	0.0030
8	4.4050×10^{-6}	7.9486×10^{-6}	0.0021

Taking the Monte Carlo simulation of a successful acquisition as an example, as shown in Figure 17, it can be observed that successful acquisition can be achieved when the acquisition threshold V_{th} is between $V_{\min}$ and V_{max}.

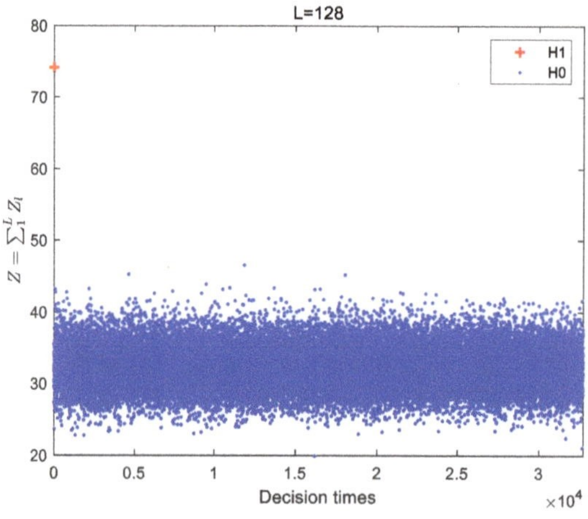

Figure 17. A successful acquisition of the Monte Carlo simulation.

6.3. Hardware Resource Consumption

The hardware of model xc7k325tffg676-2 was selected for simulation on the Vivado 2018 software platform. This paper adopts 8× clock multiplexing with two sample points per chip. As can be seen from the comparison of Tables 8 and 9, LUT decreased by 52.74%, FF decreased by 48.72%, and BRAM decreased by 32.58%.

Table 8. Non multiplexing.

Resource	Utilization	Available	Utilization%
LUT	189,470	203,800	92.97
FF	320,806	407,600	78.71
BRAM	414	445	93.03

Table 9. 8× clock multiplexing.

Resource	Utilization	Available	Utilization%
LUT	81,982	203,800	40.23
FF	125,913	407,600	30.89
BRAM	269	445	60.45

6.4. Frame Transmission Efficiency

Assuming an Eb/N0 of 3 dB, according to the above simulation, in order to achieve a reliable decision, the synchronization overhead of the spread spectrum code synchronization is 128 symbols (after coding) [1,2]. Valid data comprise 128 symbols. Frame length is (synchronization overhead + valid data =) 256 symbols. The data transfer efficiency with synchronization overhead is (valid data length/frame length=) 128/256 = 50%. After the synchronization cost is removed, the data transmission efficiency can be 100%, and the data transmission efficiency can be increased by up to 50%. The total signal length and transmission time are reduced by half, and the concealment of communication is improved.

7. Conclusions

In this paper, a signal structure based on the aperiodic long code spread spectrum is designed to improve the concealment and security of communication based on a transparent forwarding point-to-point satellite communication system. It can achieve reliable signal acquisition without special physical layer synchronization overhead, which can effectively shorten signal transmission time and improve the concealment of communication. The receiver is designed based on a matched filter with high-power clock multiplexing to reduce hardware complexity and hardware resource consumption. In addition, the decision threshold is estimated based on the auxiliary channel in real time to realize the adaptive synchronous decision in the noisy environment and ensure the acquisition success probability. The simulation results show that under typical short burst conditions, the transmission time of a burst signal can be shortened by up to 50% compared with the matching filter acquisition method with synchronization cost, and the acquisition success probability can reach more than 99.99% when Eb/N0 = 3 dB. In this paper, xc7k325tffg676-2 hardware is used to simulate on the Vivado platform, and the hardware resource consumption of the capture matching filter can be reduced by more than 40%.

In future work, we will study the acquisition technology of the burst spread spectrum satellite communication with high dynamic and large frequency offset.

Author Contributions: Conceptualization, S.P. and L.Y.; methodology, S.P. and L.Y.; software, L.Y. and S.P.; validation, L.Y.; formal analysis, S.P. and Y.W.; investigation, L.Y. and Y.T.; resources, S.P. and Y.T.; data curation, Y.T. and Y.W.; writing—original draft preparation, L.Y.; writing—review and editing, S.P., L.Y. and Y.W.; visualization, L.Y.; supervision, S.P. and Y.W.; project administration, Y.T.; funding acquisition, S.P. and Y.T. All authors have read and agreed to the published version of the manuscript.

Funding: This research was funded by an Adaptive Satellite Communication Technology Project, under grant number 2024-173ZD-040.

Data Availability Statement: The original contributions presented in the study are included in the article, and further inquiries can be directed to the corresponding authors.

Conflicts of Interest: Authors Shenfu Pan, Leyu Yin and Yan Wang were employed by the company The 54th Research Institute of China Electronics Technology Group Corporation. The remaining authors declare that the research was conducted in the absence of any commercial or financial relationships that could be construed as a potential conflict of interest.

Abbreviations

The following abbreviations are used in this manuscript:

GEO	Geosynchronous Orbit
MIMO	Multiple-Input Multiple-Output
FFT	Fast Fourier Transform
PMF-FFT	Partial Matched Filter-Fast Fourier Transform
XFAST	Extended replica Folding Acquisition Search Technique
SNR	Signal to Noise Ratio
ISI	Intersymbol Interference

Appendix A. Signal Expression

The spread spectrum signal waveform can be expressed as follows (data symbol d_0 corresponds to spread spectrum code chip $c_0 \sim c_{N-1}$):

$$s(t) = \sum_{p=-\infty}^{\infty} d_p \sum_{m=-\infty}^{\infty} c_m w(mT_c - pT) h_T(t - mT_c) \tag{A1}$$

where d_p is the data symbol sent, the chip period is T_c, the spread spectrum multiple is N and the symbol width is $T = NT_c$. $h_T(t)$ is the signal forming pulse, assumed to be the root rising cosine pulse in this paper, where α is the roll-off factor and has $0 < \alpha \leq 1$ [25]. $w(t)$ is the window function,

$$w(t) = \begin{cases} 1 & 0 \leq t < NT_c \\ 0 & else \end{cases} \tag{A2}$$

The receiver applies a matched filter with an impulse response of $T_c^{-1} h_T(t)$ to the signal in Formula (A1). Then the channel response has the raised cosine characteristic. Considering the complex baseband model in an ideal Gaussian channel, the matched filtered signal can be approximated as

$$r(t) = As(t - \tau T_c) e^{j(2\pi f t + \varphi)} + n(t) \tag{A3}$$

where $s(t - \tau T_c)$ is the baseband signal waveform and $n(t)$ is the noise after matched filtering. In addition, A is the signal amplitude, τ is the normalized signal transmission delay, f is the carrier frequency and φ is the carrier phase after matched filtering.

Sample $s(t - \tau T_c)$ is set according to the chip rate and the sampled data $s(k)$. Assuming that the local spread spectrum code is not strictly aligned with the received signal, the normalized spread spectrum code synchronization error is τT_c. When the signal of Formula (A3) is sampled at $t = kT_c$, the baseband signal after sampling can be expressed as

$$\begin{aligned} r(k) &= As(k) e^{j(2\pi f k T_c + \varphi)} + n(k) \\ &= A \sum_{p=-\infty}^{\infty} d_p \sum_{m=-\infty}^{\infty} c_m w(mT_c - pT) h(kT_c - mT_c - \tau T_c) e^{j(2\pi f k T_c + \varphi)} + n(k) \end{aligned} \tag{A4}$$

Appendix B. Correlation Operation

Using the local spread spectrum code sequence and the received signal after sampling $r(k)$ to perform correlation operations, the despreading data are obtained:

$$\begin{aligned} y_l &= \frac{1}{N} \sum_{k=lN}^{(l+1)N-1} c^*(k) r(k) \\ &= \frac{A}{N} \sum_{k=lN}^{(l+1)N-1} c^*(k) \sum_{p=-\infty}^{\infty} d_p \sum_{m=-\infty}^{\infty} c_m w(mT_c - pT) h(kT_c - mT_c - \tau T_c) e^{j(2\pi f k T_c + \varphi)} \\ &\quad + \frac{1}{N} \sum_{k=lN}^{(l+1)N-1} c^*(k) n(k) \\ &= S_l + \frac{1}{N} \sum_{k=lN}^{(l+1)N-1} c^*(k) n(k) \end{aligned} \tag{A5}$$

where $m = k - q$, q is the delay, since the spread spectrum code is an aperiodic long code, $0 < q < lN$. S_l in (A5) can be expressed as

$$\begin{aligned} S_l(q) &= \frac{A}{N}\sum_{k=lN}^{(l+1)N-1} c^*(k)\sum_{p=-\infty}^{\infty} d_p \sum_{m=-\infty}^{\infty} c_m w(mT_c - pT)h(kT_c - mT_c - \tau T_c)e^{j(2\pi fkT_c+\varphi)} \\ &= \frac{A}{N}\sum_{k=lN}^{(l+1)N-1} c^*(k)\sum_{p=-\infty}^{\infty} d_p \sum_{q=-\infty}^{\infty} c_{k-q} w(kT_c - qT_c - pT)h(qT_c - \tau T_c)e^{j(2\pi fkT_c+\varphi)} \end{aligned} \tag{A6}$$

When $q = 0$ in (A6), the corresponding signal component is

$$\begin{aligned} S_l(q=0) &= \frac{A}{N}h(-\tau T_c)\sum_{k=lN}^{(l+1)N-1} c^*(k)c(k)\sum_{p=l} d_l \sum_{m=k} w(kT_c - lT)e^{j(2\pi fkT_c+\varphi)} \\ &= \frac{A}{N}h(-\tau T_c)d_l\sum_{k=lN}^{(l+1)N-1} c^*(k)c(k)e^{j(2\pi fkT_c+\varphi)} \\ &\approx \frac{A}{NT_c}h(-\tau T_c)d_l\int_{lNT_c}^{(l+1)NT_c} e^{j(2\pi ft+\varphi)}dt \\ &= Ad_l h(-\tau T_c)\frac{\sin(\pi fT)}{\pi fT}e^{j\varphi_l} \end{aligned} \tag{A7}$$

Its average power is expressed as

$$p_s = E\left[\left(S_{m,q=0}\right)\left(S_{m,q=0}\right)^*\right] \approx A^2h^2(-\tau T_c)\left[\frac{\sin(\pi fT)}{\pi fT}\right]^2 \tag{A8}$$

$q \neq 0$ corresponds to the ISI caused by the channel band limiting effect (time domain trailing effect). If only the ISI of two adjacent symbols is considered, $|q| < N$ can be assumed. p_{J1} is the interference power when q > 0, while p_{J2} is the interference power when $q < 0$. The total interference power can be expressed as

$$p_J = p_{J1} + p_{J2} = \frac{A^2}{N}\left[\sum_{q=1}^{N-1} h^2(qT_c - \tau T_c) + \sum_{q=-N+1}^{-1} h^2(qT_c - \tau T_c)\right] \tag{A9}$$

In summary, the signal after despreading can be approximated as

$$\begin{aligned} Y_l(q) &= \frac{1}{N}\sum_{k=lN}^{(l+1)N-1} c^*(k)r(k-q) \approx Ad_l h(-\tau T_c)\frac{\sin(\pi fT)}{\pi fT}e^{j\varphi_l} + \frac{1}{N}\sum_{k=lN}^{(l+1)N-1} c^*(k)n(k) + n_{2,l}(q) \\ &= Be^{j\varphi_l} + n_{1,l}(q) + n_{2,l}(q) \end{aligned} \tag{A10}$$

where $B = Ah(-\tau T_c)\sin c(fT)$, $n \sim N(0,\sigma_n^2)$, and $n_{1,l} \sim N(0,\sigma_n^2/N)$. $n_{2,l}$ is the noise caused by ISI. The simulation results show that when $N > 32$, $|fT| < 1/4$ and $|\tau T_c| < 1/4$, the interference power of (A8) is at least 20 dB lower than the signal power of (A9), which is negligible in the context of engineering.

References

1. Stojanovic, M.; Freitag, L. Acquisition of direct sequence spread spectrum acoustic communication signals. In Proceedings of the Oceans 2003. Celebrating the Past... Teaming Toward the Future (IEEE Cat. No.03CH37492), San Diego, CA, USA, 22–26 September 2003; Volume 1, pp. 279–286. [CrossRef]
2. Le Nir, V.; Scheers, B. Robust Blind Carrier Frequency Synchronization for Direct Sequence Spread Spectrum Systems. *Electron. Lett.* **2015**, *51*, 425–427. [CrossRef]
3. Zhou, F.; Zhao, L.; Li, L.; Hu, Y.; Jiang, X.; Yu, J.; Liang, G. GNSS Signal Acquisition Algorithm Based on Two-Stage Compression of Code-Frequency Domain. *Appl. Sci.* **2022**, *12*, 6255. [CrossRef]
4. Yang, F.; Wang, G.; Meng, Z.; Guo, X.; Yan, E.; Liu, S. A High Dynamic Weak Spread Spectrum Signal Acquisition Strategy Based on Iterative Local Search. *Electronics* **2022**, *11*, 2881. [CrossRef]
5. Ling, F. Initial Acquisition and Frame Synchronization. In *Synchronization in Digital Communication Systems*; Cambridge University Press: Cambridge, UK, 2017; pp. 67–130.
6. Morelli, M.; D'Amico, A. Maximum Likelihood Timing and Carrier Synchronization in Burst-Mode Satellite Transmissions. *EURASIP J. Wirel. Commun. Netw.* **2007**, *2007*, 065058. [CrossRef]

7. Xing, Y. A New Compatible Fast Acquisition Algorithm for Parameter-varying Spread Spectrum Signals. In Proceedings of the 2018 IEEE 18th International Conference on Communication Technology (ICCT), Chongqing, China, 8–11 October 2018; pp. 1067–1072. [CrossRef]
8. Jiang, B.; Wu, W. A Algorithm of Preamble Acquisition Base on Differential Correlation Module for CDMA System. In Proceedings of the 2018 12th International Symposium on Antennas, Propagation and EM Theory (ISAPE), Hangzhou, China, 3–6 December 2018; pp. 1–4. [CrossRef]
9. Sun, Y.; Jiang, H.; Yi, L.; Hu, W.; Zhuge, Q. Preamble Design for Joint Frame Synchronization, Frequency Offset Estimation and Channel Estimation in Burst Mode Coherent PONs. In Proceedings of the 2024 Optical Fiber Communications Conference and Exhibition (OFC), San Diego, CA, USA, 24–28 March 2024; pp. 1–3.
10. Chen, H.; Qiao, K.; Tong, Y.-P. A Burst Transmission Synchronization Algorithm for Satellite Communication. *Acta Electron. Sin.* **2023**, *51*, 907–913. [CrossRef]
11. Ding, Z.; Hao, T.; Zeng, N.; Hu, L.; Li, L.; Ding, L.; Yang, F. Design of Distributed Synchronization Header for Short Burst Communication with Low Earth Orbit Satellites. In *Digital Multimedia Communications. IFTC 2023. Communications in Computer and Information Science*; Zhai, G., Zhou, J., Ye, L., Yang, H., An, P., Yang, X., Eds.; Springer: Singapore, 2024; Volume 2066.
12. Baeza, V.M.; Ha, V.N.; Querol, J.; Chatzinotas, S. Non-coherent massive MIMO integration in satellite communication. In Proceedings of the 39th International Communications Satellite Systems Conference (ICSSC 2022), Stresa, Italy, 18–21 October 2022; pp. 200–205. [CrossRef]
13. Wu, Y.; Mei, R.; Xu, J. Non pilot data-aided carrier and sampling frequency offsets estimation in fast time-varying channel. *Big Data Res.* **2024**, *36*, 100461. [CrossRef]
14. Lee, D.-U.; Kim, H.; Jones, C.R.; Villasenor, J.D. Pilotless Frame Synchronization via LDPC Code Constraint Feedback. *IEEE Commun. Lett.* **2007**, *11*, 683–685. [CrossRef]
15. Spangenberg, S.M.; Scott, I.; McLaughlin, S.; Povey, G.J.; Cruickshank, D.G.; Grant, P.M. An FFT-based approach for fast acquisition in spread spectrum communication systems. *Wirel. Pers. Commun.* **2000**, *13*, 27–55. [CrossRef]
16. Zhongyuan, B.; Bo, L.; Wen, C. A two-stage fast pseudo-code acquisition algorithm based on pmf-fft. In Proceedings of the 2020 IEEE Asia-Pacific Microwave Conference (APMC), Hong Kong, 8–11 December 2020; IEEE: New York, NY, USA, 2020; pp. 982–984.
17. Qi, J.; Luo, F.; Song, Q. Fast acquisition method of navigation receiver based on folded PMF-FFT. In Proceedings of the 2014 IEEE Computers, Communications and IT Applications Conference, Beijing, China, 20–22 October 2014; IEEE: New York, NY, USA, 2014; pp. 62–66.
18. Liu, J. Status and development of the Beidou navigation satellite system. *J. Telem. Track. Command.* **2013**, *34*, 1–8. [CrossRef]
19. Xu, H. Research on GPS P-Code Direct Acquisition Technology. Master's Thesis, National University of Defense Technology, Changsha, China, 2017.
20. Nie, Y. Using DOP algorithm to evaluate jamming efficiency in GPS jamming tests. *Electron. Inf. Warf. Technol.* **2017**, *5*, 58–62. [CrossRef]
21. Zeng, F.; Ouyang, X.; Xu, H.; Lü, D. Improved Long-code Direct Acquisition Algorithm Based on Time-frequency Fusion. *Dianzi Yu Xinxi Xuebao/J. Electron. Inf. Technol.* **2019**, *41*, 309–316. [CrossRef]
22. Yang, C. FFT Acquisition of periodic, aperiodic, puncture, and overlaid code sequences in GPS. In Proceedings of the GPS Institute of Navigation, Salt Lake City, UT, USA, 11–14 September 2011; pp. 137–148.
23. Zhao, P.; Yao, Y. An ambiguity resolving algorithm for XFAST acquisition with less computational amount in satellite navigation. *Telecommun. Eng.* **2017**, *57*, 1205–1212. [CrossRef]
24. Tang, X.; Yong, S.; Wang, F. Performance of XFAST in the presence of code Doppler. *J. Commun.* **2010**, *31*, 54–59. [CrossRef]
25. Viterbi, A.J. *CDMA Principles of Spread Spectrum Communication*; Addison-Wesley Pub.: Redwood City, CA, USA, 1995.
26. Gil, A.; Segura, J.; Temme, N.M. Algorithm 939: Computation of the Marcum Q-function. *ACM Trans. Math. Softw. (TOMS)* **2014**, *40*, 1–21. [CrossRef]
27. Khan, M.A.; Bicholia, D.; Dubey, R. Comparison of Rayleigh & Ricean Fading Channels using BPSK, QPSK & 16-QAM Modulation Techniques in terms of BER vs. SNR using RRC FIR Filter with DT OFDM Technique. *Int. J. Eng. Res. Technol.* **2015**, *4*, 620–625.
28. Buzzi, S.; Poor, H.V. Channel estimation and multiuser detection in long-code DS/CDMA systems. *IEEE J. Sel. Areas Commun.* **2001**, *19*, 1476–1487. [CrossRef]
29. Weiss, A.J.; Friedlander, B. Channel estimation for DS-CDMA downlink with aperiodic spreading codes. *IEEE Trans. Commun.* **1999**, *47*, 1561–1569. [CrossRef]
30. Iinatti, J.H.J. On the threshold setting principles in code acquisition of DS-SS signals. *IEEE J. Sel. Areas Commun.* **2000**, *18*, 62–72. [CrossRef]
31. Iinatti, J. Comparison of threshold setting rules for the probability of acquisition in a DS spread spectrum receiver. In *1998 International Zurich Seminar on Broadband Communications. Accessing, Transmission, Networking. Proceedings (Cat. No. 98TH8277)*; IEEE: New York, NY, USA, 1998; pp. 23–28.

electronics

MDPI

Article

Joint Task Offloading and Resource Scheduling in Low Earth Orbit Satellite Edge Computing Networks †

Jinhong Li [1], Rong Chai [1,*], Kangan Gui [1] and Chengchao Liang [1,2,*]

[1] School of Communications and Information Engineering, Chongqing University of Posts and Telecommunications, Chongqing 400065, China; d180101003@stu.cqupt.edu.cn (J.L.); guika01@outlook.com (K.G.)
[2] Department of Systems and Computer Engineering, Carleton University, Ottawa, ON K1S 5B6, Canada
* Correspondence: chairong@cqupt.edu.cn (R.C.); chengchaoliang@sce.carleton.ca (C.L.)
† This paper is an extended version of our paper published in 19th EAI International Conference on Communications and Networking in China, Chongqing, China, 2–3 November 2024.

Abstract: In view of the future of the Internet of Things (IoT), the number of edge devices and the amount of sensing data and communication data are expected to increase exponentially. With the emergence of new computing-intensive tasks and delay-sensitive application scenarios, terminal devices need to offload new business computing tasks to the cloud for processing. This paper proposes a joint transmission and offloading task scheduling strategy for the edge computing-enabled low Earth orbit satellite networks, aiming to minimize system costs. The proposed system model incorporates both data service transmission and computational task scheduling, which is framed as a long-term cost function minimization problem with constraints. The simulation results demonstrate that the proposed strategy can significantly reduce the average system cost, queue length, energy consumption, and task completion rate, compared to baseline strategies, thus highlighting the strategy's effectiveness and efficiency.

Keywords: low Earth orbit satellite networks; edge computing; task offloading; resource scheduling; deep Q-network

1. Introduction

Traditional terrestrial networks, like 5G, rely on base stations as their fundamental units of operation. These stations enable the delivery of real-time and rapid services [1]. Nonetheless, terrestrial network deployment faces certain challenges. Primarily, the reach of terrestrial base stations is limited, making it difficult to establish them in regions such as deserts, forests, or mountains. Additionally, terrestrial base stations are susceptible to natural disasters. When events like typhoons, tsunamis, and earthquakes damage these stations, communication systems can fail, and restoring them swiftly is challenging, which leads to severe societal and economic disruptions due to interrupted communications. Consequently, satellite communication systems, especially the low Earth orbit (LEO) satellite networks, offer a flexible deployment option with global coverage, effectively complementing terrestrial networks [2,3]. In densely populated regions, satellite communication networks can provide substantial support, utilizing their resources for communication, computing, and storage to ease the burden on terrestrial systems. In locales where terrestrial networks are unable to reach, satellite networks can assume the role of transmitting information, thus complementing the terrestrial networks [4].

In the forthcoming era of the Internet of Things (IoT), achieving global connectivity and intelligence will lead to an exponential increase in the number of edge devices and the volume of sensing and communication data. Traditional centralized cloud networks will no longer be able to efficiently process the massive amounts of data generated by the IoT [5,6]. The advent of augmented reality, enhanced reality, high-definition live streaming, and other novel application scenarios further accentuates this challenge. Limitations in terminal computing power and navigation capabilities may result in suboptimal user experiences, such as the inability to interact in real time. Mobile Edge Computing (MEC), a pivotal technology in fifth-generation mobile communications, addresses these issues by decentralizing resources from central devices to network edge devices. This enables edge devices to possess certain computing capabilities for processing user requests locally, providing users with low-latency, high-bandwidth data processing services. Consequently, MEC alleviates the pressure on core network resources and effectively resolves problems such as data traffic bottlenecks in the core network [7–9].

In traditional satellite communication systems, satellites usually act as relays for transferring data from user devices. With the emergence of new computing-intensive tasks and delay-sensitive application scenarios, terminal devices need to offload new business computing tasks to the cloud for processing. However, the traditional mode of uploading data to the cloud for processing cannot meet the current computing-intensive business needs in terms of low latency and high bandwidth transmission. Therefore, the integration of satellite communication and edge computing is the development trend of future satellite networks. Moreover, LEO satellite communication systems inherently offer low-latency communication due to their proximity to the Earth's surface and typically orbit at altitudes between 500 and 2000 km [10]. This close proximity significantly reduces the signal propagation delay compared to Geostationary Earth Orbit (GEO) satellites, which orbit at approximately 35,786 km. As a result, LEO satellites can achieve round-trip signal latencies as low as 20–30 ms, making them suitable for real-time and delay-sensitive applications [10,11]. The deployment of LEO satellite constellations enhances communication reliability through redundant coverage and seamless handover capabilities.

These low-latency and high-reliability characteristics of satellite communication systems, combined with edge computing and storage resources deployed on satellites, can provide users with strong computing and storage resource support [12–14]. This can significantly reduce data transmission between the edge and the central cloud, reduce delay, and lower energy consumption. In addition, with the increasing demand for computing and storage resources in new application scenarios, satellite edge computing can provide strong computing and storage resource support for new applications.

By integrating MEC technology with LEO satellite communication networks, satellites not only can receive and forward tasks to ground stations but also deploy MEC servers onboard, enabling LEO satellites to possess computing and storage resources. This is crucial for expanding service capabilities. Considering the characteristics of the LEO satellite communication networks and MEC technology, their integration can fully utilize the flexible satellite chain routes of LEO communication networks for efficient transmission [14]. Additionally, by deploying satellite servers, we can meet the needs of end users for faster, better, and safer applications. Therefore, building a LEO satellite edge computing network is an inevitable trend in the development of communication networks [15]. In our initial result [16], integrating edge computing technology with LEO networks can significantly reduce the service response time of LEO satellites, enhance the autonomous task processing capabilities of satellites, and improve network performance. Moreover, dynamic task offloading and transmission scheduling are critical for optimizing performance in satellite networks as they must adapt to various dynamic factors such as fluctuating user demand, the movement of satellites, changes in network topology, and environmental conditions.

This adaptability ensures that tasks are efficiently offloaded to the most suitable resources in real time, thereby enhancing overall system performance and responsiveness.

The existing challenges of latency issues and bandwidth allocation must be addressed to improve system optimization. Although satellite networks are known for their low latency advantages, certain applications may still experience latency issues, especially when bandwidth is constrained during peak usage periods. Effective bandwidth allocation is crucial as it directly influences task processing speed and response times. Furthermore, optimizing bandwidth is essential in enhancing resource utilization efficiency, which ultimately contributes to better overall system performance. The LEO satellite systems integrated with edge computing need to communicate through satellites and ground stations, and there is a certain signal transmission delay between satellites and ground stations. This delay may adversely affect real-time applications and tasks that require rapid data processing; the bandwidth of satellite communications is relatively limited and insufficient for supporting large-scale data transmission and processing needs. This means that bottlenecks may be encountered when processing large-scale data sets and performing complex computing tasks; the LEO satellite systems integrated with edge computing require a large amount of energy supply to support their computing and communication needs. However, current solar battery technology and battery capacity limitations restrict the energy supply capacity of satellite systems. This may lead to the system being unable to meet the needs of high-intensity computing tasks. Additionally, satellite systems still face difficulties in maintenance and updates, as well as high costs.

The major contributions of this paper are as follows:

- This paper proposes a novel integration of Mobile Edge Computing (MEC) technology with LEO satellite communication networks while comprehensively considering constraints such as onboard computing and transmission, satellite–ground and inter-satellite links, and traffic conservation.
- This paper formulates the problem of efficient task offloading in the integrated LEO satellite edge computing network. It defines a cost function based on caching and energy consumption and models the dynamic task offloading and transmission scheduling problem as a constrained long-term cost function minimization problem.
- The original optimization problem is divided into two key components: data task transmission scheduling and computation task offloading. The data task transmission scheduling problem focuses on determining the most efficient routes for data transmission from observation satellites to ground stations, utilizing Dijkstra's algorithm to identify the shortest paths while accounting for energy consumption. In contrast, the computation task offloading and computation scheduling problem is modeled as a Markov Decision Process (MDP), where dynamic factors such as the arrival of IoT device tasks and the status of relay satellite buffers are considered. This approach enables the application of reinforcement learning techniques, specifically the Deep Q-Network (DQN) algorithm, to develop optimal strategies for task offloading and computation scheduling, which minimize long-term system costs under varying conditions.
- Extensive simulations are performed to evaluate the proposed scheme compared with the existing peers.

The rest of the paper is organized as follows. Section 2 presents an overview of related work. Section 3 describes the system model, and Section 4 formulates the joint transmission, task offloading, and computation scheduling problem. The proposed schemes are introduced in Section 5. Simulation results are presented in Section 6. Finally, we conclude this paper in Section 7.

2. Related Works

This section outlines the current research related to task offloading and resource scheduling in LEO satellite networks.

2.1. Task Offloading in LEO Satellite Networks

In scenarios involving task offloading in LEO satellite networks, references [17,18] suggest energy-efficient algorithms for task offloading. The work in [17] examines a ground network supported by LEO satellites and offers a cloud–edge collaborative offloading strategy that accounts for service quality constraints, facilitating efficient energy offloading within a satellite network. It provides separate solutions utilizing deep reinforcement learning and game theory. Meanwhile, Ref. [18] explores a hybrid cloud–edge LEO satellite computing network with a three-layer structure, focusing on reducing total energy consumption for ground users. This research applies game theory to devise task offloading strategies for each ground user.

Studies [19–21] introduce algorithms that optimize delay in task offloading. Ref. [19] presents a network architecture integrating space and ground components to deliver edge computing services to LEO satellites and UAVs, focusing on reducing task execution delays through deep reinforcement learning for offloading decisions. Ref. [20] examines task offloading challenges within modern small satellite systems, proposing a strategy that optimizes delay while offloading and dismantling tasks. For satellite relay computing, Ref. [21] suggests a collaborative computing approach between satellite and ground to specifically reduce task response delays.

Other works [22–25] focus on creating energy-efficient and delay-optimized strategies for task offloading. Ref. [22] analyzes the time delays inherent in satellite–ground communications within relay computing scenarios, proposing a heuristic search algorithm to balance the computational load and reduce system energy consumption and task execution delay. Ref. [23] introduces a task offloading method employing a heuristic search algorithm specifically to lower both energy consumption and task execution delay for ground users. Ref. [24] also focuses on modern small satellite systems, suggesting a delay-optimized strategy for both offloading and dismantling tasks. Ref. [25] proposes a local search matching algorithm to determine the optimal task offloading strategy, targeting a reduction in energy usage and delay for ground users.

2.2. Resource Scheduling Algorithms for LEO Satellite Networks

The works of [26–28] examine the challenge of satellite transmission delays. Specifically, Ref. [26] explores the use of ground stations, high-altitude platforms, and LEO satellites to offer data offloading services to ground users. Meanwhile, Ref. [27] investigates using satellite chains for data offloading to ground stations, suggesting optimization algorithms to enhance the download throughput of data at these stations and for data transfer among co-operative satellites and from satellites to ground stations. Ref. [28] develops a transmission scheduling algorithm to maximize bandwidth usage in satellite constellations by enhancing inter-satellite link capacity. Furthermore, Ref. [29] examines the integration of LEO satellite networks with terrestrial networks, proposing a matching algorithm aimed at maximizing data transmission rates and user access, factoring in various capacity constraints.

Additionally, Refs. [30–33] delve into the scheduling of computing and storage resources in LEO satellite networks. For instance, Ref. [30] introduces a hybrid satellite–ground network utilizing software-defined networking, focusing on optimizing the long-term usage efficiency of communication, storage, and computing resources despite variable network, storage, and computing states, employing deep Q-learning algorithms for joint resource allocation optimization. Ref. [31] considers LEO satellites equipped with MEC platform computing and storage resources, proposing a service request scheduling approach to optimize resource use and service quality while reducing system costs. Ref. [32] suggests a strategy for satellite network storage resource scheduling to maintain user experience quality, and Ref. [33] explores the joint optimization of wireless resource allocation and task offloading in satellite networks to leverage the computing power of edge nodes effectively.

2.3. Task Offloading and Resource Scheduling Algorithms for LEO Networks

In addressing the challenges of task offloading and resource scheduling within LEO network environments, Ref. [34] examines the varying aspects of task segmentation and the energy priorities of edge devices. The study introduces an energy-efficient task offloading and resource scheduling algorithm to minimize both overall delay and energy usage.

Studies [35,36] put forward joint task offloading and resource scheduling algorithms leveraging deep reinforcement learning. Specifically, Ref. [35] employs a distributed algorithm grounded in convex optimization to tackle the task offloading and resource scheduling issue, recasting it as a direct optimization problem of computing resource scheduling for LEO satellites and simulating the system to maintain high energy efficiency. Meanwhile, Ref. [36] utilizes a deep reinforcement learning approach for resource allocation, addressing task offloading to the edge device and task segmentation issues, and applies a deep reinforcement learning strategy to identify the optimal offloading approach to minimize delay and energy consumption at the edge device.

Moreover, Refs. [37,38] suggest joint task offloading and resource scheduling frameworks using deep reinforcement learning. For instance, Ref. [37] proposes an integrated strategy for computing resource allocation and task offloading that adapts to dynamic changes in the network environment to enhance energy efficiency long-term, employing deep reinforcement learning to modulate the offloading strategy dynamically. Meanwhile, Ref. [38] evaluates LEO network architecture with edge computing, focusing on minimizing user task processing delay and energy use at the edge node, and proposes a comprehensive strategy for task offloading and computing resource scheduling to achieve long-term reductions in delay and energy consumption. Furthermore, Ref. [39] considers dynamic shifts in the locations of edge devices, satellites, and the network setting, using a deep reinforcement learning algorithm to optimize user association, offloading, computing, and communication resource allocation strategy for minimal long-term delay and energy use. Additionally, Ref. [40] presents a joint task offloading and resource scheduling strategy tailored for LEO edge computing aimed at minimizing system costs, with simulation outcomes indicating that the proposed strategy can notably decrease the average system cost.

Despite significant advancements in LEO satellite task offloading research, there remains a substantial gap in fully leveraging satellite collaboration for efficient resource utilization. Existing studies, summarized in the Table 1, primarily focus on optimizing task execution delay and energy efficiency without accounting for the dynamic nature of LEO networks and the potential of satellite cooperation. To address these gaps, our research makes the following unique contributions. We propose a groundbreaking integration of Mobile Edge Computing (MEC) technology with LEO satellite communication networks. This integration considers unique constraints such as onboard computing capabilities, satellite–ground and inter-satellite link capacities, and traffic conservation laws, which have not been collectively addressed in prior works.

Table 1. Comparison of related works on task offloading in LEO satellite networks.

References	Focus Area	Methodology	Key Contributions	Limitations
[17,18]	Task offloading with energy optimization	Game theory, energy-efficient algorithms	Proposes energy-efficient offloading strategies with service quality constraints	Does not address joint resource scheduling
[19–21]	Delay-optimized task offloading	Deep reinforcement learning	Reduces task execution delays through offloading optimization	Limited focus on computational resource scheduling
[22–25]	Energy and delay optimization	Heuristic algorithms	Balances computational load and reduces system energy consumption and delays	Lacks adaptability to dynamic network conditions
[26–29]	Satellite resource scheduling for communication	Optimization algorithms	Enhances inter-satellite bandwidth usage and throughput	Focuses only on data transmission scheduling, not computation
[29–33]	Resource scheduling in hybrid networks	Software-defined networking, Q-learning	Improves communication, storage, and computing efficiency in hybrid networks	Assumes static resource demands and lacks joint optimization
[34–36]	Joint task offloading and resource scheduling	Deep reinforcement learning, convex optimization	Proposes energy-efficient strategies for task segmentation and scheduling	Limited handling of dynamic network environments
[37–40]	Long-term optimization in LEO networks	Deep reinforcement learning, joint optimization	Provides comprehensive strategies for delay and energy optimization in dynamic scenarios	Assumes fixed task arrival rates and limited scalability
This work	Joint task offloading and dynamic resource scheduling	Deep Q-Network-based optimization	Long-term cost minimization, real-time adaptability to dynamic task arrivals and link variability	Requires high computational resources for reinforcement learning model training

3. System Model

This section considers the system model comprising the network model, task and computation model, transmission service model, and link transmission rate model.

3.1. Network Model

The proposed LEO satellite network includes N observation satellites, K relay satellites, M IoT devices, and a ground station. Observation satellites equipped with sensors and instruments to collect data from the Earth's surface, such as imagery or environmental measurements, focus on data acquisition and initial data processing [41,42]. Relay satellites are defined as satellites that facilitate communication by receiving data from observation satellites or IoT devices [43,44] and forwarding it to other satellites, ground stations, or processing nodes. They possess greater computational and communication capabilities to handle data relay and task offloading.

In this paper, we distinguish between two types of information handled by satellites: data and tasks. Data refer to raw information transmitted or forwarded without additional processing, such as images relayed from an observation satellite to a ground station. Tasks, on the other hand, involve computational workloads where data need to be processed to extract useful insights. For example, an IoT sensor may collect temperature readings, which require analysis to detect anomalies. In our model, if data only require transmission, we classify it as 'data'; if computation is required, we classify it as a 'task'.

When direct communication between an observation satellite and the ground station is not possible due to their relative positions, observation satellites gather data from the Earth's surface and transmit it to relay satellites, which then relay the data to the ground station cache. For example, the data can be raw or processed information, such as images, sensor readings, and measurements that need to be transmitted to ground stations. Computational tasks may originate from IoT devices and require offloading to relay satellites for processing due to limited device capabilities. IoT devices create computational tasks and transmit them either to relay satellites or directly to the ground station for processing. Computational tasks encompass processing activities generated by IoT devices or satellites, requiring computational resources to execute functions like data analysis, image processing, and complex calculations. Accordingly, IoT devices establish connections with relay satellites through an association process, where each device selects an appropriate relay satellite for communication and task offloading based on factors such as signal quality, available resources, and connectivity status.

Let OS_n denote the n-th observation satellite, RS_k denote the k-th relay satellite with a computational capacity of ϕ_k, and ID_m denote the m-th IoT device. Each relay satellite's sub-channel offers a bandwidth of B, allowing multiple IoT devices to connect with relay satellites via Orthogonal Frequency Division Multiple Access (OFDMA) technology. The specific system layout is depicted in Figure 1.

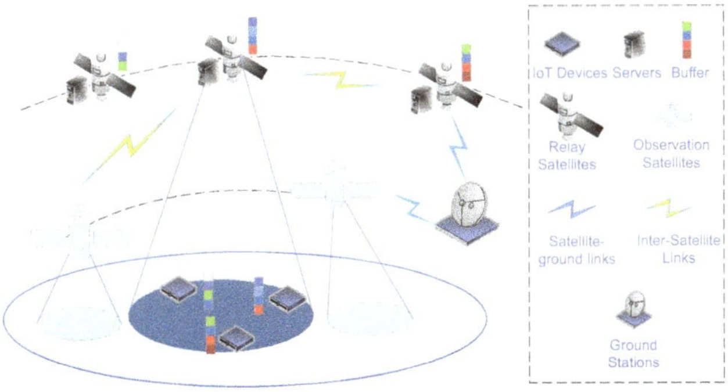

Figure 1. System model.

In the dynamic LEO satellite scenario explored in this paper, to more clearly illustrate the system's link dynamics, we consider that the system time T is divided into continuous intervals of duration τ. Given the nature of inter-satellite links in the LEO network, each LEO satellite is equipped with four interconnections with other satellites. The links between satellites within the same orbit are comparatively stable, while those between satellites in adjacent orbits achieve relative stability through steerable beam technology. Therefore, the inter-satellite network topology is assumed to remain stable. To describe the connection status of the inter-satellite communication links, we define $x^o_{n,k} \in \{0,1\}$ as the physical link indicator between the observation satellite OS_n and the relay satellite RS_k. If $x^o_{n,k} = 1$, a physical link exists between OS_n and RS_k; otherwise, $x^o_{n,k} = 0$. Similarly, $x^r_{k,k'} \in \{0,1\}$ represents the link indicator between relay satellites RS_k and $RS_{k'}$, where $x^r_{k,k'} = 1$ indicates that a link exists, and $x^r_{k,k'} = 0$ indicates no link. The satellite-to-ground network topology is assumed to change with each time slot, yet it remains constant within a single time slot. To describe the satellite-to-ground link status, $x^i_{m,k,t} \in \{0,1\}$ is used to indicate the existence of a physical link between the IoT device ID_m and the relay satellite RS_k at time slot t. Here, $x^i_{m,k,t} = 1$ indicates a link exists, otherwise $x^i_{m,k,t} = 0$. Likewise, $x^g_{k,t} \in \{0,1\}$ serves as the indicator for a link between the relay satellite RS_k and the ground station at time t, where $x^g_{k,t} = 1$ symbolizes an existing link, and $x^g_{k,t} = 0$ symbolizes its absence.

3.2. Task and Computation Model

Consider that the computational tasks for IoT devices arrive unpredictably during each time slot and conform to a Poisson distribution. Let $\theta_{m,t}$ signify the computational task that arrives at IoT device ID_m during time slot t. This task $\theta_{m,t}$ is characterized as $\theta_{m,t} = (D_{m,t}, F_{m,t})$, where $D_{m,t} = \lambda_m \tau$ denotes the size of the task $\theta_{m,t}$, and $F_{m,t}$ indicates the CPU cycle count necessary to process each bit of data in the task $\theta_{m,t}$, where λ_m signifies the typical task arrival rate for device ID_m.

Considering that IoT devices often have limited computational capabilities and constrained energy resources due to their compact size and design, they are not suited for processing computationally intensive tasks locally. Performing such tasks would result in significant delays and quickly deplete the device's battery life. Therefore, rather than executing computational tasks themselves, IoT devices offload these tasks to relay satellites. Relay satellites possess enhanced processing power and energy availability, enabling them to handle tasks efficiently either through onboard computation or by forwarding tasks to ground stations for processing. This offloading strategy not only conserves the limited resources of IoT devices but also improves overall system performance by leveraging the capabilities of more powerful network nodes. Define $z_{m,t,k,t'} \in \{0,1\}$ as the variable indicating the transmission of task $\theta_{m,t}$. A value of $z_{m,t,k,t'} = 1$ means that the task $\theta_{m,t}$ is sent to the relay satellite RS_k during time slot t', while $z_{m,t,k,t'} = 0$ means it is not. Furthermore, let $\lambda^g_{m,t,k,t'} \in \{0,1\}$ denote the decision variable for selecting the ground station offloading mode for task $\theta_{m,t}$. If $\lambda^g_{m,t,k,t'} = 1$, it signifies that the relay satellite RS_k chooses the ground station mode for task $\theta_{m,t}$ during time slot t', whereas $\lambda^g_{m,t,k,t'} = 0$ indicates otherwise.

3.3. Transmission Service Model and Transmission Model

In the scenario outlined in this paper, observation satellites are tasked with continuously monitoring the ground and collecting data transmission services at the outset of system operations. Denote by D^o_n the data collected by the observation satellite OS_n. Assume that, following data collection, an observation satellite OS_n is required to relay the data to the ground station through relay satellites in the following time slots. Define $\gamma^o_{n,k,t} \in \{0,1\}$ as the selection variable for satellite transmission relays concerning observation satellites. A value of $\gamma^o_{n,k,t} = 1$ indicates that the observation satellite OS_n sends data to relay satellite RS_k in time slot t, whereas $\gamma^o_{n,k,t} = 0$ indicates no transmission. Let $\gamma^r_{n,k,k',t} \in \{0,1\}$ represent the transmission selection variable between relay satellites. If $\gamma^r_{n,k,k',t} = 1$, relay satellite RS_k forwards data D^o_n to relay satellite $RS_{k'}$ in time slot t; otherwise, $\gamma^r_{n,k,k',t} = 0$. Additionally, let $\gamma^g_{n,k,t} \in \{0,1\}$ be the variable for selecting the ground station transmission from relay satellites. If $\gamma^g_{n,k,t} = 1$, the relay satellite RS_k transmits data D^o_n to the ground station in the time slot t; if not, then $\gamma^g_{n,k,t} = 0$.

3.4. Link Transmission Model

Define $R^i_{m,k,t}$ as the data rate between the IoT device ID_m and the satellite RS_k at time t:

$$R^i_{m,k,t} = x^i_{m,k,t} B \log_2\left(1 + \frac{P^i_m G_m G^r_k L^i_{m,k,t} L^{pt}_{m,k,t}}{\sigma^2}\right) \tag{1}$$

Here, P^i_m is the transmission power, G_m and G^r_k are antenna gains, $L^i_{m,k,t}$ is path loss, $L^{pt}_{m,k,t}$ is rain attenuation, and σ^2 is noise power. The antenna gains and free space path loss are critical components in the link budget analysis. They determine the signal-to-noise ratio (SNR) at the receiver, which in turn affects the achievable data rate as per Shannon's capacity formula. Higher antenna gains improve the SNR, while larger path losses deteriorate it. According to Equation (23) in [45], $L^i_{m,k,t} = \left(4\pi d^i_{m,k,t} f/c\right)^2$ where c is the speed of light, $d^i_{m,k,t}$ is the distance, and f is the frequency.

Let $R^g_{k,t}$ represent the link rate between the relay satellite RS_k and the ground station in time slot t. It is given by the following:

$$R^g_{k,t} = x^g_{k,t} B \log_2 \left(1 + \frac{P^{rg}_k G^{rg}_k G^g_r L^g_{k,t} L^{pt}_{k,t}}{\sigma^2}\right) \tag{2}$$

where P^{rg}_k is the relay satellite's transmission power to the ground station, G^{rg}_k is the transmission antenna gain, G^g_r is the receiving antenna's gain, $L^{pt}_{k,t}$ denotes the rain attenuation, and $L^g_{k,t}$ indicates the free space loss, which is defined as follows: $L^g_{k,t} = \left(4\pi d^g_{k,t} f/c\right)^2$ where $d^g_{k,t}$ is the distance between RS_k and the ground station during time slot t.

Let $R^o_{n,k}$ represent the link rate between the observation satellite OS_n and the relay satellite RS_k, which is given by Equation (4) in [46]:

$$R^o_{n,k} = x^o_{n,k} \frac{G^o_n G^r_k P_n L^o_{n,k}}{k_s T_s E_b / N_0} \tag{3}$$

where G^o_n is the observation satellite's transmission gain, G^r_k is the relay satellite's receiving gain, P_n is the power of OS_n, $L^o_{n,k} = \left(4\pi d^o_{n,k} f/c\right)^2$ is the free space loss between OS_n and RS_k, k_s is the Boltzmann constant, T_s is the system noise temperature, E_b is the energy per bit, and N_0 is the noise power spectral density. $d^o_{n,k}$ is the distance between OS_n and RS_k.

$R^r_{k,k'}$ specifies the rate between the relay satellites RS_k and $RS_{k'}$, which is given by the following:

$$R^r_{k,k'} = x^r_{k,k'} \frac{G^r_k G^r_{k'} P^r_k L^r_{k,k'}}{k_s T_s E_b / N_0} \tag{4}$$

where G^r_k and $G^r_{k'}$ are the relay satellite k's transmission gain and k''s receiving gain, P^r_k is the power for inter-satellite transmission, $L^r_{k,k'} = \left(4\pi d^r_{k,k'} f/c\right)^2$ is the free space loss between RS_k and $RS_{k'}$. $d^r_{k,k'}$ is the distance between RS_k and $RS_{k'}$.

4. Optimization Problem Modeling

This section first models the system's cache queue and energy consumption then defines the system cost function based on the cache queue and energy consumption. Next, it considers constraints such as association and computation limitations to model the joint task offloading and resource scheduling problem as a long-term cost function minimization problem under these constraints.

The dynamic nature of LEO networks requires a flexible approach to resource scheduling. As the number of IoT devices and their corresponding tasks vary, our system dynamically schedules the transmission and computation resources to ensure efficient processing and transmission. Continuous assessment of the operational status of satellites and ground stations allows our framework to respond promptly to changes in network conditions. The quality of communication channels is monitored to optimize data transmission paths, ensuring reliable connectivity. The relationships between satellites are considered, allowing for adaptive scheduling based on the current inter-satellite network topology.

To effectively manage these dynamic factors, our resource scheduling strategy is designed to adapt in real time. By modeling the task offloading and computation scheduling processes as a Markov Decision Process (MDP), we enable our system to make informed decisions based on the current state of the network. This approach ensures that both the data transmission and computational tasks are scheduled efficiently, minimizing latency and maximizing throughput.

4.1. System Cache Queue Modeling

As the computational tasks for IoT devices appear unpredictably and are sent to relay satellites in successive time slots, it is necessary to model the computation task cache queue at the devices. Given that relay satellites might handle data transmission services, accept and transfer computational tasks, and perform onboard tasks in each time slot, it is important to model a queue at the relay satellites to illustrate the caching scenario. Inspired by the analysis and modeling in [47–49], we model the queueing as follows.

4.1.1. IoT Device Queues

Let $Q^i_{m,t}$ represent the length of the task queue for computations at the IoT device ID_m during time slot t. The formula for updating the queue $Q^i_{m,t}$ is given by the following:

$$Q^i_{m,t+1} = \min\left(Q^{i,\max}_m, Q^i_{m,t} - \sum_{k=1}^{K} \sum_{t'=1}^{t} z_{m,t,k,t'} D_{m,t'} + D_{m,t} \right) \tag{5}$$

where $Q^{i,\max}_m$ denotes the maximum allowable length of the cache queue for the IoT device ID_m and $\sum_{k=1}^{K} \sum_{t'=1}^{t} z_{m,t,k,t'} D_{m,t'} + D_{m,t}$ denotes the incoming data.

4.1.2. Relay Satellite Queues

Define $Q_{k,t}$ as the queue size for the relay satellite RS_k at time t. The update equation for the queue at the satellite RS_k is given by the following:

$$\begin{aligned} Q_{k,t+1} = \min \Bigg\{ & Q^{\max}_k, Q_{k,t} + \sum_{n=1}^{N} \gamma^o_{n,k,t} D^o_n - \sum_{k' \neq k}^{K} \gamma^r_{n,k',t} D^o_n \\ & - \sum_{n=1}^{N} \gamma^g_{n,k,t} D^o_n + \sum_{m=1}^{M} \sum_{t'=1}^{t-1} \lambda^g_{m,t',t} D^g_{m,t',t} \\ & - \sum_{m=1}^{M} \sum_{t'=1}^{t-1} y_{m,t',k,t} \min\left(\frac{\varphi_k \tau}{D_{m,t'} F_{m,t'}}, D_{m,t',k,t} \right) \Bigg\} \end{aligned} \tag{6}$$

Here, $Q^{\max}_k$ denotes the upper limit of the queue for RS_k; $y_{m,t',k,t} \in \{0,1\}$ is a binary task indicator for the satellite's task execution. A value of 1 signals that RS_k processes the task $\theta_{m,t'}$ at time t. The symbol φ_k stands for the computational capacity of RS_k, and $D_{m,t',k,t}$ indicates the workload associated with task $\theta_{m,t'}$ for RS_k at t. The workload update is defined as follows:

$$D_{m,t',k,t+1} = D_{m,t',k,t} - \min\left(\frac{\varphi_k \tau}{D_{m,t} F_{m,t}}, D_{m,t',k,t} \right) \tag{7}$$

Initially, $D_{m,t',k,t}$ for satellite RS_k is determined by the following:

$$D_{m,t',k,t} = \begin{cases} 0, & t < t'', \text{ and } z_{m,t',k,t''} = 1 \\ D_{m,t'}, & z_{m,t',k,t''} = 1 \text{ and } \sum_{\bar{t}=t''}^{t} y_{m,t',k,\bar{t}}, \ t'' \leq t \leq \bar{t}. \end{cases} \tag{8}$$

Equation (8) implies that the task is considered empty when there are no incoming data from the IoT devices, and it reflects the total task size once task execution begins.

4.2. System Energy Consumption Model

Let E_t^c denote the energy consumption for executing tasks at time t, comprising energy spent transmitting tasks from IoT devices to relay satellites, relay satellites computing tasks, and transmitting tasks to ground stations. The energy consumed by ground stations is neglected due to their ample power. E_t^c is given by the following:

$$E_t^c = \sum_{m=1}^{M}\sum_{k=1}^{K} E_{m,t,k}^t + \sum_{m=1}^{M}\sum_{k=1}^{K} E_{m,t,k}^c + \sum_{m=1}^{M}\sum_{k=1}^{K} E_{m,t,k}^g \tag{9}$$

In Equation (9),

- $E_{m,t,k}^t = z_{m,t,k} D_{m,t}^o P_{m,t}^i / R_{m,t,k}^i$ is the energy to transmit a task $\theta_{m,t}$ from IoT device ID_m to relay RS_k at t;
- $E_{m,t,k}^c = y_{m,t,k}\epsilon_k (D_{m,t}^o/\varphi_k)^2$ is the energy for RS_k to compute $\theta_{m,t}$ at t where ϵ_k is the energy coefficient for RS_k;
- $E_{m,t,k}^g = \lambda_{m,t,k} D_{m,t}^o P_{m,t}^g / R_{m,t,k}^g$ is the energy to send $\theta_{m,t}$ from RS_k to ground stations at t.

Define E_t^o as the energy needed for data transmission at time t, incorporating data transfer from observation satellites to relay satellites, among relay satellites, and from relay satellites to ground stations. The formula for E_t^o is

$$E_t^o = \sum_{n=1}^{N}\sum_{k=1}^{K} E_{n,k,t}^o + \sum_{k=1}^{K}\sum_{k'=1,k'\neq k}^{K} E_{n,k,k',t}^r + \sum_{n=1}^{N}\sum_{k=1}^{K} E_{n,k,t}^g \tag{10}$$

where the following holds:

- $E_{n,k,t}^o = \gamma_{n,k,t}^o D_n^o P_n / R_{n,k}^o$ denotes the energy for observation satellite OS_n to send the data D_n^o to the relay satellite RS_k at t;
- $E_{n,k,k',t}^r = \gamma_{n,k,k',t}^r D_n^o P^r / R_{k,k'}^r$ quantifies the energy for relay satellite RS_k to transmit the data D_n^o to another relay satellite $RS_{k'}$ at t;
- $E_{n,k,t}^g = \gamma_{n,k,t}^g D_n^o P^{rg} / R_{k,g}^g$ defines the energy for RS_k to transmit the data D_n^o to ground stations at t.

4.3. System Cost Model

The system cost function, denoted by U, takes into account both the task queue length and energy usage and is represented as follows:

$$U = \sum_{t=1}^{T} U_t = \sum_{t=1}^{T} \mu_1 E_t + \mu_2 Q_t \tag{11}$$

Here, U_t indicates the cost at each time t. The energy consumed, $E_t = E_t^c + E_t^o$, includes computation and data transmission at time t, while $Q_t = \sum_{m=1}^{M} Q_{m,t}^o + \sum_{k=1}^{K} Q_{k,t}$ reflects the system's queue length at the same time. The parameters μ_1 and μ_2 prioritize energy and queue length, respectively.

4.4. Optimization Constraints

Considering transmission links, link availability, flow conservation, and other constraints, the optimization problem needs to satisfy the following constraints.

4.4.1. Transmission Link Constraints

This study assumes that, for any time t, each IoT device can link to only one relay satellite, which is expressed as follows:

$$\sum_{k=1}^{K} z_{m,t',k,t}^i \leq 1, \quad \forall m, t' < t \tag{12}$$

Similarly, at any given t, each observation satellite is restricted to a connection with a single relay satellite, which is expressed as follows:

$$\sum_{k=1}^{K} \gamma^{o}_{n,k,t} \leq 1, \quad \forall n, t \tag{13}$$

For the relay satellites, at any moment t, data transfer is limited to transmitting from one observation satellite. This is denoted by the following:

$$\sum_{k'=1,k'\neq k}^{K} \gamma^{r}_{n,k,k',t} \leq 1, \quad \forall n, k, t \tag{14}$$

Further, at any point in time t, each relay satellite can only facilitate the transmission of data from one observation satellite within a single relay transfer:

$$\sum_{n=1}^{N} \gamma^{r}_{n,k,k',t} \leq 1, \quad \forall k, k' \neq k, t \tag{15}$$

Finally, each relay satellite is also constrained to transmitting data from a single observation satellite to the ground station. This is defined by the following:

$$\sum_{n=1}^{N} \gamma^{g}_{n,k,t} \leq 1, \quad \forall k, t \tag{16}$$

4.4.2. Link Availability

IoT devices can transmit computational tasks to relay satellites only when connectivity is established, which is defined by

$$\sum_{t'=1}^{t-1} z^{i}_{m,t',k,t} \cdot 1_{\{x^{i}_{m,k}=0\}} = 0, \quad \forall m, k, t \tag{17}$$

where $1_{\{x\}}$ is an indicator function that returns 1 if x is true and 0 otherwise. Similarly, relay satellites can relay data to ground stations only with an active link:

$$\sum_{n=1}^{N} \gamma^{g}_{n,k,t} \cdot 1_{\{x^{g}_{k,g}=0\}} = 0, \quad \forall k, t \tag{18}$$

$$\sum_{t'=1}^{t-1} \lambda^{g}_{m,t',k,t} \cdot 1_{\{x^{g}_{k,g}=0\}} = 0, \quad \forall k, t \tag{19}$$

For data transfer from observation satellites to relay satellites, a valid connection is necessary:

$$\sum_{t=1}^{T} \gamma^{o}_{n,k,t} \cdot 1_{\{x^{o}_{n,k}=0\}} = 0, \quad \forall n, k \tag{20}$$

Finally, data can only be exchanged between relay satellites if they are interconnected:

$$\sum_{t=1}^{T} \gamma^{r}_{n,k,k',t} \cdot 1_{\{x^{r}_{k,k'}=0\}} = 0, \quad \forall k, k', k \neq k' \tag{21}$$

The term 'x' serves as an indicator of link availability between IoT devices and satellites. When $x = 0$, it signifies that the link is not available; thus, the transmission variables must be set to zero, reflecting the inability to transmit data. This is supported by the indicator function, which will yield a value of 1 under these circumstances. Conversely, if $x = 1$, which indicates an available link, the transmission variables can assume non-zero values, allowing for data communication. This logical structure ensures that our model accurately represents the conditions of link availability within the network.

4.4.3. Task Computation

Given that the relay satellite initiates computing tasks only when all tasks have been transmitted and that they process just one task in each time slot, we have

$$y_{m,t',k,t} \cdot 1_{\{t<t'' \cdot z^{i}_{m,t',k,t''}\}} = 0, \quad \forall m,t',k,t,t'',t' < t'' < t \tag{22}$$

$$\sum_{m=1}^{M} \sum_{t'=1}^{T} y_{m,t',k,t} \leq 1, \quad \forall k,t,t > t' \tag{23}$$

4.4.4. Flow Conservation

In the transmission of user data via inter-satellite and satellite-to-ground communications, it is essential to adhere to flow conservation laws for both observation and relay satellites. The flow conservation constraint for an observation satellite OS_n is defined by the following:

$$\sum_{k=1}^{K} \sum_{t=1}^{T} \gamma^{o}_{n,k,t} = 1, \quad \forall n \tag{24}$$

Equation (24) ensures that each observation satellite OS_n (where n is the index of the satellite) must transmit its data to exactly one relay satellite RS_k within the time frame $[1,T]$. The summation over k (relay satellites) and t (time slots) ensures that the total number of transmissions initiated by each observation satellite OS_n equals 1. This reflects the principle that each observation satellite must transmit its data to exactly one relay satellite during the defined time horizon, maintaining data flow from observation satellites to relay satellites.

For a relay satellite RS_k handling the data D^{o}_{n}, the constraint is formulated as follows:

$$\sum_{t=1}^{T} \gamma^{o}_{n,k,t} + \sum_{k'=1,k'\neq k}^{K} \sum_{t=1}^{T} \gamma^{r}_{n,k,k',t} = \sum_{k'=1,k'\neq k}^{K} \sum_{t=1}^{T} \gamma^{r}_{n,k',k,t} + \sum_{t=1}^{T} \gamma^{g}_{n,k,t'}, \quad \forall n,k \tag{25}$$

Equation (25) enforces the principle of flow conservation at each relay satellite RS_k for data originating from a given observation satellite OS_n. It ensures that the total incoming data flow into the relay satellite RS_k equals the total outgoing data flow from RS_k, preserving the continuity of the data flow within the network. The left-hand side represents the total incoming flow into the relay satellite RS_k, which includes data directly transmitted from the observation satellite and relayed from other relay satellites. The right-hand side represents the total outgoing flow from the relay satellite RS_k, which includes data relayed to the other relay satellite and transmitted to the ground station.

Equations (24) and (25) embody the principle of flow conservation, stating that for any given node in the network, the total data inflow must equal the total data outflow. This principle is vital for maintaining the integrity and efficiency of data transmission within the satellite network. By applying this conservation principle, we ensure that our model accurately tracks data movement and resource utilization across the network, thereby facilitating optimal routing and task offloading strategies.

4.5. Optimization Model

Considering constraints such as link availability and computation limitations, the optimization model for the joint dynamic task offloading and transmission scheduling problem based on the long-term cost function minimization under the constraints is the following:

$$\begin{aligned} \min_{X} \quad & \lim_{T\to\infty} \frac{1}{T}\mathbb{E}\left[\sum_{t=1}^{T} U_t\right] \\ s.t. \quad & (12)\text{–}(25) \end{aligned} \tag{26}$$

where $X = \{\gamma^o_{n,k,t}, \gamma^r_{n,k,k',t}, \gamma^g_{n,k,t}, z^i_{m,t',k,t}, \lambda^g_{m,t',k,t}, y_{m,t',k,t}\}$ is the set of all variables.

5. Proposed Schemes for Joint Task Offloading and Transmission Scheduling

Solving the long-term cost minimization problem $P1$ modeled in (26) involves dynamic task offloading and computation scheduling strategies for observation satellite data tasks, IoT device computation tasks, and inter-satellite network topology stability. The original optimization problem is divided into data task transmission scheduling and computation task offloading and scheduling problems. For the data task transmission scheduling problem, a transmission scheduling strategy based on Dijkstra's algorithm [50] is proposed. For the dynamic task offloading and scheduling problem, considering the Markov decision process, a deep learning method is adopted to propose a dynamic task offloading and scheduling strategy based on DQN.

5.1. Data Task Transmission Scheduling

This section models and addresses the data task transmission scheduling issue, aiming to devise a transmission strategy. Assuming a static inter-satellite network topology, the problem is simplified to a directed shortest path with weighted edges. Dijkstra's algorithm is utilized to find the shortest routes from observation satellites to the ground station. To solve the capacity limitations of relay satellites, an enhanced scheduling strategy is introduced. Excluding conflicts among relay satellites, a scheduling strategy for observation satellites maps the data transmission route to ground stations. The link weights, determined by energy consumption, help model this as a directed shortest path problem. Let W_n be the cumulative energy expenditure from OS_n to the ground station, which is represented by the following:

$$W_n = \sum_{k=1}^{K}\sum_{t=1}^{T} \gamma^o_{n,k,t} W^o_{n,k} + \sum_{k=1}^{K}\sum_{k'=1,k'\neq k}^{K}\sum_{t=1}^{T} \gamma^r_{n,k,k',t} W^r_{n,k,k'} + \sum_{k=1}^{K}\sum_{t=1}^{T} \gamma^g_{n,k,t} W^g_{n,k,t} \tag{27}$$

Here, $W^o_{n,k}$ denotes the link weight between the observation satellite OS_n and relay satellite RS_k, $W^r_{n,k,k'}$ the link weight for the data D^o_n between the relay satellites RS_k and $RS_{k'}$, and $W^g_{n,k,t}$ the link weight for the data D^o_n from the relay satellite RS_k to the ground station at time t. These weights are computed as follows.

$$W^o_{n,k} = \begin{cases} \frac{D^o_n P_n}{R^o_{n,k}}, & x^o_{n,k} = 1 \\ +\infty, & x^o_{n,k} = 0 \end{cases}, \quad W^r_{n,k,k'} = \begin{cases} \frac{D^o_n P^r_k}{R^r_{k,k'}}, & x^r_{k,k'} = 1 \\ +\infty, & x^r_{k,k'} = 0 \end{cases}, \quad W^g_{n,k,t} = \begin{cases} \frac{D^o_n P^g_k}{R^g_{k,g}}, & x^g_{k,g} = 1 \\ +\infty, & x^g_{k,g} = 0 \end{cases} \tag{28}$$

From (28), it is observed that greater transmission energy reduces link weight. The shortest path problem between the observation satellite OS_n and the ground station is modeled as follows:

$$\min_{\gamma^o_{n,k,t}, \gamma^r_{n,k,k',t}, \gamma^g_{n,k,t}} W_n, \quad s.t. \quad (13)\text{–}(16), (18), (20), (21), (24), (25) \tag{29}$$

The inter-satellite link's state over time is modeled by representing the ground station as T nodes, GS_t, with $1 \leq t \leq T$. At each time t, link weights to relay satellites depend on link conditions. Using Dijkstra's algorithm [50] for shortest path calculations, super nodes are employed, and virtual ground station nodes have link weights set to zero. Specifically, to model the time-varying nature of the satellite–ground links, we represent the ground station as a set of virtual nodes GS_t for each time slot t. Each virtual ground station node GS_t corresponds to the ground station at time t, capturing the link availability and conditions at that specific time. To facilitate the application of Dijkstra's algorithm, we introduce a 'super node'. The super node serves as a single common destination node and is connected to all virtual ground station nodes GS_t with edges of zero weight. This construction ensures that the shortest path algorithm can effectively consider all possible paths reaching the ground station over different time slots.

The graph $G_n = (V, E, W_n)$ is constructed, where V includes observation satellites OS_n, relay satellites, and virtual ground station nodes; E consists of satellite–ground station links, and W_n contains link weights. All link weights from ground stations to super nodes are zero. By constructing the time-expanded graph $G_n = (V, E, W_n)$, we capture the dynamic topology of the inter-satellite network over time. The vertices V include the observation satellite OS_n, relay satellites RS_k at different time slots, virtual ground station nodes GS_t representing the ground station at time t, and the super node serving as the common destination. Edges E represent possible transmissions between nodes at consecutive time slots, with weights Wn indicating the transmission costs. We apply Dijkstra's algorithm to this graph to find the shortest path from OS_n to the super node. This path denotes the optimal sequence of transmissions through relay satellites and time slots, minimizing the overall transmission cost to the ground station.

Figure 2 illustrates a schematic of the link formed by one observation satellite OS_n and six relay satellites. Applying Dijkstra's algorithm to the augmented graph G_n identifies the shortest path between OS_n and the virtual ground station nodes. Let $\tilde{\pi}^o_{n,t} = \{\tilde{\gamma}^o_{n,k,t'}, \ldots, \tilde{\gamma}^r_{n,k,k',t'}, \ldots, \tilde{\gamma}^g_{n,k,t'}, 1 \leq t' \leq t\}$ represent the data set D^o_n from OS_n to GS_t. Due to multiple hops and the dynamic nature of the inter-satellite link, the original transmission strategy may become infeasible, requiring adjustments. The strategy set $\tilde{\pi}^o_{n,t}$ is checked for conflicts. If the maximum conflict time $t_1 = \arg\max\{\tilde{\gamma}^r_{n,k,k',t}\}$ of the relay satellite transmission strategy exceeds t, the strategy is invalidated; if $t_1 < t$, the strategy set $\tilde{\pi}^o_{n,t}$ is retained. The strategy with the minimum weight is selected as the local transmission scheduling strategy set $\tilde{\pi}^o_{n,t*}$ for OS_n.

Figure 2. Diagram of link state.

In the context of independently designing transmission scheduling strategies for observation satellites, instances may arise where two distinct data tasks commence transmission from a similar origin and are destined for the same endpoint concurrently. Such scenarios precipitate a transmission scheduling conflict. To address this issue, the following section introduces an algorithm capable of adjusting transmission strategies through prioritization. For illustrative purposes, consider the transmission strategies $\tilde{\gamma}^{r}_{n,k,k',t}$ and $\tilde{\gamma}'^{r}_{n',k,k',t}$ associated with the conflicting data D^{o}_{n} and D'^{o}_{n}, respectively. The algorithm's fundamental concept is developed as follows:

1. Assess the Priority of Present Data Tasks:At the current time, evaluate the transmission energy expended by both conflicting data tasks, assuming that the task with greater accumulated transmission energy is accorded higher priority. Define $E_{n,t'-1}$ as the total transmission energy accumulated by the data task D^{o}_{n} preceding time t'. The expression for $E_{n,t'-1}$ is given by the following:

$$E_{n,t'-1} = \sum_{k=1,t=1}^{K,t'-1} E^{o}_{n,k,t} + \sum_{\substack{k=1,k'=1\\k'\neq k,t=1}}^{K,K,t'-1} E^{r}_{n,k,k',t} \tag{30}$$

The accumulated energy $E^{o}_{n,t'-1}$ in Equation (30) quantifies the total energy expended for transmitting data task D^{o}_{n} up to time $t'-1$, considering both observation-to-relay and relay-to-relay transmissions. Specifically, for conflicting data tasks D^{o}_{n} and D'^{o}_{n}, if $E^{o}_{n,t'-1} > E_{n',t'-1}$, the data task D^{o}_{n} is granted higher priority (i.e., the one that has already consumed more energy up to time t'). This metric is crucial for prioritizing tasks during scheduling conflicts as tasks with higher accumulated energy are considered to have greater investment and are, therefore, prioritized to minimize system cost and resource waste.

2. Modify Transmission Strategy in Accordance with Priority:The aforementioned data task D^{o}_{n}, enjoying higher precedence, will implement the strategy encapsulated in $\tilde{\pi}^{o}_{n,t^{*}}$. Here, $\tilde{\pi}^{o,*}_{n,t}$ denotes the designated set of strategies for transmitting the data task D^{o}_{n}. Consequently, the optimal scheduling strategy set is $\tilde{\pi}^{o,*}_{n} = \tilde{\pi}^{o}_{n}$. For the data task D'_{n} possessing a subordinate priority, it is imperative to revert to the observation satellite OS_{n} and reformulate the transmission scheduling strategy, as detailed in the Dijkstra-based algorithm developed above. To circumvent subsequent conflicts, careful consideration must be given to arranging the set containing strategy $\tilde{\gamma}^{r}_{n,k,k',t'}$, ultimately achieving the refined strategy set $\tilde{\pi}^{o,*}_{n',t^{*}}$.

5.2. Dynamic Task Offloading and Computation Scheduling Problem Modeling and Solving

With the transmission scheduling strategy for data tasks established, the relay satellite's buffer queue status is known for each time slot. Due to the relay satellite's limited buffer capacity, dynamic satellite–ground links, and the random arrival of IoT device computation tasks, the task offloading and computation scheduling problem becomes highly complex and dynamic.

This complexity makes traditional optimization methods less effective, as they may not efficiently handle the high-dimensional and stochastic nature of the problem. Therefore, we model the problem as a Markov Decision Process (MDP). By formulating the problem as an MDP, we can employ reinforcement learning techniques to find optimal policies. Specifically, we utilize the Deep Q-Network (DQN) algorithm, which integrates deep learning with Q-learning. DQN is adept at estimating the state–action value function in high-dimensional spaces, making it suitable for our model. The algorithm leverages deep neural networks to approximate the optimal Q-values and incorporates experience replay and target networks to enhance learning stability and efficiency.

The application of DQN enables the agent (relay satellite) to learn optimal task offloading and computation scheduling strategies through interactions with the environment, ultimately minimizing the long-term system cost in the face of uncertainty and dynamic conditions.

Consider the function $\tilde{U}_t$ as indicative of the cost associated with the relay satellite's activities in managing the offloading of tasks and scheduling of computational procedures at time instance t. The function $\tilde{U}_t$ is defined mathematically as follows:

$$\tilde{U}_t = \mu_1 E_t + \mu_2 Q_t \quad (31)$$

The challenge surrounding the optimization model for dynamic task offloading and computation scheduling is formulated by the following:

$$\min_{z^i_{m,t',k,t}, \lambda^g_{m,t',k,t}, y_{m,t',k,t}} \lim_{T\to\infty} \frac{1}{T}\mathbb{E}\left[\sum_{t=1}^{T} \tilde{U}_t\right], \quad s.t. \quad (12), (17), (19), (22), (23) \quad (32)$$

Through solving the above-stated Equation (32), it becomes possible to determine the strategic approach for selecting computation modes by the relay satellite, denoted as $z^i_{m,t',k,t}$. Additionally, this leads to the identification of strategies for offloading mode selection by the ground station, represented by $\lambda^g_{m,t',k,t}$, along with determining the computation scheduling strategy denoted by $y_{m,t',k,t}$.

5.2.1. Markov Decision Process Modeling

When we closely examine Equation (32), it becomes apparent that we are dealing with a long-term optimization challenge that incorporates randomness through its variables. To effectively address this problem, it is structured as a Markov decision process, which is mathematically characterized by the tuple (S, A, r_t). Here, S signifies the array of system states, $A = \{a_t\}$ signifies the suite of potential actions, and r_t stands for the immediate reward function associated with the process [51]. The constituents of the state space, action space, and reward function within this context are further detailed as follows. Let $S = \{s_t\}$ denote the array of system states at a given time t, with each state s_t conceptualized as follows:

$$s_t = \{Q_t, Q^i_t, L^{pt}_r, x^i_t\} \quad (33)$$

In this expression, $Q_t = \{Q_{1,t}, Q_{2,t}, \ldots, Q_{k,t}, \ldots, Q_{K,t}\}$ captures the buffer queue conditions for each relay satellite at time t; $Q^i_t = \{Q^i_{1,t}, Q^i_{2,t}, \ldots, Q^i_{m,t}, \ldots, Q^i_{M,t}\}$ reflects the buffer queue status for each IoT device during the same period; $L^{pt}_t = \{L^{pt}_{m,k,t}, 1 \le m \le M, 1 \le k \le K\}$ indicates the rain attenuation coefficient relevant to the communication link connecting each IoT device and each relay satellite at time t; and $x^i_t = \{x^i_{m,k,t}, 1 \le m \le M, 1 \le k \le K\}$ depicts the status of logical link existence between each IoT device and each relay satellite at this time instance. At time t, IoT devices have the capability to transfer computation tasks stored within their buffer queues to relay satellites. Accordingly, let $A = \{a_t\}$ denote the comprehensive set of actions available at time t, where each action a_t encapsulates the strategies for selecting task offloading modes by IoT devices at that specific moment and can be expressed as follows:

$$a_t = \{z_{m,t',k,t}, \lambda^g_{m,t',k,t}, y_{m,t',k,t}, 1 \le m \le M, 1 \le t' < t, 1 \le k \le K\} \quad (34)$$

Under the scenario wherein IoT devices occupy state s_t and undertake action a_t, the resulting immediate reward r_t is meticulously designed as follows:

$$r_t = -\mu_1\left(\sum_{k=1}^{K} Q_{k,t} + \sum_{m=1}^{M} Q^i_{m,t}\right) - \mu_2 E_t \quad (35)$$

5.2.2. Determining Satellite Dynamic Link Strategy Based on DQN

Figure 3 illustrates the developed framework for task offloading and scheduling. The algorithm of DQN, introduced by the DeepMind group, is a reinforcement learning algorithm integrating deep learning principles widely used in current complex satellite networks design [19,30,52–54]. Unlike traditional Q-learning, DQN is adept at managing complex state spaces characterized by high dimensionality. In the realm of Q-learning, such state spaces often pose dimensionality challenges, impeding effective learning. DQN circumvents these limitations by employing deep neural networks as function approximators, which allows for the efficient handling of high-dimensional spaces by representing states as continuous variables, thus enhancing learning efficacy. Moreover, conventional Q-learning mandates a discrete action space, thereby restricting its utility in scenarios necessitating continuous action spaces. DQN addresses this limitation by leveraging deep neural networks to estimate the Q-values for each possible action, thus enabling the handling of continuous action spaces and the selection of the optimal action via optimization techniques [55]. The integration of an experience replay mechanism permits the storage of the agent's experiences in a replay buffer, from which samples are randomly extracted for training. This approach alleviates the correlation between successive training data, thereby mitigating the sample correlation issue and bolstering training efficiency and stability. Consequently, the DQN method is proficient at maximizing interactions between intelligent agents and stochastic environments. The 'Environment' encapsulates the temporal dynamics that influence the state, allowing the model to maintain a distinction between the current conditions and the actions taken, until such action results in a transition to a 'Next State'.

Figure 3 illustrates the iterative process of the Deep Q-Network (DQN) learning cycle. Although the figure does not explicitly show an entry point, the process naturally begins with the first observation of the environment, denoted as the initial state s_0. This initial state is used to determine the first action a_0, initiating the sequence of interactions between the agent and the environment. From this point forward, the system continuously updates states and actions based on learned policies. This design aligns with conventional DQN representations, where the emphasis is on the continuous reinforcement learning process rather than a discrete starting point.

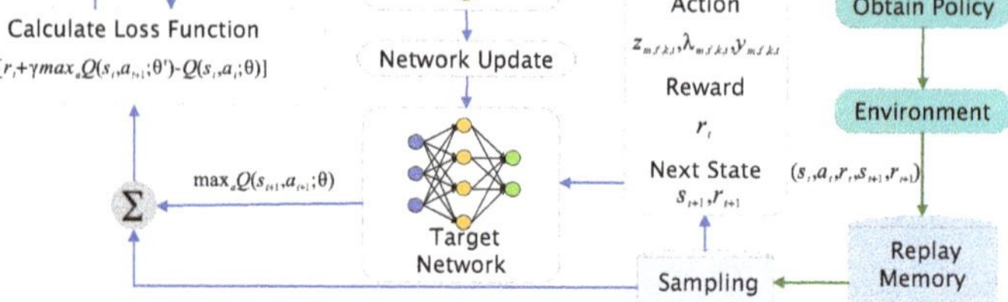

Figure 3. Proposed DQN-based task offloading algorithm framework.

During the training phase of the DQN (Deep Q-Network) algorithm, the strengths of both convolutional and recurrent neural networks are harnessed to proficiently estimate the state–action value function, known as the Q-function. The architecture of DQN comprises two distinct networks: the prediction network tasked with generating experience replay data, and the target network, which computes the state–action value, referred to as the Q-value. To expedite convergence, DQN employs an experience replay strategy, which involves the random sampling of accumulated data for training, alongside a smoothing technique for managing the data effectively. When IoT devices execute actions during the state s_t, the associated action value function Q is updated as follows:

$$Q_{t+1}(s,a) = Q(s,a) + \alpha\left[r(a_t|s_t) + \gamma \max_{a'} Q_t(s',a') - Q_t(s,a)\right] \quad (36)$$

In Equation (36), α denotes the learning rate, and γ represents the discount factor. To tackle the computational intricacies of Equation (36), and prevent divergence of Q-values, DQN introduces a prediction network $Q(s,a;\theta)$ and a target network $Q(s,a;\theta')$, which are responsible for predicting and assessing the Q-values, respectively. Herein, θ and θ' are the parameters of the prediction network and target network. The parameters θ are updated by reducing the loss function:

$$\theta \leftarrow \theta + \nu \nabla F_t(Q(s,a|*)) \quad (37)$$

where ν signifies the step size, and $F_t(Q(s,a|*))$ denotes the Q-value loss function, which is expressed as follows:

$$F_t(Q(s,a|*))) = \mathbb{E}_{(s_t,a_t,r_t,s_{t+1})\sim U(D)}\left[(y_t - Q(s,a;\theta))^2\right] \quad (38)$$

In Equation (38), $(s_t, a_t, r_t, s_{t+1}) \sim U(D)$ symbolizes samples randomly extracted from the experience replay buffer D, and y_t is the output generated by the target network at time t, which is given by the following:

$$y_t = r_t + \gamma \max_{a'} Q'(s',a';\theta') \quad (39)$$

The detailed procedures for the computation task offloading and scheduling algorithm, rooted in the DQN learning algorithm for interactions between satellite and IoT devices, are outlined in Algorithm 1. Begin by initializing the prediction and target networks alongside the state, action, and parameters within the experience replay buffer. Within the state s_t, implement the ϵ-greedy policy to identify the optimal action a_t for the target network, facilitating the transition of the system state from s_t to s_{t+1}. Record the experience tuple (s_t, a_t, r_t, s_{t+1}) in the experience replay buffer, whose maximum capacity is denoted as $D^{\max}$. Select a random batch of samples with size D^{batch} from this buffer and feed them into the prediction network to generate predicted Q-values. The target Q-values are computed in line with Equation (36). Then, determine the discrepancy between the predicted and target Q-values to update the parameters of the prediction network via Equation (37). The target network parameters are periodically revised utilizing the prediction network's parameters. Iterate these steps until convergence is reached.

We define 'resource scheduling' in the context of our study as the strategic scheduling and management of computational and communication resources within the satellite network. This encompasses both data scheduling, which prioritizes data packet transmission, and task offloading, which involves distributing computational tasks among various network nodes. Although our primary focus is on optimizing data scheduling paths and task offloading strategies, these actions are intrinsically linked to efficient resource utilization, ensuring that the network operates optimally in conjunction with its available resources.

Algorithm 1 DQN-Based Task Offloading and Scheduling

Initialize prediction network $Q(s, a; \theta)$ and target network $Q'(s, a; \theta')$
Initialize experience replay buffer D with capacity $D^{\max}$
for each episode **do**
  Initialize state s_t
  for each step in episode **do**
    Select action a_t using ϵ-greedy policy
    Execute action a_t and observe reward r_t and next state s_{t+1}
    Store (s_t, a_t, r_t, s_{t+1}) in D
    Sample random batch of size D^{batch} from D
    Compute target $y_t = r_t + \gamma \max_{a'} Q'(s_{t+1}, a'; \theta')$
    Compute loss $F_t = \mathbb{E}[(y_t - Q(s_t, a_t; \theta))^2]$
    Update prediction network parameters $\theta \leftarrow \theta + \nu \nabla_\theta F_t$
    if step mod update_frequency == 0 **then**
      Update target network parameters $\theta' \leftarrow \theta$
    end if
    $s_t \leftarrow s_{t+1}$
  end for
end for

6. Simulation Results and Analysis

This section uses the STK simulation software version 6.0 to build a system scenario in order to obtain real satellite and inter-satellite physical link information, apply Pytorch 2.2.2 and MATLAB 2024a to simulate the algorithms proposed in this paper and the algorithm in the literature [40], and analyze the performance differences between the two. This paper presents a novel cost model for an LEO satellite edge computing system, addressing the challenges of computation offloading. It introduces a joint computation offloading and resource allocation strategy, which decomposes the problem into two sub-problems: optimizing computation offloading using game theory and allocating communication resources via the Lagrange multiplier method. Ref. [40] is selected as a baseline due to its relevance to IoT applications and its use of energy consumption as a cost metric, which aligns with our study's objectives. However, our approach differs significantly in methodology, as we employ a DQN-based optimization rather than the game-theoretic and Lagrange multiplier methods used in [40]. To enable a fair comparison, we replace our proposed DQN-based optimization algorithm with the JCORA method from [40] while keeping all other simulation parameters unchanged. This allows us to directly compare the performance impact of different optimization strategies under the same system conditions.

Simulation Scenario

In the simulation scenario of this paper, a LEO satellite network is built using the STK satellite tool (version 6.0), which includes 24 medium-orbit satellites distributed in three orbital planes with an altitude of 1000 km, 10 observation satellites, and one ground station. This section uses MATLAB (version 2024a) the Python programming language (version 3.8) and a Pytorch (version 2.1.2) simulation environment based on the Gym reinforcement learning environment to simulate the DQN network. During the simulation process, the number of training steps is set to 4000 steps, and the number of training episodes is set to 30 h. The parameters of the DQN network and other parameters are shown in Table 2. This section averages the simulation results of 600 independent experiments and compares the proposed algorithm with the algorithm in the literature.

Table 2. Parameter settings for DQN network and simulation

Parameter	Value
Learning Rate α	0.001
Discount Factor γ	[0.1, 0.5, 0.9]
Experience Replay Buffer Capacity $D^{\max}$	10,000
Sample Size D^{batch}	64
Exploration Value ϵ	[0.1, 1]
Link Bandwidth B	20 MHz
Arrival Rate	[0.8, 2] Mbps
μ_1 and μ_2	[1,3]
Rain Attenuation $L^{\text{rain}}_{m,k,t}$	[10, 20] dB
Boltzmann Constant k_s	1.380649×10^{-23} J/K
Thermal Noise Temperature T_s	300 K
Signal-to-Noise Ratio E_b/N_0	17 dB
Relay Satellite Transmission Power P^r	40 W
Relay Satellite Transmission Antenna Gain G^{rg}_t	27 dBi
Relay Satellite Reception Antenna Gain G^{rg}_r	24 dBi
Observation Satellite Transmission Power P_n	40 W
IoT Device Transmission Power P^i_m	0.1 W
Time Slot Length T	0.2 s

Figures 4 and 5 show the impact of different learning rates and discount factors on long-term rewards. As shown in the figure, a learning rate of 0.1 is very clear that the DQN network does not converge and fluctuates significantly. This is because the learning rate is too large, and the network jumps directly to the local optimal value, resulting in suboptimal results. When the learning rate is 0.0001, the network converges slowly, and even after 400 training steps, there is still a large fluctuation. This is because, during the learning process, the network converges slowly and requires more training steps to reach the final reward. When the learning rate is 0.001, the network can reach the optimal reward after 1000 steps and maintain the reward through continuous training.

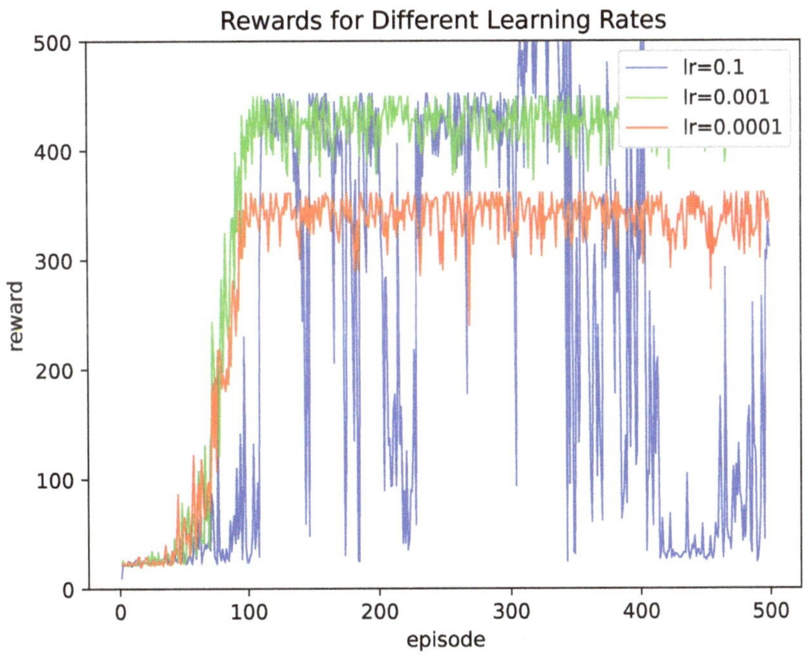

Figure 4. Long-term reward versus number of training steps (with different learning rates).

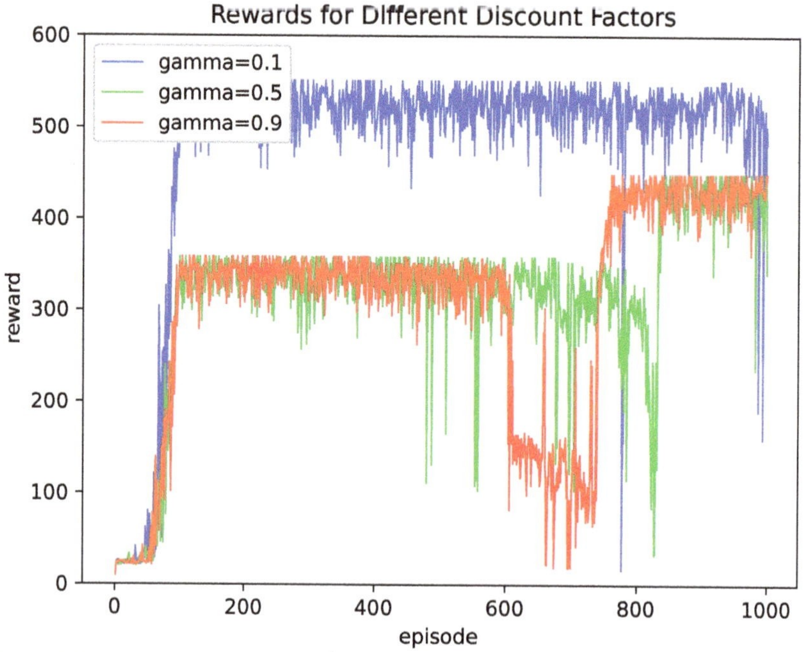

Figure 5. Long-term reward versus number of training steps (with different discount factors).

Figure 5 illustrates the impact of different discount factors γ on the long-term rewards during the training of our DQN-based task offloading and computation scheduling algorithm. The discount factor is a critical parameter in reinforcement learning algorithms, determining how future rewards are valued relative to immediate rewards. When $\gamma = 0.1$, the agent exhibits slow convergence with high fluctuations in rewards. This low discount factor implies that the agent undervalues future rewards, leading to shortsighted decisions that do not contribute to long-term system optimization. For $\gamma = 0.5$, the agent shows improved convergence and reduced fluctuations. The balance between immediate and future rewards allows the agent to learn better policies but still falls short of optimal performance. With $\gamma = 0.9$, the agent achieves rapid convergence and higher steady-state rewards. The high emphasis on future rewards enables the agent to learn strategies that significantly minimize the long-term system cost. This result indicates that a higher discount factor is beneficial in the context of our satellite edge computing network, where long-term planning is essential due to the dynamic and continuous nature of tasks and resources.

In Figure 6, we illustrate the connection between the relay satellite's maximum queue length and the system cost, alongside analysis of varied scenarios concerning task computation difficulty. The figure reveals that as the relay satellite's maximum queue length grows, the system cost declines. This occurs because a longer queue allows the satellite to manage more data transmission tasks and adjust computation task scheduling, enhancing the system's capacity to prevent data loss and stabilize cost fluctuations. Notably, the system cost achieved by our proposed method in this study is lower than that reported in prior research, and with a task computation difficulty of 1000 cycles, the cost reduction using our approach is slower compared to the method from the literature [40].

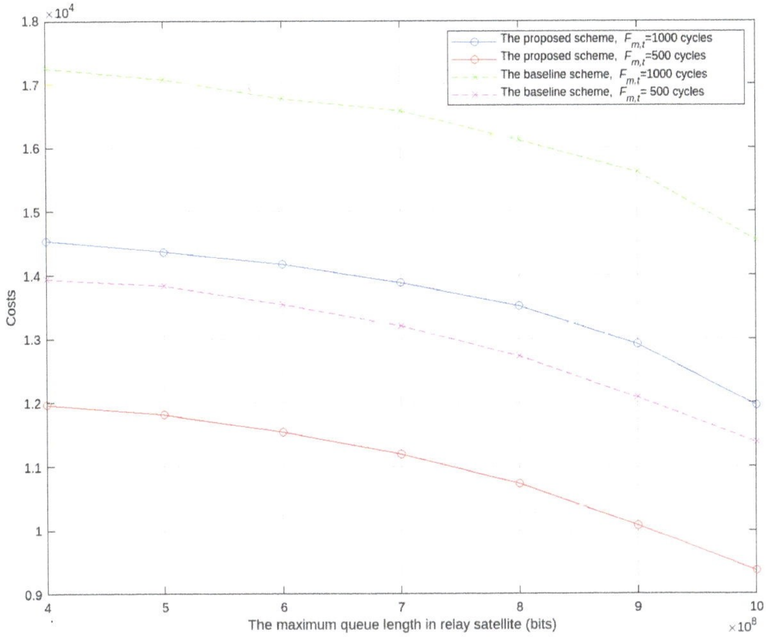

Figure 6. System cost versus maximum queue length of relay satellites.

Figure 7 presents the correlation between the relay satellite's computation capability and system cost. It demonstrates that as the satellite's computation capability increases, the system cost decreases unevenly across various methods. Figure 7 demonstrates the relationship between the computational capability of relay satellites and the system cost in the proposed LEO satellite edge computing network. As the computational capability of relay satellites increases, the system cost decreases significantly due to improved task processing efficiency and reduced communication overhead. However, the curve exhibits a diminishing return effect, where further increases in computational capability yield only marginal reductions in system cost. This highlights the importance of balancing computational resources to achieve cost-effective performance. The figure underscores the efficiency of the proposed joint task offloading and resource scheduling strategy in optimizing system performance while minimizing costs.

Furthermore, an increase in IoT device numbers also results in uneven decreases in system costs. This is due to the reduction in task computation time as satellite computation improves, allowing the task queue to remain stable within a reasonable timeframe, thus lowering costs. Moreover, a rise in device data quantity leads to more tasks for the system to process, adding to the costs. The cost via our proposed methodology notably surpasses that of previous approaches [40], and with enhanced satellite computation, our method's cost increase is slower than that of existing methods, underscoring the benefits of our approach.

Figure 8 evaluates the relationship between the average task arrival rate and system cost under different numbers of IoT devices for the proposed algorithm in this chapter and the algorithm from [40]. As shown in the figure, as the task arrival rate increases, the system cost decreases. This is because a higher task arrival rate leads to greater utilization of the network's computational, communication, and caching resources, resulting in increased system costs. Moreover, the system cost achieved by the proposed algorithm in this chapter is lower than that of the algorithm in [40]. Furthermore, as the task arrival rate increases, the cost under the proposed algorithm increases at a relatively slower rate.

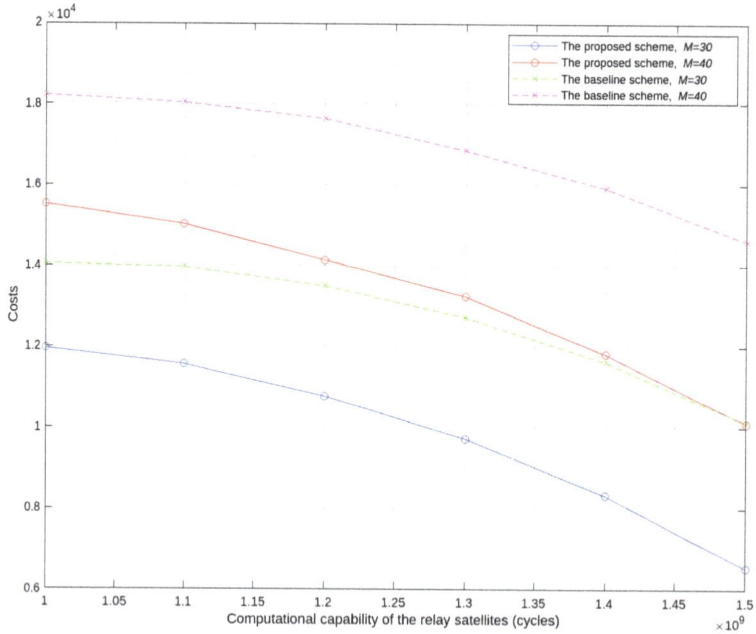

Figure 7. System cost versus computing capability of relay satellites.

The system cost metric used in this study incorporates energy consumption, task processing delays, and other operational factors. By minimizing system costs, the proposed method achieves significant energy savings in addition to reducing delays and improving task completion rates. This holistic evaluation provides a comprehensive assessment of the system's performance.

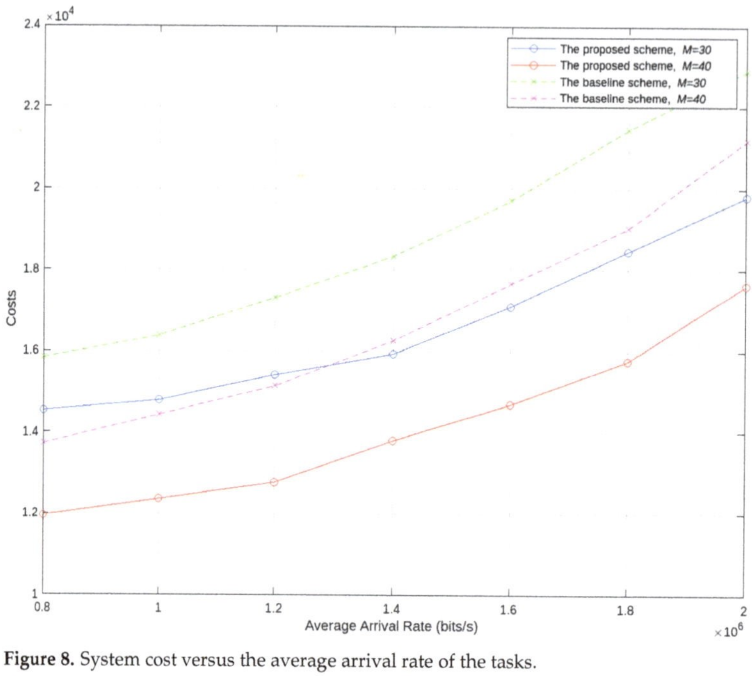

Figure 8. System cost versus the average arrival rate of the tasks.

7. Conclusions

This paper has comprehensively addressed the joint dynamic task offloading and resource scheduling problem in LEO satellite edge computing networks. The proposed system model incorporates both data service transmission and computational task offloading and is framed as a long-term cost function minimization problem with constraints. Key contributions include the development of a priority-based policy adjustment algorithm for handling transmission scheduling conflicts and a DQN-based algorithm for dynamic task offloading and computation scheduling. These methods are integrated into a joint scheduling strategy that optimizes overall system performance. Simulation results demonstrate significant improvements in average system cost, queue length, energy consumption, and task completion rate, compared to baseline strategies, highlighting the strategy's effectiveness and efficiency. Future work will extend the framework to more complex network scenarios and explore the integration of advanced machine learning techniques to further enhance security performance [56,57].

Author Contributions: Conceptualization, J.L. and R.C.; methodology, J.L. and K.G.; software, K.G. and R.C.; formal analysis, J.L.; writing—original draft preparation, J.L.; writing—review and editing, R.C. and C.L.; project administration, C.L.; funding acquisition, C.L. All authors have read and agreed to the published version of the manuscript.

Funding: This research was funded by Science and Technology Innovation Key R&D Program of Chongqing grant number : CSTB2023TIAD-STX0025.

Data Availability Statement: Access to the experimental data presented in this article can be obtained by contacting the corresponding author.

Conflicts of Interest: The authors declare no conflicts of interest.

References

1. Chen, M.; Li, J.; Chai, R. 5G Multi-Service-Oriented User Association and Routing Algorithm for Integrated Terrestrial-Satellite Networks. In Proceedings of the 13th International Conference on Wireless Communications and Signal Processing (WCSP), Changsha, China, 20–22 October 2021; pp. 1–5.
2. De Gaudenzi, R.; Ottersten, B.; Perez-Neira, A.; Yanikomeroglu, H.; Heyn, T.; Lichten, S.M. Guest Editorial Space Communications New Frontiers: From Near Earth to Deep Space. *IEEE J. Sel. Areas Commun.* **2024**, *42*, 1023–1028. [CrossRef]
3. Huang, Y.; Cui, H.; Hou, Y.; Hao, C.; Wang, W.; Zhu, Q.; Li, J.; Wu, Q.; Wang, J. Space-based electromagnetic spectrum sensing and situation awareness. *Space: Sci. Technol.* **2024**, *4*, 0109. [CrossRef]
4. Bai, L.; Zhu, L.; Zhang, X.; Zhang, W.; Yu, Q. Multi-satellite relay transmission in 5G: Concepts, techniques, and challenges. *IEEE Netw. Mag.* **2018**, *32*, 38–44. [CrossRef]
5. De Cola, T.; Bisio, I. QoS Optimisation of eMBB Services in Converged 5G-Satellite Networks. *IEEE Trans. Veh. Technol.* **2020**, *69*, 12098–12110. [CrossRef]
6. Li, B.; Fei, Z.; Zhou, C.; Zhang, Y. Physical-layer Security in Space Information Networks: A Survey. *IEEE Internet Things J.* **2019**, *7*, 33–52. [CrossRef]
7. Ranaweera, P.; Jurcut, A.D.; Liyanage, M. Survey on Multi-Access Edge Computing Security and Privacy. *IEEE Commun. Surv. Tutorials* **2021**, *23*, 1078–1124. [CrossRef]
8. Abbas, N.; Zhang, Y.; Taherkordi, A.; Skeie, T. Mobile Edge Computing: A Survey. *IEEE Internet Things J.* **2017**, *5*, 450–465. [CrossRef]
9. Pan, J.; McElhannon, J. Future Edge Cloud and Edge Computing for Internet of Things Applications. *IEEE Internet Things J.* **2018**, *5*, 439–449. [CrossRef]
10. Al-Hraishawi, H.; Chougrani, H.; Kisseleff, S.; Lagunas, E.; Chatzinotas, S. A survey on nongeostationary satellite systems: The communication perspective. *IEEE Commun. Surv. Tutor.* **2022**, *25*, 101–132. [CrossRef]
11. Huang, Y.; Dong, M.; Mao, Y.; Liu, W.; Gao, Z. Distributed Multi-Objective Dynamic Offloading Scheduling for Air–Ground Cooperative MEC. *IEEE Trans. Veh. Technol.* **2024**, *73*, 12207–12212. [CrossRef]
12. Liu, J.; Shi, Y.; Fadlullah, Z.M.; Kato, N. Space-Air-Ground Integrated Network: A Survey. *IEEE Commun. Surv. Tutor.* **2018**, *20*, 2714–2741. [CrossRef]

13. Xie, R.; Tang, Q.; Wang, Q.; Liu, X.; Yu, F.R.; Huang, T. Satellite-Terrestrial Integrated Edge Computing Networks: Architecture, Challenges, and Open Issues. *IEEE Netw.* **2020**, *34*, 224–231. [CrossRef]
14. Huang, C.; Chen, G.; Xiao, P.; Xiao, Y.; Han, Z.; Chambers, J.A. Joint offloading and resource allocation for hybrid cloud and edge computing in SAGINs: A decision assisted hybrid action space deep reinforcement learning approach. *IEEE J. Sel. Areas Commun.* **2024**, *42*, 1029–1043. [CrossRef]
15. Tang, Q.; Xie, R.; Fang, Z.; Huang, T.; Chen, T.; Zhang, R.; Yu, F.R. Joint service deployment and task scheduling for satellite edge computing: A two-timescale hierarchical approach. *IEEE J. Sel. Areas Commun.* **2024**, *42*, 1063–1079. [CrossRef]
16. Li, J.; Gui, A.; Rong, C.; Liang, C. Cost-Optimized Dynamic Offloading and Resource Scheduling Algorithm for Low Earth Orbit Satellite Networks. In Proceedings of the 19th EAI International Conference on Communications and Networking in China, Chongqing, China, 2–3 November 2024; p. 1.
17. Tang, Z.; Zhou, H.; Ma, T.; Yu, K.; Shen, X.S. Leveraging LEO assisted cloud-edge collaboration for energy efficient computation offloading. In Proceedings of the 2021 IEEE Global Communications Conference (GLOBECOM), Madrid, Spain, 7–11 December 2021; pp. 1–6.
18. Tang, Q.; Fei, Z.; Li, B.; Han, Z. Computation offloading in LEO satellite networks with hybrid cloud and edge computing. *IEEE Internet Things J.* **2021**, *8*, 9164–9176. [CrossRef]
19. Xu, F.; Yang, F.; Zhao, C.; Wu, S. Deep reinforcement learning based joint edge resource management in maritime network. *China Commun.* **2020**, *17*, 211–222. [CrossRef]
20. Chen, Y.; Ai, B.; Niu, Y.; Zhang, H.; Han, Z. Energy-constrained computation offloading in space-air-ground integrated networks using distributionally robust optimization. *IEEE Trans. Veh. Technol.* **2021**, *70*, 12113–12125. [CrossRef]
21. Zhang, Z.; Zhang, W.; Tseng, F.-H. Satellite mobile edge computing: Improving QoS of high-speed satellite-terrestrial networks using edge computing techniques. *IEEE Netw.* **2019**, *33*, 70–76. [CrossRef]
22. Wang, Y.; Yang, J.; Guo, X.; Qu, Z. A game-theoretic approach to computation offloading in satellite edge computing. *IEEE Access* **2019**, *8*, 12510–12520. [CrossRef]
23. Wang, Y.; Zhang, J.; Zhang, X.; Wang, P.; Liu, L. A computation offloading strategy in satellite terrestrial networks with double edge computing. In Proceedings of the 2018 IEEE International Conference on Communication Systems (ICCS), Chengdu, China, 19–21 December 2018; pp. 450–455.
24. Pang, B.; Gu, S.; Zhang, Q.; Zhang, N.; Xiang, W. CCOS: A coded computation offloading strategy for satellite-terrestrial integrated networks. In Proceedings of the 2021 International Wireless Communications and Mobile Computing (IWCMC), Harbin , China, 28 June–2 July 2021; pp. 242–247.
25. Zhang, L.; Zhang, H.; Guo, C.; Xu, H.; Song, L.; Han, Z. Satellite-aerial integrated computing in disasters: User association and offloading decision. In Proceedings of the ICC 2020—2020 IEEE International Conference on Communications (ICC), Dublin, Ireland, 7–11 June 2020; pp. 554–559.
26. Alsharoa, A.; Alouini, M.S. Improvement of the global connectivity using integrated satellite-airborne-terrestrial networks with resource optimization. *IEEE Trans. Wirel. Commun.* **2020**, *19*, 5088–5100. [CrossRef]
27. Zhang, M.; Zhou, W. Energy-efficient collaborative data downloading by using inter-satellite offloading. In Proceedings of the 2019 IEEE Global Communications Conference (GLOBECOM), Waikoloa, HI, USA, 9–13 December 2019; pp. 1–6.
28. Yarr, N.; Ceriotti, M. Optimization of intersatellite routing for real-time data download. *IEEE Trans. Aerosp. Electron. Syst.* **2018**, *54*, 2356–2369. [CrossRef]
29. Di, B.; Zhang, H.; Song, L.; Li, Y.; Li, G.Y. Ultra-dense LEO: Integrating terrestrial-satellite networks into 5G and beyond for data offloading. *IEEE Trans. Wirel. Commun.* **2018**, *18*, 47–62. [CrossRef]
30. Qiu, C.; Yao, H.; Yu, F.R.; Xu, F.; Zhao, C. Deep Q-learning aided networking, caching, and computing resources allocation in software-defined satellite-terrestrial networks. *IEEE Trans. Veh. Technol.* **2019**, *68*, 5871–5883. [CrossRef]
31. Li, C.; Zhang, Y.; Hao, X.; Huang, T. Jointly optimized request dispatching and service placement for MEC in LEO network. *China Commun.* **2020**, *17*, 199–208. [CrossRef]
32. Jiang, D.; Wang, F.; Lv, Z.; Mumtaz, S. QoE-aware efficient content distribution scheme for satellite-terrestrial networks. *IEEE Trans. Mob. Comput.* **2021**, *22*, 443–458. [CrossRef]
33. Wang, G.; Zhou, S.; Niu, Z. Radio resource allocation for bidirectional offloading in space-air-ground integrated vehicular network. *J. Commun. Inf. Netw.* **2019**, *4*, 24–31. [CrossRef]
34. Song, Z.; Hao, Y.; Liu, Y.; Sun, X. Energy-efficient multiaccess edge computing for terrestrial-satellite Internet of Things. *IEEE Internet Things J.* **2021**, *8*, 14202–14218. [CrossRef]
35. Ding, C.; Wang, J.-B.; Zhang, H.; Lin, M.; Li, G.Y. Joint optimization of transmission and computation resources for satellite and high altitude platform assisted edge computing. *IEEE Trans. Wirel. Commun.* **2021**, *21*, 1362–1377. [CrossRef]
36. Cui, G.; Long, Y.; Xu, L.; Wang, W. Joint offloading and resource allocation for satellite assisted vehicle-to-vehicle communication. *IEEE Syst. J.* **2020**, *15*, 3958–3969. [CrossRef]

37. Cheng, N.; Lyu, F.; Quan, W.; Zhou, C.; He, H.; Shi, W.; Shen, X. Space/aerial-assisted computing offloading for IoT applications: A learning-based approach. *IEEE J. Sel. Areas Commun.* **2019**, *37*, 1117–1129. [CrossRef]
38. Wei, K.; Tang, Q.; Guo, J.; Zeng, M.; Fei, Z.; Cui, Q. Resource scheduling and offloading strategy based on LEO satellite edge computing. In Proceedings of the 2021 IEEE 94th Vehicular Technology Conference (VTC2021-Fall), Norman, OK, USA, 27–30 September 2021; pp. 1–6.
39. Cui, G.; Li, X.; Xu, L.; Wang, W. Latency and energy optimization for MEC enhanced SAT-IoT networks. *IEEE Access* **2020**, *8*, 55915–55926. [CrossRef]
40. Wang, B.; Feng, T.; Huang, D. A joint computation offloading and resource allocation strategy for LEO satellite edge computing system. In Proceedings of the 2020 IEEE 20th International Conference on Communication Technology (ICCT), Nanning, China, 28–31 October 2020; pp. 649–655.
41. Tani, S.; Hayama, M.; Nishiyama, H.; Kato, N.; Motoyoshi, K.; Okamura, A. Multi-carrier relaying for successive data transfer in earth observation satellite constellations. In Proceedings of the 2017 IEEE Global Communications Conference, Singapore, 4–8 December 2017; pp. 1–5.
42. He, C.; Dong, Y. Multi-Satellite Observation-Relay Transmission-Downloading Coupling Scheduling Method. *Remote. Sens.* **2023**, *15*, 5639. [CrossRef]
43. Qu, Z.; Zhang, G.; Cao, H.; Xie, J. LEO satellite constellation for Internet of Things. *IEEE Access* **2017**, *5*, 18391–18401. [CrossRef]
44. Cui, G.; Duan, P.; Xu, L.; Wang, W. Latency optimization for hybrid GEO–LEO satellite-assisted IoT networks. *IEEE Internet Things J.* **2022**, *10*, 6286–6297. [CrossRef]
45. Saeed, N.; Elzanaty, A.; Almorad, H.; Dahrouj, H.; Al-Naffouri, T.Y.; Alouini, M.-S. CubeSat communications: Recent advances and future challenges. *IEEE Commun. Surv. Tutor.* **2020**, *22*, 1839–1862. [CrossRef]
46. Golkar, A.; i Cruz, I.L. The federated satellite systems paradigm: Concept and business case evaluation. *Acta Astronaut.* **2015**, *111*, 230–248. [CrossRef]
47. Han, D.; Chen, W.; Fang, Y. Joint channel and queue aware scheduling for latency sensitive mobile edge computing with power constraints. *IEEE Trans. Wirel. Commun.* **2020**, *19*, 3938–3951. [CrossRef]
48. Liu, J.; Zhang, Q. Computation resource allocation for heterogeneous time-critical IoT services in MEC. In Proceedings of the 2020 IEEE Wireless Communications and Networking Conference (WCNC), Virtually, 25–28 May 2020; pp. 1–6.
49. Hu, H.; Song, W.; Wang, Q.; Hu, R.Q.; Zhu, H. Energy efficiency and delay tradeoff in an MEC-enabled mobile IoT network. *IEEE Internet Things J.* **2022**, *9*, 15942–15956. [CrossRef]
50. Zheng, F.; Wang, C.; Zhou, Z.; Pi, Z.; Huang, D. LEO laser microwave hybrid inter-satellite routing strategy based on modified Q-routing algorithm. *Eurasip J. Wirel. Commun. Netw.* **2022**, *2022*, 36. [CrossRef]
51. Kallus, N.; Uehara, M. Double reinforcement learning for efficient off-policy evaluation in Markov decision processes. *J. Mach. Learn. Res.* **2020**, *21*, 6742–6804.
52. Lyu, Y.; Liu, Z.; Fan, R.; Zhan, C.; Hu, H.; An J. Optimal computation offloading in collaborative LEO-IoT enabled MEC: A multi-agent deep reinforcement learning approach. *IEEE Trans. Green Commun. Netw.* **2022**, *7*, 996–1011. [CrossRef]
53. Wu, M.; Guo, K.; Li, X.; Lin, Z.; Wu, Y.; Tsiftsis, T.A.; Song, H. Deep reinforcement learning-based energy efficiency optimization for RIS-aided integrated satellite-aerial-terrestrial relay networks. *IEEE Trans. Commun.* **2024**, *72*, 4163–4178. [CrossRef]
54. Zhang, H.; Zhao, H.; Liu, R.; Kaushik, A.; Gao, X.; Xu, S. Collaborative Task Offloading Optimization for Satellite Mobile Edge Computing Using Multi-Agent Deep Reinforcement Learning. *IEEE Trans. Veh. Technol.* **2024**, 1–16. *early access*. [CrossRef]
55. Chen, T.; Liu, J.; Tang, Q.; Huang, T.; Liu, Y. Deep reinforcement learning based data offloading in multi-layer Ka/Q band LEO satellite-terrestrial Networks. In Proceedings of the 2021 IEEE 21st International Conference on Communication Technology (ICCT), Tianjin, China, 13–16 October 2021; pp. 1417–1422.
56. Wang, J.; Du, H.; Liu, Y.; Sun, G.; Niyato, D.; Mao, S.; Kim, D.I.; Shen, X. Generative AI based Secure Wireless Sensing for ISAC Networks. *arXiv* **2024**, arXiv:2408.11398.
57. Lin, Y.; Gao, Z.; Du, H.; Niyato, D.; Kang, J.; Xiong, Z.; Zheng, Z. Blockchain-based Efficient and Trustworthy AIGC Services in Metaverse. *IEEE Trans. Serv. Comput.* **2024**, *17*, 2067–2079. [CrossRef]

electronics

MDPI

Article

A Machine Learning Evaluation of the Impact of Bit-Depth for the Detection and Classification of Wireless Interferences in Global Navigation Satellite Systems

Gianmarco Baldini [1,*] and Fausto Bonavitacola [2]

[1] European Commission, Joint Research Centre, 21027 Ispra, Italy
[2] Piksel, 20126 Milan, Italy; fausto.bonavitacola@ext.ec.europa.eu
* Correspondence: gianmarco.baldini@ec.europa.eu

Abstract: The performance of the services provided by Global Navigation Satellite Systems (GNSSs) can be seriously degraded by the presence of wireless interferences, and Machine Learning (ML) has been applied to address this problem using the digital artifacts generated by the GNSS receiver. While such an application is not novel in the literature, the analysis of the impact of the bit-depth at which the GNSS signal is recorded has not received significant attention. The type and power level of the wireless interference are also important factors to investigate in this context. This paper addresses this gap by performing an extensive analysis of the impact of these factors on a data set of GNSS signals subject to three different types of wireless interferences with ML and DL algorithms. The analysis is a combination of a pre-processing phase where the Carrier-to-Noise Ratio (CNR) values of different satellites are evaluated, the extraction of relevant features for ML, and the application of a Convolutional Neural Network (CNN) with a multi-head attention layer. The results show that the proposed approach is able to detect the presence of interference with great accuracy (e.g., 99%) but the type of interference and bit-depth can decrease the performance.

Keywords: GNSS; machine learning; deep learning; wireless interference

1. Introduction

Global Navigation Satellite System (GNSS) technologies are used in many applications in the modern world from the automotive domain to the provision of Location-Based Services (LBSs), synchronization of cellular networks, and so on. Such applications set the need for delivering high-performance GNSS solutions in terms of performance parameters, such as accuracy, availability, continuity, and integrity [1].

On the other hand, GNSS signals may be vulnerable to either intentional or unintentional Radio Frequency Interference (RFI) due to their weak received power and the wide availability of low-cost jammers [2,3]. In particular, jamming is an intentional RFI aimed to disrupt GNSS-based services by deliberately transmitting powerful signals in GNSS bands. It would be useful to characterize the type of jamming attack to apply appropriate mitigation measures.

As in other fields, Machine Learning (ML) and Deep Learning (DL) have been increasingly used in GNSSs to address a number of problems including the detection of ionospheric disturbances, scenario classification [1], and so on. In parallel, ML and DL were used to detect and classify types of disturbances like interference in wireless communications [4] or power quality disturbances in power transmission systems [5].

The underlying concept of this paper is to apply ML and DL algorithms to the detection and classification of interferences of GNSSs taking into consideration the specific aspects of the GNSS context: (1) the digital observables identified with the Carrier-to-Noise Ratio (CNR) from the GNSS receivers are the result of a pre-processing step where discriminating information of the wireless interferences may be lost, (2) GNSS satellites may not be fully visible at the time of the data acquisition and processing, which can also obfuscate important information on the interferences, and (3) the signal may be processed using different bit-depths as there is a trade-off between processing time and storage.

The aim of this paper is to investigate the application both of 'shallow' (e.g., Decision Tree) and neural networks 'deep' ML algorithms to detect and classify different types of wireless interferences in GNSS data collected from real measurements and processed at different power levels and bit-depths. Additional details on the novelty and advancement of this study in comparison to the state of the art in the research literature are provided at the end of Section 2.

The structure of this paper is the following: Section 2 provides a literature review on the application of ML and DL to the problem of detection and classification of wireless interference in GNSS signals. Section 3 outlines the overall methodology used for the implementation of data collection, pre-processing, and the application of the ML and DL algorithms. In particular, Section 3.2 describes the ML and DL algorithms used to perform the detection and classification including the Convolutional Neural Network (CNN) architecture and the features used in the ML application. Section 4 describes the data set used to evaluate the approach. The data set was generated by the authors in their laboratory and it will be available to the research community after the publication of the paper. Section 5 provides the results of the comprehensive analysis of the application of the proposed approach on the data set for different values of the hyper-parameters (e.g., number of considered satellites) and the types of wireless interference. Finally, Section 6 concludes the paper and outlines potential future developments.

2. Related Work

This section provides an overview on the related work on the detection and identification of wireless interferences in GNSS signals. Section 2.1 gives an overview on the application of ML to this problem while Section 2.2 reports on the literature where neural networks including DL are applied.

2.1. Related Work on Machine Learning

An extensive survey on the application of ML and DL to GNSS problems is given in [1], where the detection of interference is one of the considered sub-cases among others (estimate of ionospheric effects, signal detection, positioning), and it is often associated with jamming detection and classification. One of the first studies, which applied ML to jamming identification and classification is [6] where a twin Support Vector Machine (SVM) algorithm (TWSVM) was used for real-time interference monitoring. The TWSVM is the preferred SVM for improved classification performance. In contrast, DL algorithms were not used and the impact of the bit-depth was not considered at all. Another paper, which used shallow Machine Learning algorithms was [7], which adopted a combination of SVM and Principal Component Analysis (PCA) to reduce the feature space. The goal of the authors of [7] was to detect both jamming and spoofing attacks. However, DL was not used. Several recent papers have adopted DL algorithms, demonstrating in general a superior performance to shallow Machine Learning algorithms in this context as discussed in the next sub-section.

2.2. Related Work on Neural Networks and Deep Learning

The authors in [8] adopted a CNN (as in this paper) to classify different types of jamming signals present in GNSS bands. In particular, pulse jamming, narrowband, and wideband chirp jamming were used to generate the jamming signals. As in this study, different power levels were also used. The main difference of this paper with [8] is that the analysis with CNNs is performed directly on the radio frequency space in [8], while in this study the CNR values are used. The approach in this paper is more realistic because the classification of the interferences in the original radio frequency space would require an additional spectrum analyzer component, while the CNR values (as C/N_0 time series) are provided even by consumer-grade GNSS receivers. Apart from this significant difference, the authors of [8] do not take into consideration the impact of the bit-depth. The authors of [9] have also used a CNN with a similar limitation that the DL algorithm is applied directly to the spectrograms of the combined GNSS and jamming signals, which can also be computing-intensive due to the need to generate the spectrograms from the signal in space. The jamming signals used in [9] are chirp-based, pulse, or narrowband signals, which are similar to the ones used in this paper. The results of [9] show that SVM achieves better accuracy than the CNN, whereas this study obtained opposite results. Furthermore, it should be taken in consideration that this study used a more sophisticated CNN architecture with a multi-head attention layer rather than the baseline two-layer CNN architecture proposed in [9]. A similar approach was also followed in [10] with a CNN architecture and [11], where DL algorithms (i.e., ResNet18 and Transformers) were applied both to the time domain and spectral domain representation of the combination of the GNSS signal and the interferences. Pulse, chirp, and modulated signals were used to generate the interference. An evolution of these approaches is to use combinations of time frequency transforms as in [12], where the Wigner-Ville transform (WVT) and the spectrograms were used in combination to identify and classify interferences in GNSSs using a CNN. The studies [10–12] have the same limitation as the previous papers because the application of the CNN to the original signal or its spectral domain (either with Fast Fourier Transform (FFT) or with Time Frequency (TF)) does not take into consideration the processing (e.g., signal correlation) of the GNSS receiver and it would require additional components. The bit-depth aspect was also not considered. A similar use of spectrogram and WVT and CNN was adopted in [13], while the authors of [3] applied more sophisticated DL algorithms and architectures than a vanilla CNN for the classification of interference, but again the spectrogram was used and no analysis on bit-depth was performed. Regarding the use of C/N_0 instead of spectrograms, the authors in [14] analyzed the impact of jamming signals with different levels of power on the C/N_0 but the bit-depth was not considered and no classification attempt with ML or DL was made.

We summarize in the following bullet list the key contributions and advancements of this paper in comparison to the reviewed literature:

- With a good degree of novelty, the CNN is applied to the C/N_0 digital artifacts created by the GNSS receiver instead of the spectral domain representation of the signal in space. Even if the application of the CNN to the C/N_0 observables may be more challenging because the original Radio Frequency (RF) signal is pre-processed and some information could be lost, the approach is more practical and realistic because it relies only on the output of the GNSS receiver and no additional components (e.g., spectrum analyzer) are needed.
- For the first time in the literature, the impact of the bit-depth storage and reproduction of the original GNSS signal is evaluated in combination with CNN and the C/N_0 artifacts for the problem of detection and classification of interferences in GNSSs.

- A CNN with a multi-head attention layer is used for classification, which is more sophisticated than the CNN architectures used in the literature so far.
- The authors have produced a novel (because it is based on C/N_0 data) and comprehensive data set with different types of interference, levels of attenuation, and bit-depths, which was not made available before to the research community. This data set will be available after the publication of the manuscript.

3. Methodology

This section describes the approach used in this study to detect and classify the different types of interference. In particular, Section 3.1 describes the overall set of procedures and how they are integrated among them. Section 3.2 describes the ML algorithms while Section 3.3 describes the CNN architecture and related hyper-parameters. Section 3.4 describes the metrics of evaluation.

3.1. Main Flow and Procedures of the Proposed Approach

The overall methodology is pictorially described in Figure 1. On each of the C/N_0 time series generated by the GNSS receiver for each satellite in the constellation, an analysis of the quality of the data is performed. On the overall set of 47 satellites (24 GPS and 23 Galileo satellites), roughly half were in conditions of visibility and not all of them provided significant information for the detection of the interference because the GNSS signal did not have enough strength for the whole measurement duration. The consequence is that the generated C/N_0 would not have enough C/N_0 points for further processing or its level would be below an acceptable threshold. For this reason, a pre-processing step is performed, where only the satellites with a significant mean value are considered. The application of a threshold-based approach for filtering the appropriate satellites may not be suitable in this context for the following reasons: (1) It is difficult to set a specific threshold level and the consequent steps in the methodology could be biased by a wrong threshold value. On the other hand, an optimization of the threshold would require the repeated execution of the entire set of procedures described in this methodology, which would be quite time consuming. (2) Due to the presence of interference, the C/N_0 series are different for each interference condition or level of attenuation or bit-depth in the data set. As a consequence, a filtering step with the threshold may create unbalanced data sets across the different conditions, which may complicate the application of ML/DL more. This approach is based on a sorting of the average C/N_0 values for each satellite time series where only the first 8 satellites are considered. The number 8 was chosen as a trade-off between having enough satellites for the application of ML/DL and the need to have C/N_0 time series with a significant level of quality (C/N_0 over 36 dB-Hz). As a consequence, this first pre-processing step for the GNSS satellite selection is data driven and it does not require hyper-parameters.

The subsequent step is to apply a sliding window with an overlapping factor on the C/N_0 time series for each of the considered 8 satellites. While both the overlapping factor and the length of the sliding window could be considered hyper-parameters in a generic time series analysis, their values can be defined in this specific context because the GNSS receiver will process the GNSS signal in a limited number of samples and a small value of overlapping factor should be used. The length of the sliding window should also be long enough to support the estimate of statistical features in the ML approach and to support the application of the CNN in the DL approach. In addition, the overlapping factor should be small enough to produce a significant number of samples in the C/N_0 time series with a length of 17,448 samples. Based on these considerations, a value of 4 was chosen for the overlapping factor and a value of 32 for the sliding window size to produce a data set of 433 samples for each satellite. In the next phase, two different branches of the methodology are adopted. For the ML approach, a set of features were applied to each sliding window to create a feature space for each satellite. The feature spaces were concatenated on the 8 satellites to create the final feature space on which the shallow Machine Learning algorithms described in Section 3.2 are applied. For the DL approach, the segments based on the sliding window length were used as input to the CNN. However, the size of the data set (433 samples) may be too small for the application of the CNN. To address this aspect, a simple data augmentation based on the application of Additive White Gaussian Noise (AWGN) with values ± 1 dB was used to increase it to three times the size of the original data set. See [15,16] for a discussion on the advantages of data augmentation for CNNs. Finally, both the ML and DL algorithms were applied to each of the 15 interference conditions (3 levels of attenuation $\times$ 5 levels of bit-depth) for the different tasks of interference detection and interference identification (see Section 4 for a description of the interferences). The detection task is a binary classification problem (identify samples with the presence of interference from the ones with the absence of interference), while the identification of the type of interference is a multi-class (the three classes are Wide, Gaussian, and Narrow interference) problem.

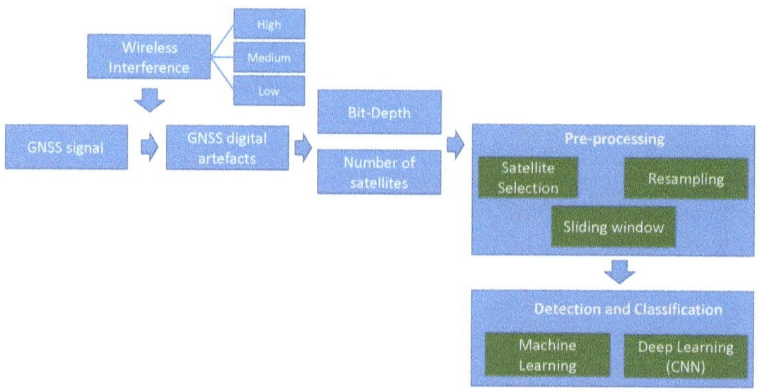

Figure 1. Methodology of the proposed approach.

3.2. Feature-Based Approach with Machine Learning Algorithms

A feature-based approach with three different shallow ML algorithms was used as a comparison to the CNN. The features are applied to the windows of the C/N_0 time series and they are selected due to the specific characteristics of this context: the presence or absence of interference will, respectively, decrease or increase the C/N_0 values. Features that provide an estimate of the values or a change in values are preferred. Ultimately, the nine features listed here were selected: minimum value, maximum value, Root Mean Square (RMS), skewness, kurtosis, standard deviation, and quantile with proportion 0.7, 0.8, and 0.9.

Three different classifiers were used in this study: the Decision Tree (DT) algorithm, the Random Forest (RaF) algorithm, and the Extremely Randomized Tree (ERT) algorithm. The ERT algorithm, which is an evolution of tree-based ensemble methods, is described by the authors of [17]. The DT algorithm is based on the MATLAB 2023a implementation using the *templateTree* and *fitecoc* functions configured with the automatic hyper-parameters optimization. The RaF algorithm is based on the MATLAB 2023a implementation using the *templateTree* and *fitcensemble* functions with the Adaptive Boosting algorithm (i.e., AdaBoost) and the automatic hyper-parameters optimization.

3.3. CNN Architecture and Hyper-Parameters

A two-layer CNN with multi-head attention and the Adaptive moment estimation (Adam) solver is used to implement the DL branch of the proposed approach.

A visual representation of the CNN architecture is shown in Figure 2, while the specific values of the CNN parameters are shown in Table 1.

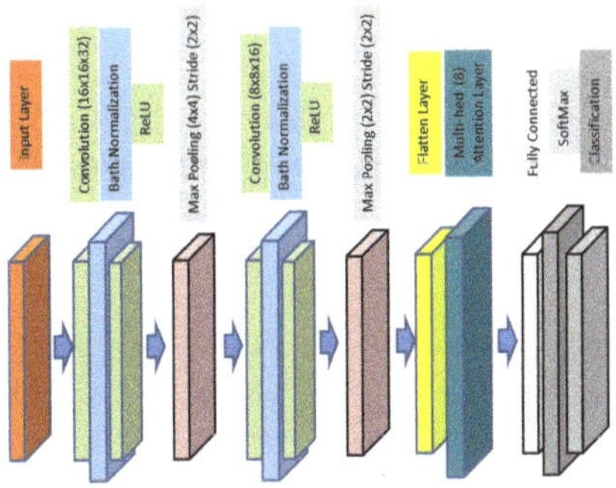

Figure 2. Architecture of the Convolutional Neural Network.

These values were identified as optimal for the classification problem. The relatively low value for the number of epochs was set to 30 to mitigate the risk of overfitting. For the same reason, a 3-fold approach was used and the CNN classification was repeated 10 times (for a total of 30 executions of the ML algorithm). The resulting metrics were averaged. The cross-entropy loss was used as loss function.

Table 1. List of the CNN parameters used in this study.

CNN Parameter	Value
Width, filter size, and number of filters of the 1st convolutional layer	16, 16, 32
Width, filter size, and number of filters of the 2nd convolutional layer	8, 8, 16
1st pooling layer	Max pooling (4,4) with Stride (2,2)
2nd pooling layer	Max pooling (2,2) with Stride (2,2)
Activation functions	REctified Linear Unit (RELU)
Number of heads in the attention layer	8
Number of channels for keys and queries in the attention layer	64
Maximum number of epochs	30

3.4. Metrics of Evaluation

The evaluation metrics used in this study are accuracy, recall, and precision.

The *accuracy* is defined as

$$Accuracy = \frac{TP + TN}{TP + TN + FP + FN}. \tag{1}$$

where TP and TN are the numbers of True Positives and Negatives, respectively, FP and FN are the numbers of False Positives and Negatives, respectively.

The *precision* is defined by the following equation:

$$Precision = \frac{TP}{TP + FP}. \tag{2}$$

The *recall* is defined by the following equation:

$$Recall = \frac{TP}{TP + FN}. \tag{3}$$

F1 score is defined by the following equation:

$$F1score = 2 \times \frac{Precision \times Recall}{Precision + Recall}. \tag{4}$$

Confusion matrices are also used a classification metrics. This paper adopts the convention that column of the matrix represents the instances in an actual class while each row represents the instances in a predicted class,.

3.5. Computing Platform

The computing platform used to conduct the experimental evaluation is a workstation equipped with an Intel I9-990KF Central Processing Unit (CPU) (Intel is based in Santa Clara, CA, USA) with a clock speed of 3.6 GHz, 32 Gbytes of Random Access Memory (RAM), and the Graphic Processing Unit (GPU) NVIDIA RTX4000. NVIDIA is based in Santa Clara, CA, USA. MATLAB R2023a from MathWorks was used to perform the scientific computations with the Signal Processing toolbox and the Machine Learning and Deep Learning toolbox. MathWorks is based in Natick, MA, USA.

4. Data Set Generation and Processing

This section describes the data set used to evaluate the proposed approach with a breakdown in Section 4.1 for the description of the test bed and equipment used to generate the data set, Section 4.2 for the GNSS signal model and the description of the types of interference, Section 4.3 for the playback configuration with different bit-depths, Section 4.4 for the GNSS receiver configuration, and Section 4.5 for the description of the data set structure and format.

4.1. Test Bed Setup

The test bed is composed of the following equipment and the overall setup of the components. The transformation of the GNSS signal in space to the GNSS receiver's artifacts (i.e., C/N_0) values) is shown in Figure 3.

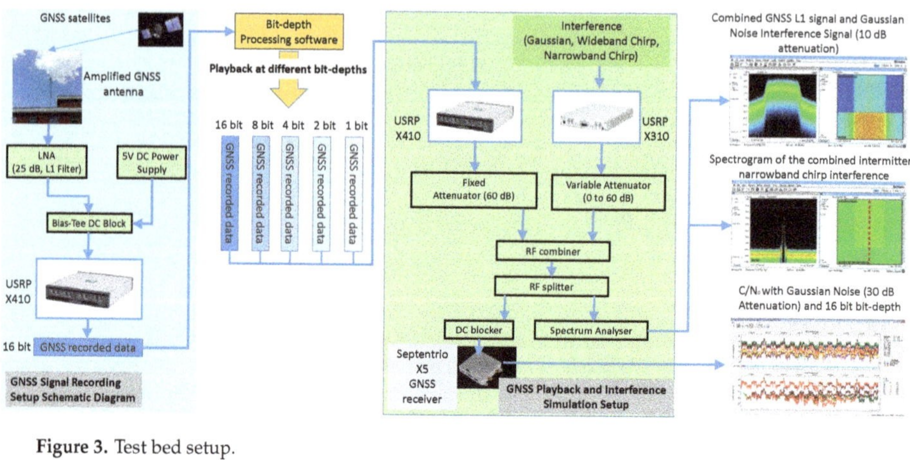

Figure 3. Test bed setup.

The system includes the following:

- Hardware component USRP X410, which is the primary GNSS recording and playback system, configured with a custom LabVIEW-based application. Labview is a graphical system design and development platform produced and distributed by National Instruments. National Instruments is based in Austin, TX, USA.
- Hardware component X310 (interferent signal generation) is utilized with GNU's Not Unix (GNU) Radio on Ubuntu to generate diverse interferent signals, including Gaussian noise, narrowband chirp, and wideband chirp.
- Variable attenuator precisely controls the power level of the interferer.
- Software component Windows Platform LabVIEW-based application with standard LabVIEW driver based on USRP Hardware Driver (UHD) was chosen for its compatibility with the X410 Universal Software Radio Peripheral (USRP) and for the implementation of bit-depth reduction.
- Software component Ubuntu Environment GNU Radio is used for generating the interference signals.

The Software-Defined Radio (SDR) platforms are key components of the test bed and they are briefly described here. As described above, two SDR platforms are used in the test bed: the NI-Ettus X410 USRP and X310. The X410 is a new high-performance platform that builds on the success of the previous generation of USRPs, such as the N210, N310, X300, X310, and NI-2944r. This model is implemented with a new FPGA-based architecture that provides up to 400 MHz of instantaneous bandwidth per channel, which is twice that of the previous generation of USRPs. In addition, the X410 communication element includes two QSFP28 ports, which are able to support data rates of up to 4×25 Gbps, which provides significant support for high-speed data transfer and communication. The X410 analog front-end includes the ZBX two-channel superheterodyne transceiver, which supports frequencies ranging from 1 MHz to 8 GHz. It is an improvement in comparison to its older USRP counterparts thanks to the the implementation of a double intermediate frequency (IF) architecture, encompassing both intermediate (IF1) and baseband (IF2) stages. This design mitigates the DC offset problem associated with older zero-IF devices. In other words, unlike the X310's zero-IF approach, the X410 eliminates the potential issue of an undesired carrier component acting as an unintentional GNSS jammer. This, combined with its improved sensitivity and dynamic range, makes the X410 ideal for faithful recording and playback of GNSS signals, ensuring the integrity of our study. In this setup, the X410 is used for the collection of the GNSS signal with 16 bit and the consequent playback with different bit-depths as described in Section 4.3.

The X310 is a high-performance, scalable SDR, which combines two extended-bandwidth daughterboard slots covering DC 6 GHz with up to 160 MHz of baseband bandwidth. In this study, only one of the ports is used. The X310 is used to generate the interference signals described in Section 4.2 using its zero-IF architecture. The X310 output gain is kept constant throughout the experiment to ensure that the interfering signal always maintains a consistent power level not greater than -50 dB over the entire 10 MHz bandwidth.

4.2. GNSS Signal Model and Types of Interference

At the GNSS receiver front-end, the GNSS signal from the different satellites is down-converted to Intermediate Frequency (IF), sampled into a digital sequence, and finally sent to the acquisition and tracking blocks. The received signal is expressed as follows:

$$r(t) = \sum_{i=1}^{N} s(i) + w(t) + q(t) \tag{5}$$

where N is the overall number of satellites considered in this study, i indicates the ith satellite signal, $w(t)$ is the zero-mean white Gaussian noise with variance σ^2, and $q(t)$ is the interference signal. The desired GNSS signal from the i_{th} satellite is given by

$$s_i(t) = \sqrt{P_i} D_i(t - \tau_i) \cdot C_i(t - \tau_i) \cdot e^{(j(2\pi(f_{IF} - f_{d,i}t + \phi_i)} \quad (6)$$

where i denotes the index of in-view satellites; P_i is the received power; $C_i()$ and $D_i()$ are the ±1-valued pseudorandom spreading code and navigation message, respectively; and τ_i, f_{IF}, $f_{d,i}$, and ϕ_i are the code delay introduced by the channel, *IF* used by the receiver front-end, carrier Doppler frequency, and carrier phase, respectively. Three type of interferences are considered. Wideband and narrowband chirp interferences are represented by

$$q(t) = \sqrt{P_j e^{j2\pi f_{init} t + \pi \frac{f_{max} - f_{min}}{T_{swp}} t^2 + \phi_j}} \quad (7)$$

where for wideband f_{max} and f_{min} are 1.57542 GHz (i.e., f_{init}) $\pm$ 5 MHz and T_{swp} is 5 µs. For narrowband interference, f_{max} and f_{min} are 1.57542 GHz (i.e., f_{init}) $\pm$ 20 KHz and T_{swp} is 1 ms.

The Gaussian band interference (random noise characterized by its Gaussian normal distribution) is represented by

$$q(t) = \sqrt{P_j} \sum_i a_i g(t - iT_a) cos(2\pi f_j t + \phi_j) \quad (8)$$

where $g(t)$ represents the impulse response of shaping filter, a_i is the pseudorandom code that takes the value -1, +1, and T_a is the code duration time.

These types of interference have been used because they are adopted in [3] and for the following reasons. Gaussian noise is introduced to simulate the background noise present in real-world environments, such as electromagnetic interference from natural and man-made sources. Wideband chirps mimic interference sources such as radar systems or broadband communication signals. Narrowband chirp interference poses a unique threat to GNSS receivers by potentially disrupting the timing synchronization process.

4.3. GNSS Signal and Playback with Specific Bit-Depth

The data set is created using recorded GNSS signals from an amplified roof antenna (to ensure adequate reception strength) installed in the JRC premises. Then, real-world signal and not simulated signals are used for the study. A 25 dB Low Noise Amplifier (LNA), which includes a GPS L1 band RF filter as well, is introduced to further bolster the signal level, mitigating potential noise issues. We consider the GPS L1 band even if the data set includes Galileo E1. To power this external LNA, a bias-tee and a 5V Direct Current (DC) power supply are incorporated, ensuring its functionality while simultaneously DC blocking the RF to the output. This step prevents any unwanted DC components from reaching the subsequent component in the signal processing chain (e.g., X410's RF input), maintaining signal integrity. In the X410 SDR, the RF analog signal undergoes conversion to the digital domain with a sampling rate of 10 MSamples/Sec using a bandwidth of 10 MHz and bit-depth of 16 bits. This sampling rate and bit-depth provide sufficient resolution to capture the relevant information within the chosen GPS L1 and Galileo E1 bands and keep the recorded files small enough. See [18,19] for an analysis of the bit-depth impact in GNSS recording and playback. The center frequency is set to 1.57542 GHz, encompassing both the L1 and E1 bands and the maximum gain of 59 dB was used in the LNA of the X410 (to ensure adequate signal strength for optimal recording). An extended recording duration of 30 min was employed.

This timeframe allows for the observation of signal variations arising from environmental fluctuations, satellite movements, and potential interference events.

The next step is to create a representation of the GNSS signal for playback with different bit-depths, which is one of the main objectives of the study presented in this paper. This was achieved with a custom LabVIEW application designed specifically for this purpose to synthesize signal files at various bit-depths: 8, 4, 2, and 1 bit-depths starting from the original 30 min file recorded at 16 bits. The algorithm implemented with LabVIEW discards less significant bits to achieve the desired bit-depth. For example, transitioning from 16 to 8 bits involves eliminating the 8 least significant bits while retaining the most significant 8 bits in the output file. The algorithm was designed to preserve negative numbers with two's complement: the bit reduction may impact negative values, essential for In-phase and Quadrature component (IQ) signals. By using this representation, discarding the Least Significant Bits (LSBs) preserves the sign bit and, consequently, the negative values embedded within the signal. This ensures accurate representation of both positive and negative components, which is crucial for faithful signal reconstruction. On the other hand, discarding LSBs effectively reduces the signal's dynamic range. To maintain accurate representation after this reduction, the original signal must possess sufficient power in its Most Significant Bits (MSBs), which is the reason why the maximum gain of 59 dB was used.

Some background information on the USRP digital representation and the two's complement is provided here because it is a core part of the bit-depth aspect of this study. USRPs handle numbers in integer binary format, using 16-bit sequences of 0 s and 1 s. An I16 integer (signed 16-bit integer) can represent values from −32,768 to 32,767. Each bit has a specific weight based on its position, starting with 2^0 for the least significant bit (LSB) and increasing by a power of 2 for each bit to the left. The positive numbers are directly represented by their binary equivalent (e.g., the decimal number 10 in 16-bit binary is 0000 0000 0000 1010 with $0 \times 2^0 + 1 \times 2^1 + 0 \times 2^2 + 1 \times 2^3 + 0 \times 2^4 + 0 \times 2^5 + 0 \times 2^6 + 0 \times 2^7 + 0 \times 2^8 + 0 \times 2^9 + 0 \times 2^{10} + 0 \times 2^{11} + 0 \times 2^{12} + 0 \times 2^{13} + 0 \times 2^{14} + 0 \times 2^{15} = 10$ dec). The negative numbers are represented with the two's complement by taking the absolute value of the number (e.g., 10), converting it to 16-bit binary (e.g., 0000 0000 0000 1010), inverting each bit (e.g, 1111 1111 1111 0101), and adding 1 to generate the final binary representation (e.g., 1111 1111 1111 0110). The key point is that the Most Significant Bit (MSB) of a two's complement number determines its sign: 0 for positive, 1 for negative. Therefore, discarding less significant bits in two's complement representation preserves the sign bit, maintaining both positive and negative values. In the algorithm implemented in this study, to achieve bit-depths lower than 16 (e.g., 4), the discarding of the Least Significant Bit (LSB)s does not impact the sign information. If we take, for example, the decimal number −15,234, represented in binary as 1100 0100 0111 1110, discarding the four LSBs generates 1100 0100 0111. During playback, the digital content is converted back to I16 by treating the discarded bits as zeros: 1100 0100 0111 becomes 1100 0100 0111 0000. This results in the decimal value −15,248, which is very close to the original one, as expected. Although the precision is reduced, the sign bit (1) is preserved, reflecting the negative nature of the original value. In general, the error introduced by discarding the four LSBs ranges from 0000 to 1111 (15 dB), causing a minor loss in precision.

To enhance the realism of the interference simulation, the continuous interference signals generated in the L1 band were transformed into burst transmissions. This intermittent behavior, where the jammer alternates between one-minute 'on' and 'off' periods, better reflects the characteristics of certain real-world interference sources. This intermittent effect was created on the continuous data stream generated by the chirp signal process (involving the Signal Source and Voltage-Controlled Oscillator (VCO) blocks) by segmentation using the stream to tag the stream block (see Figure 4). The burst shaper block was employed to selectively manipulate the tagged packets. This block was configured to create alternating one-minute intervals where the interference signal is passed through ('on' state) and suppressed ('off' state). These intermittent values were also used to define the label information.

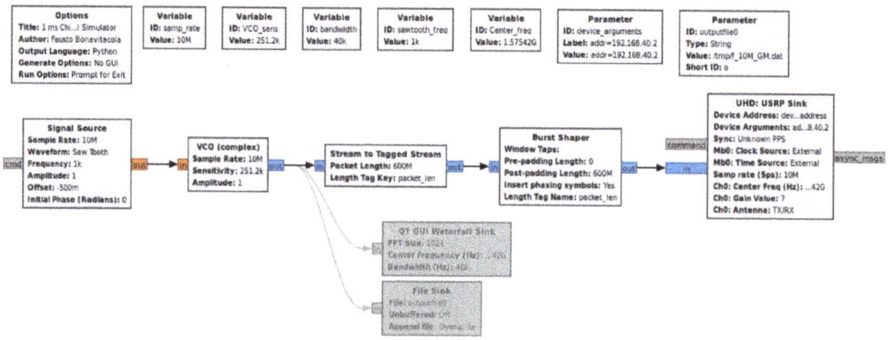

Figure 4. GNU radio flow graphs for the narrow chirp generation (signal source frequency set to 1 KHz).

Each interference condition is generated using a specific power level of the interferer, which may change due to the type of interference in the attempt to make the impact visible in the C/N_0 equivalent (e.g., a wideband chirp interferer has a larger impact than a narrowband chirp interferer at the same level of dB). The power levels of the interference are defined by setting the attenuation (using the variable attenuator shown in Figure 3) to the values presented in Table 2 and represented in the rest of this paper as L_I = L (low), L_I = M (medium), and L_I = H (high),

Table 2. Definition of the power levels of interference and related attenuation level in the variable attenuator.

Interference Type	L_I = H (High) Attenuation Value	L_I = M (Medium) Attenuation Value	L_I = L (Low) Attenuation Value
Gaussian Noise	10 dB	20 dB	30 dB
Wideband Chirp	10 dB	20 dB	30 dB
Narrowband Chirp	0 dB	5 dB	10 dB

4.4. GNSS Receiver Configuration and Generated Digital Artifacts

The experiment utilizes a Septentrio Mosaic X5 GNSS receiver for real-time data acquisition and logging, which generates the C/N_0 time series per satellite at a frequency of 1 Hz. This metric measures the signal strength of each tracked satellite relative to the background noise level. C/N_0 indicates a strong capability to accurately track and decode satellite data.

An example of the impact of the interference on the GNSS C/N_0s artifacts is shown in Figure 5.

The presence of spikes can be noted in Figure 5. These spikes are due to the Automatic Gain Control (AGC) algorithm employed in the Septentrio Mosaic X5 GNSS receiver. Septentrio is based in Leuven, Belgium. This algorithm dynamically adjusts the gain level to maintain a consistent signal strength, leading to temporary increases when the signal weakens corresponding to the observed C/N_0 spikes. Other mechanisms implemented in the GNSS receiver, which may influence the C/N_0 generation, are the carrier notch filters (which attenuate narrowband interference around the GNSS carrier frequency), integration time (affects smoothing of C/N_0 plots), and GNSS selection (allows focusing on specific constellations). We use the default settings for the GNSS receiver. The selected timing source is the GPS constellation.

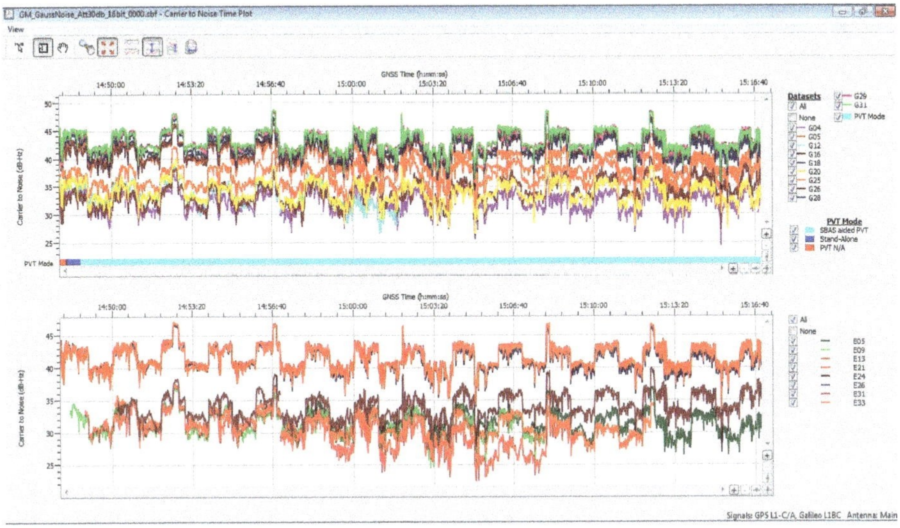

Figure 5. C/N_0 for the satellite considered in this study with the impact of the Gaussian noise (moderate level: 30 dB attenuation) for the 16 bit playback.

4.5. Data Set Structure and Format

The generated data set is composed of the C/N_0 data generated by the Septentrio receiver for almost 30 min of signal playback at a rate of 1 Hz. Thus, a time series composed of 1748 C/N_0 values was generated for each of the considered satellites (32) for each level of bit-depth (i.e., five bit-depths), three power levels (e.g., $L_I = M$), and four types of interference scenarios (i.e., three types of interference and the absence of interference). Taking into consideration that the chosen step is 4 and the window size is 16 C/N_0 value, 433 window samples were generated on which the feature and the CNN were applied. As mentioned before, an analysis of the C/N_0 samples from the different satellites showed that not all the satellites are suitable for the analysis because of periods of signal unavailability, which led to missing samples or samples degraded to low values of C/N_0. The rule was to consider only satellite, where the number of generated values of C/N_0 was not less than 90% of the 1748 samples and value of the C/N_0 was higher than 34 dB-Hz for two-thirds of the data samples. This was an empirical approach, which can be further refined in future developments, and it is based on the consideration that there should be enough valid samples (i.e., C/N_0 values corresponding to the signals of GNSS satellites with LOS propagation conditions of satellite fix) to support the ML classification. The threshold of 90% was inspired by similar studies where such threshold values were used on the C/N_0 to distinguish between propagation conditions [20]. The result of this filtering step identified only eight satellites that were suitable for the subsequent phases.

Even these eight satellites could have significant variability in the number of samples, and a resampling step was performed with the MATLAB *interp1* function and two different algorithms: linear interpolation and nearest neighbor interpolation. These algorithms were chosen on the basis of the characteristics of the C/N_0 time series with steep transients, which prompted the authors to discard other algorithms (i.e., cubic spline). The result from the two interpolation algorithms were basically equivalent and the linear interpolation was chosen for its computational speed.

Because the number of samples for each type of interference and bit-depth could be limited to train the CNN algorithm, a simple augmentation step was implemented to increase the number of samples. Two other sets of 433 samples were generated with AWGN of 1 dB higher and lower than the measured SNR of the original C/N_0 time series. Considering the case of interference classification, this resulted in a DL problem with 1299 samples $\times$ 3 types of interference to have 3897 samples in total as input to the CNN. In addition, to provide an input wide enough for the CNN classifier, the C/N_0 time series was resampled by a factor of 2 after the interpolation (i.e., the window given in input to the CNN has length 32).

5. Results and Discussion

This section presents the results of the application of the proposed approach described in Section 3 on the data set described in Section 4. This section is structured as follows: Section 5.1 focuses on the detection of interference, which is implemented as a binary classification problem. Section 5.2 focuses on the identification (i.e., multi-class classification) of the different types of interference considered in this study: wideband, Gaussian, and narrowband interference to the GNSS signal. Finally, Section 5.3 discusses the limitations of the proposed approach.

5.1. Detection of Interference

Figure 6 and the related sub-figures show the detection results using the feature-based approach for the metric of accuracy, Figure 7 and sub-figures show the precision, and Figure 8 and sub-figures show the recall for different levels of attenuation. In particular, Figure 6a, Figure 6b, and Figure 6c show, respectively, the detection accuracy for low, medium, and high levels of interference for the three types of interference and different levels of bit-depth (called B_D in the rest of this section) as shown in the legend. In a similar way, Figure 7a–c show the precision values and Figure 8a–c show the recall values. Regarding B_D, the results are generally consistent across the metrics and the types of interference: a higher value of B_D (e.g., 16 bits in comparison to 2 bits) provides a higher value of the metric (e.g., accuracy). The differences are more significant for low levels of interference, where the detection algorithm has more difficulty in detecting the interference. This is to be expected because a high value of B_D provides richer and more discriminating information to the detection algorithm, which supports a better detection of the interference. However, the figures seem to indicate that $B_D = 16$ bits are not needed to achieve the optimal results. Values of 8 bits also provide a detection performance comparable to 16 bits. These results can be used to reduce the amount of data needed to perform the detection of interference.

The higher the level of interference, the higher the value of the performance (e.g., higher accuracy). This is also to be expected because the higher the level is the more significant the impact on the GNSS signal, which translates to a higher variation in the C/N_0 time series.

The results for the different types of interferences are somewhat varied and they depend on the level of interference. This may also be related to the different interference power levels defined in Table 2. Narrow interference can be detected more easily than Wide and Gaussian interference for low levels of interference, but the Gaussian interference is detected with higher accuracy with the higher power level of interference.

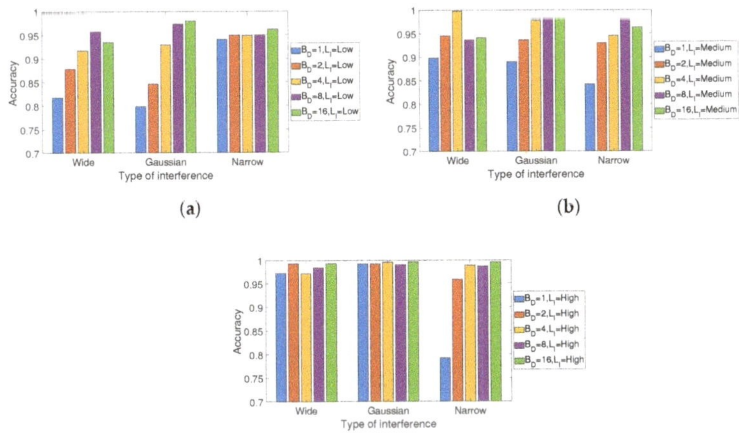

(a) (b) (c)

Figure 6. Comparison of the detection with the accuracy metric using the feature-based approach. (**a**) low interference level, (**b**) medium interference level, and (**c**) high interference level.

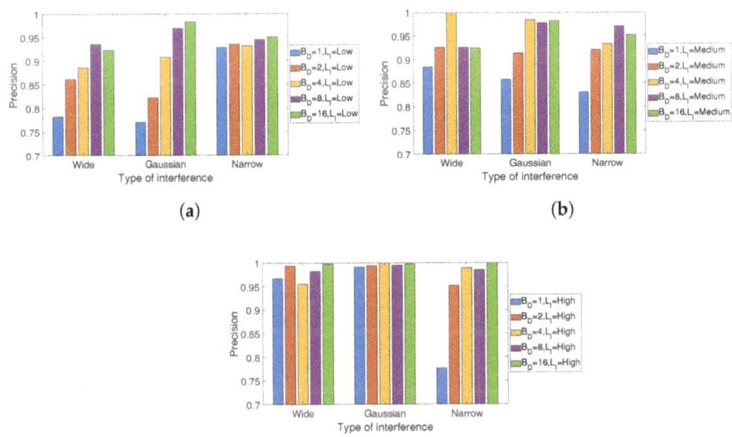

(a) (b) (c)

Figure 7. Comparison of the detection with the precision metric using the feature-based approach. (**a**) low interference level, (**b**) medium interference level, and (**c**) high interference level.

Similar results are achieved using the CNN algorithm. To conserve space, only the accuracy results are shown because the precision and recall graphs present similar results. The comparison of Figure 9 with Figure 6a–c shows that the CNN is able to obtain a superior detection performance at the cost of a higher computing complexity. In particular, for higher values of B_D, the CNN algorithm can obtain perfect detection (i.e., 100%) or near perfection. For the value of $B_D = 1$, even the CNN algorithm has a significant number of misclassification errors because the lower bit-depth degrades significantly discriminating features, which can be used by the CNN to detect the interference. As expected (and coherently with the feature-based approach), a low level of interference causes a lower detection accuracy, especially for the Wide and Narrow types of interference, while the CNN reaches almost perfect detection for values of B_D higher or equal than 4.

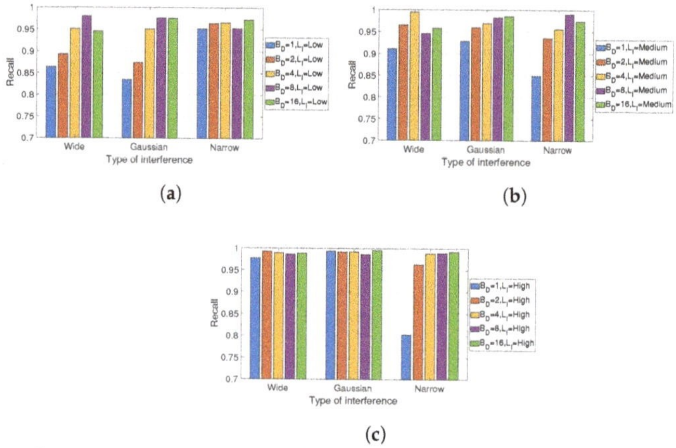

Figure 8. Comparison of the detection with the recall metric using the feature-based approach. (**a**) low interference level, (**b**) medium interference level, and (**c**) high interference level.

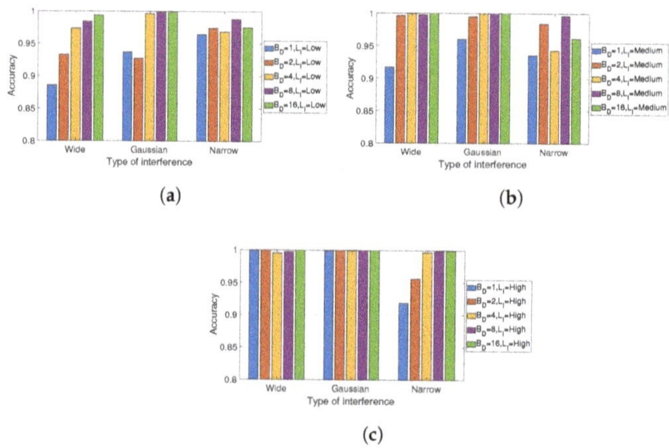

Figure 9. Comparison of the detection with the accuracy metric using the CNN-based approach. (**a**) low interference level, (**b**) medium interference level, and (**c**) high interference level.

5.2. Classification of Interference

Figure 10 and related sub-figures show the comparison of the accuracy and F1-score using the CNN and the shallow ML algorithms using the feature-based approach. These results are obtained using the C/N_0 data obtained from all eight GNSS satellites obtained after the filtering step described in Section 4.5.

In particular, Figure 10a, Figure 10b, Figure 11a, and Figure 11b present, respectively, the accuracy, F1 score, precision, and recall for different levels of bit-depth and interference level (x-axis). The figures show that the four metrics are generally coherent among themselves. The CNN algorithm generally demonstrates a superior performance to the shallow ML algorithms with the exception of the ERT, because the ERT equals CNN for some cases or it is even better with $B_D = 16$ and $L_I = M$. Considering the ERT has a significantly lower computing complexity than the CNN, ERT could be the preferred choice for computing efficient deployments of this approach. ERT mostly outperforms RaF and DT, which justifies the proposal of ERT in this study. It can be noted that for values of $B_D = 1$ and $B_D = 2$, the classification performance drops significantly for the RaF and

DT algorithms, especially at a low level of the interference power $L_I = L$. This is not surprising because the classifiers have difficulty in exploiting the discriminating content due to the limited available information (the reduction in the B_D) and the minor impact of interference signals with low levels of power because the variations of the C/N_0 are less prominent than for a high level of power. Even in the challenging conditions of low B_D and L_I, the performance of the CNN classifier is remarkable as it has almost 100% accuracy apart from the case of $B_D = 2$ and $L_I = M$, where it achieves 97% accuracy.

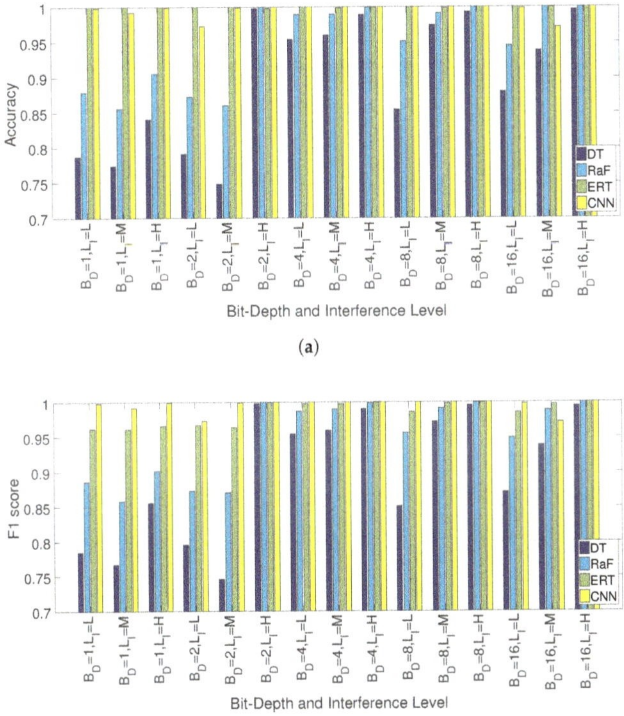

(b)

Figure 10. Comparison of the classification performance for the different approaches (feature- and CNN-based): (**a**) accuracy, (**b**) F1 score.

It is also important to evaluate the performance of the proposed approach for different numbers of satellites N_S. Figure 12 and related sub-figures show the accuracy and F1 score while Figure 13 and related sub-figures present the precision and recall for different bit-depths and levels of interference using different numbers of satellites. These figures were obtained by averaging the results for repeated execution of the CNN algorithm for the potential combinations of satellites, respectively, in sets of one (8 executions), two (28 executions), four (70 executions), six (28 executions), and eight (1 executions). It can be seen that a larger number of satellites increases the classification accuracy across the different levels of B_D and interference level. This is to be expected because a larger number of satellites can provide more (and more discriminating) information to the CNN classifier to identify the specific type of interference. On the other hand, for medium and high levels of interference, six satellites or even four satellites are enough to obtain an accuracy/precision and recall equal or similar to the one obtained with all eight satellites. Both Figures 12 and 13 show that the proposed approach is able to obtain almost perfect accuracy/F1 score/precision/recall with B_D higher than 4 and more than four satellites, which is evidence of the robustness of the proposed approach. The drop in classification performance is particularly significant between a number of satellites equal to one or two, especially for $L_I = L$ and $L_I = M$. The analysis of these figures stresses the importance of the pre-processing phase to collect data from as many satellites as possible, because a lower number of satellites may significantly degrade the classification performance.

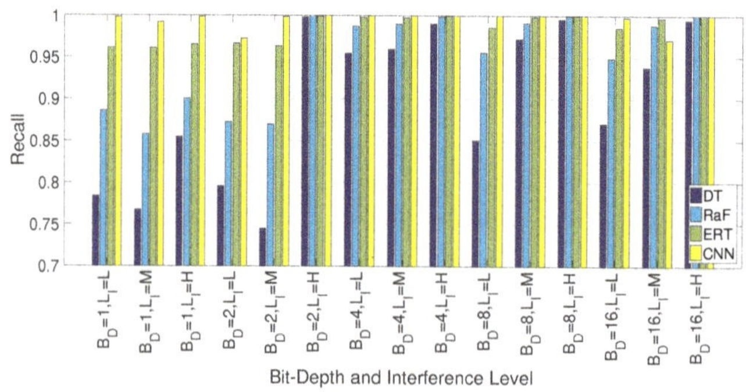

(**a**)

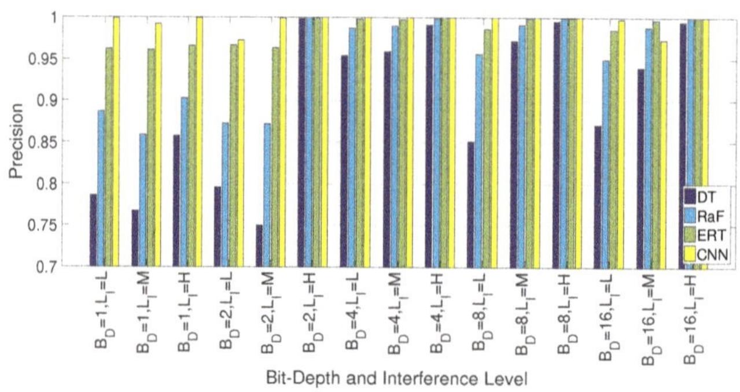

(**b**)

Figure 11. Comparison of the classification performance for the different approaches (feature- and CNN-based): (**a**) recall, (**b**) precision.

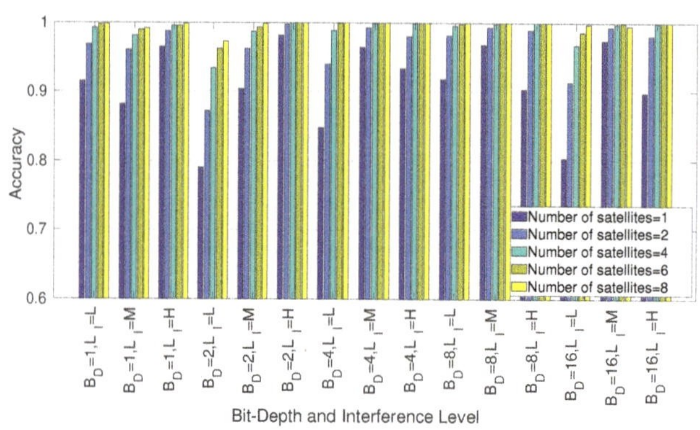

(**a**)

Figure 12. *Cont.*

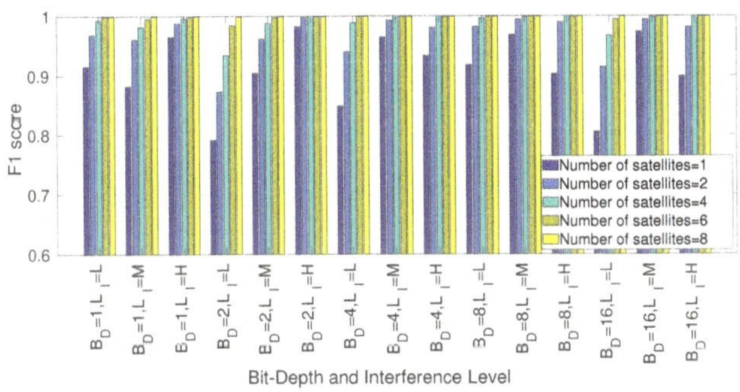

(**b**)

Figure 12. Comparison of the classification performance for different numbers of satellites (N_S) with CNN: (**a**) accuracy, (**b**) F1 score.

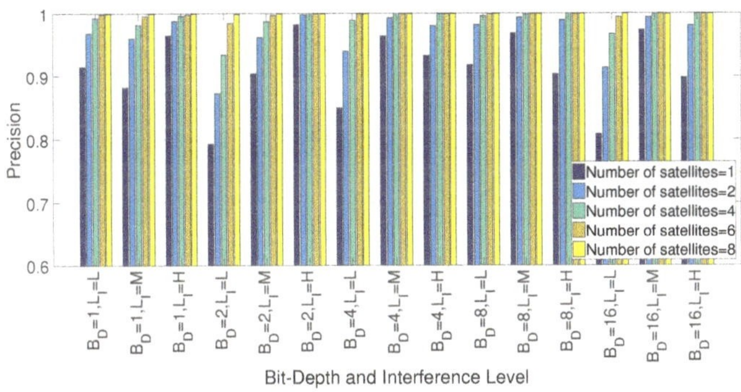

(**a**)

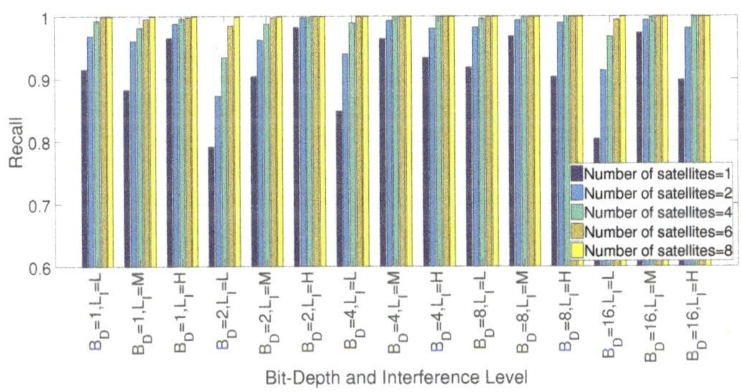

(**b**)

Figure 13. Comparison of the classification performance for different numbers of satellites (N_S) with CNN: (**a**) precision, (**b**) recall.

Figure 14 gives a more detailed view of the classification accuracy with $L_I = M$ for different values of the B_D and for one satellite and all eight satellites. As shown in the previous figures, both the decrease in B_D and N_S lead to a decrease in classification accuracy.

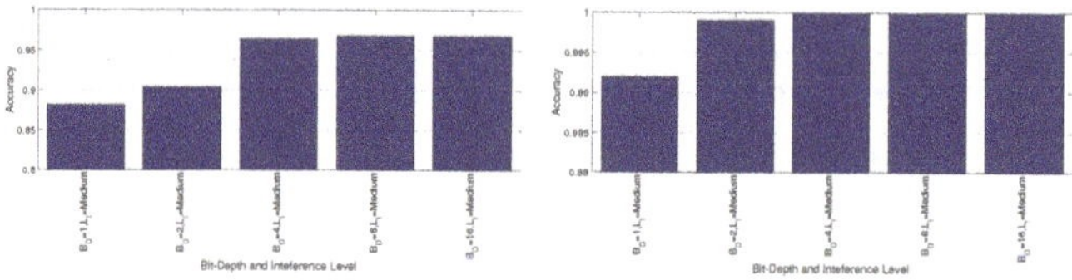

(a) (b)

Figure 14. Comparison of the classification accuracy for different numbers of satellites with CNN and L_I: (**a**) $N_S = 1$, (**b**) $N_S = 8$.

To complement the previous results for accuracy, precision, and recall, we provide in the following Figures 15 and 16 and related sub-figures the confusion matrices for different values of B_D, L_I, and N_S. Predicted values are on the y-axis and True values are on the x-axis. The figures were generated using the CNN with $L_I = L$, which is the most challenging for classification.

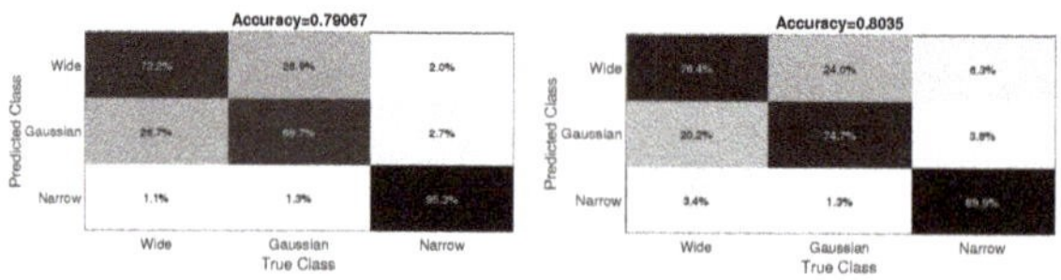

(a) (b)

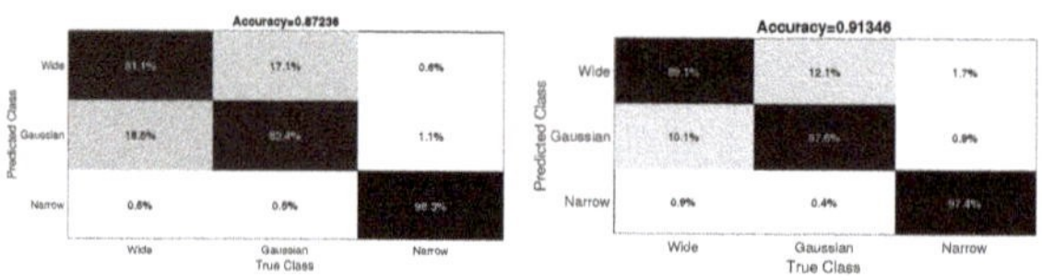

(c) (d)

Figure 15. Confusion matrices with different number of satellites N_S (range from 1 to 2) and bit-depth B_D with low level of interference: (**a**) $N_S = 1$, $B_D = 2$, (**b**) $N_S = 1$, $B_D = 16$, (**c**) $N_S = 2$, $B_D = 2$, (**d**) $N_S = 2$, $B_D = 16$.

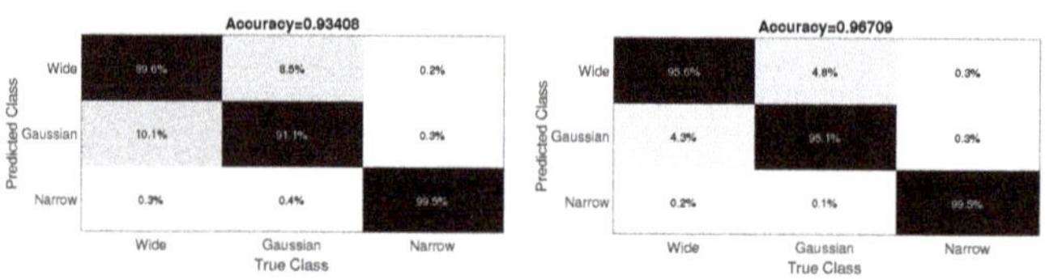

(a) (b)

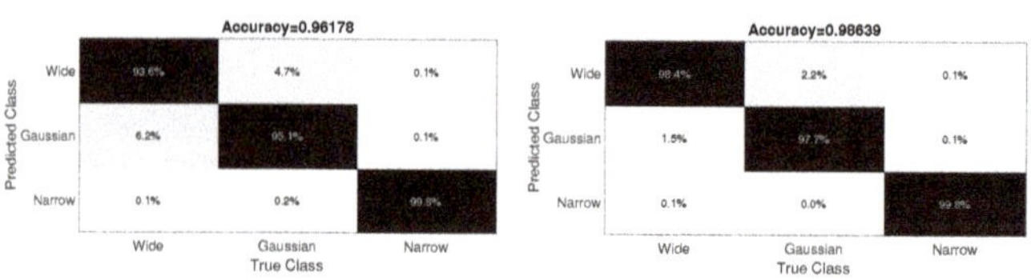

(c) (d)

Figure 16. *Cont.*

(e) (f)

Figure 16. Confusion matrices with different number of satellites N_S (range from 4 to 8) and bit-depth B_D with low level of interference: (**a**) $N_S = 4$, $B_D = 2$, (**b**) $N_S = 4$, $B_D = 16$, (**c**) $N_S = 6$, $B_D = 2$, (**d**) $N_S = 6$, $B_D = 16$, (**e**) $N_S = 8$, $B_D = 2$, (**f**) $N_S = 8$, $B_D = 16$.

The figures show that the Wide and Gaussian types of interference are more difficult to classify in comparison to the Narrow type of interference. As expected, the number of FPs and FNs is lower in percentage (values outside the diagonal) when more satellites or a higher value of B_D is used.

5.3. Discussion on the Limitations of the Proposed Approach

The main limitation of the approach presented in this paper is related to the data set and the filtering process where the data from the GNSS satellites is collected. Even if real-world signals were collected through a GNSS antenna located in a JRC premises, the position of the antenna was static and in a limited time frame of 30 min. On the basis of these data, an empirical filtering step was implemented to select a number of satellites, X, using a pre-defined threshold with the number of generated values of C/N_0 not less than 90% of the 1748 samples and values of the C/N_0 higher than 34 dB-Hz for two-thirds of the data samples. This filtering step indicated the X=8 satellites used in the analysis. In a dynamic context, where there could be significant variations in the propagation environment for the GNSS signal (e.g., a vehicle driving in a urban environment), the filtering step would benefit of a dynamic threshold setting where the number of satellites is adjusted according to the quality of C/N_0 generated by the GNSS receiver. It is also noted from the results presented in Section 5.2 that even with a number of satellites equal to four, the CNN and ERT classifiers manage to obtain a very high classification accuracy (i.e., 99% and higher). Then, the proposed approach may be robust even in a dynamic environment because even a limited set of GNSS satellites would be enough. Potential approaches to implement dynamic threshold, which have to be timely and computing-efficient, could be based on a optimization step linked to the threshold values (e.g., range of the percentage or C/N_0 threshold) where one of the classification metrics (e.g., accuracy) could be used as a cost function. Due to the need of timely execution, a shallow ML algorithm like the DT used in this paper could be selected for its computing efficiency, even if DT demonstrated a low performance in comparison to the CNN in Section 5.2. The creation of a dynamic data set with the related collection and generation of interference signals would be much more complex than the the current data set used in this paper, and this task is deferred to future developments as described in Section 6.

Another limitation of the proposed study is the use of only the C/N_0 data from the GNSS receiver. Even if they are related, other measures like Horizontal Dilution Of Precision (HDOP), Vertical Dilution Of Precision (VDOP), clock bias, the carrier phase, pseudorange metrics, and even the satellite elevation features could be used taking inspiration from the studies focused on GNSS LOS/NLOS classification [21,22]. This is also reserved to future developments.

Finally, even if the CNN classifier has demonstrated an excellent detection and classification performance, more recent and sophisticated neural networks models could be evaluated and adopted, like Transformers used for GNSS NLOS identification in [23]. On the other hand, especially in a dynamic environment, there is an important trade-off between computing efficiency and accuracy and it has to be seen if DL can be applicable or other neural network algorithms (e.g., feed-forward neural networks) could be more suitable.

6. Conclusions and Future Developments

This paper describes an approach to detect and classify different types of interferences on GNSS signals based on the application of ML/DL algorithms, where detection means the capability to reveal the presence of an interference condition in the processed GNSS signals, while classification refers to the capability to distinguish among different types of interference (i.e., wideband, Gaussian, and narrowband in this study). In particular, the impact of three main parameters was considered in the study: (1) the bit-depth at which the GNSS signal was played back before it was processed by the GNSS receiver, (2) the power level of the interference signal, and (3) the number of different GNSS satellites used to detect and classify the interference. This study used ML algorithms and neural networks/Deep Learning (DL) algorithms. The ML algorithms were a Decision Tree (DT), Random Forest (RaF), and Extremely Randomized Trees (ERT). The DL was a Convolutional Neural Network (CNN). The results show that the CNN is able to obtain almost perfect detection accuracy with high and medium levels of interference power and bit-depth equal or superior to 4 bits with the wideband and Gaussian interference and eight GNSS satellites. However, the accuracy drops with a bit-depth equal to 1 or 2, low interference level, and narrowband interference because the impact of the interference is less distinguishable in the GNSS receiver and because the bit-depth reduction removes discriminating information for the CNN classifier. Similar results are obtained for the interference classification task where perfect (100%) or almost perfect (greater than 99%) accuracy is generally obtained for bit-depth greater and including 4 bits and interference levels are medium or high, but it drops significantly for a low interference level and bit-depth of 1 or 2. The DT and RaF generally perform worst than the CNN. It is noted that the ERT algorithm often achieves a similar level of accuracy to the CNN with less computing resources, which could support its use in operational scenarios, which requires computational efficiency. The use of the number of satellites also has a significant impact. The classification performance degrades significantly when only the information from one or two satellites is used. The results show that the use of six satellites or eight satellites is quite similar, supporting the concept that with this specific data set, six satellites would be enough to distinguish with high accuracy the different types of interference. Then, there is a trade-off between the need to process more information from a larger number of satellites with the need for a high classification accuracy.

This paper has also discussed the limitations of the proposed approach. Even if real-data from GNSS constellations was used in the study in a significant amount of time, a dynamic environment where the signal quality from the different satellites changes frequently (e.g., a GNSS receiver mounted on a vehicle driving in an urban environment) may be challenging for the proposed approach where a fixed number of satellites is used. Then, an adaptive algorithm based on the quality of the received signal could be used to determine in a dynamic way the optimal number of satellites and address the trade-off mentioned above. The evaluation of the approach proposed in this paper to a vehicular dynamic context will be the objective of future extensions of this study.

Future developments could go in different directions. One direction would be to investigate the proposed approach on a dynamic environment (e.g., vehicle moving in an urban environment) where the propagation conditions change frequently. Then, new filtering steps should be investigated to determine the optimal thresholds in the data pre-processing phase. As suggested in the main body of this paper, a hybrid approach of shallow and deep ML algorithms could address the need for timely results and computing efficiency. Another direction may include the application of more sophisticated neural networks algorithms like a Transformer architecture for improved accuracy or feed-forward neural networks for higher computing efficiency. Taking in consideration the excellent performance of the ERT in this study, similar algorithms like Rotation Forest or Deep Forest could also be used. Finally, this study was mostly based on the use of Carrier-to-Noise Ratio information (e.g., C/N_0) but other relevant information provided by the GNSS receiver could also be used like carrier phase, pseudorange features, Horizontal Dilution Of Precision (HDOP) or Vertical Dilution Of Precision (VDOP).

Author Contributions: Conceptualization, G.B.; Methodology, G.B.; Software, G.B. and F.B.; Resources, G.B. and F.B.; Data Curation, G.B. and F.B.; Writing—Original Draft Preparation, G.B.; Writing—Review and Editing, G.B. and F.B. All authors have read and agreed to the published version of the manuscript.

Funding: This research received no external funding.

Institutional Review Board Statement: Not applicable.

Informed Consent Statement: Not applicable.

Data Availability Statement: The data presented in this study are available on request from the corresponding author. The data are not publicly available due to lack of approval of the data set publication by the JRC. The data set will be submitted to the internal JRC approval process for data publication after the acceptance of the publication.

Conflicts of Interest: Author Fausto Bonavitacola was employed by the company Piksel. The remaining authors declare that the research was conducted in the absence of any commercial or financial relationships that could be construed as a potential conflict of interest.

Abbreviations

The following abbreviations are used in this manuscript:

Adam	Adaptive moment estimation
AGC	Automatic Gain Control
AWGN	Additive White Gaussian Noise
CNN	Convolutional Neural Network
CNR	Carrier-to-Noise Ratio
DC	Direct Current
DL	Deep Learning
ERT	Extremely Randomized Tree
FPGA	Field-Programmable Gate Array
GPS	Global Positioning System
GNSS	Global Navigation Satellite System
HDOP	Horizontal Dilution Of Precision
IF	Intermediate Frequency
IQ	in-phase (I) and quadrature (Q)
JRC	Joint Research Centre
LNA	Low Noise Amplifier
ML	Machine Learning
RFI	Radio Frequency Interference
SVM	Support Vector Machine
TF	Time Frequency

UHD	USRP Hardware Driver
USRP	Universal Software Radio Peripheral
VCO	Voltage-Controlled Oscillator
VDOP	Vertical Dilution Of Precision

References

1. Siemuri, A.; Selvan, K.; Kuusniemi, H.; Valisuo, P.; Elmusrati, M.S. A systematic review of machine learning techniques for GNSS use cases. *IEEE Trans. Aerosp. Electron. Syst.* **2022**, *58*, 5043–5077. [CrossRef]
2. Dimc, F.; Bažec, M.; Borio, D.; Gioia, C.; Baldini, G.; Basso, M. An Experimental Evaluation of Low-Cost GNSS Jamming Sensors. *Navig. J. Inst. Navig.* **2017**, *64*, 93–109. [CrossRef]
3. Zhong, W.; Xiong, H.; Hua, Y.; Shah, D.H.; Liao, Z.; Xu, Y. TSFANet: Temporal-Spatial Feature Aggregation Network for GNSS Jamming Recognition. *IEEE Trans. Instrum. Meas.* **2024**, *73*, 3375975. . [CrossRef]
4. Baldini, G.; Bonavitacola, F.; Chareau, J.M. Wireless Interference Identification with Convolutional Neural Networks based on the FPGA implementation of the LTE Cell Specific Reference Signal (CRS). *IEEE Trans. Cogn. Commun. Netw.* **2023**, *10*, 48–63. [CrossRef]
5. Ma, J.; Zhang, J.; Xiao, L.; Chen, K.; Wu, J. Classification of power quality disturbances via deep learning. *IETE Tech. Rev.* **2017**, *34*, 408–415. [CrossRef]
6. Li, W.; Huang, Z.; Lang, R.; Qin, H.; Zhou, K.; Cao, Y. A real-time interference monitoring technique for GNSS based on a twin support vector machine method. *Sensors* **2016**, *16*, 329. [CrossRef] [PubMed]
7. Semanjski, S.; Muls, A.; Semanjski, I.; De Wilde, W. Use and validation of supervised machine learning approach for detection of GNSS signal spoofing. In Proceedings of the 2019 IEEE International Conference on Localization and GNSS (ICL-GNSS), Nuremberg, Germany, 4–6 June 2019; pp. 1–6. [CrossRef]
8. Wu, Z.; Zhao, Y.; Yin, Z.; Luo, H. Jamming signals classification using convolutional neural network. In Proceedings of the 2017 IEEE International Symposium on Signal Processing and Information Technology (ISSPIT), Bilbao, Spain, 18–20 December 2017; pp. 62–67. [CrossRef]
9. Morales Ferre, R.; de la Fuente, A.; Lohan, E.S. Jammer classification in GNSS bands via machine learning algorithms. *Sensors* **2019**, *19*, 4841. [CrossRef]
10. Pérez, A.; Querol, J.; Park, H.; Camps, A. Radio-frequency interference location, detection and classification using deep neural networks. In Proceedings of the IGARSS 2020—2020 IEEE International Geoscience and Remote Sensing Symposium, Waikoloa, HI, USA, 26 September–2 October 2020; pp. 6977–6980. [CrossRef]
11. Brieger, T.; Raichur, N.L.; Jdidi, D.; Ott, F.; Feigl, T.; van der Merwe, J.R.; Rügamer, A.; Felber, W. Multimodal learning for reliable interference classification in GNSS signals. In Proceedings of the International Technical Meeting of the Satellite Division of The Institute of Navigation (ION GNSS+ 2022), Denver, CO, USA, 19–23 September 2022; pp. 19–23. [CrossRef]
12. Guo, C.; Tu, W. GNSS interference signal recognition based on deep learning and fusion time-frequency features. In Proceedings of the 34th International Technical Meeting of the Satellite Division of The Institute of Navigation (ION GNSS+ 2021), St. Louis, MI, USA, 20–24 September 2021; pp. 855–863. [CrossRef]
13. Mehr, I.E.; Dovis, F. Detection and classification of GNSS jammers using convolutional neural networks. In Proceedings of the 2022 IEEE International Conference on Localization and GNSS (ICL-GNSS), Tampere, Finland, 7–9 June 2022; pp. 1–6. [CrossRef]
14. Ding, M.; Chen, W.; Ding, W. Performance analysis of a normal GNSS receiver model under different types of jamming signals. *Measurement* **2023**, *214*, 112786. [CrossRef]
15. Taylor, L.; Nitschke, G. Improving deep learning with generic data augmentation. In Proceedings of the 2018 IEEE Symposium Series on Computational Intelligence (SSCI), Bangalore, India, 18–21 November 2018; pp. 1542–1547. [CrossRef]
16. Hernández-García, A.; König, P. Further advantages of data augmentation on convolutional neural networks. In Proceedings of the Artificial Neural Networks and Machine Learning–ICANN 2018: 27th International Conference on Artificial Neural Networks, Rhodes, Greece, 4–7 October 2018; Proceedings, Part I 27; Springer: Berlin/Heidelberg, Germany, 2018; pp. 95–103. [CrossRef]
17. Geurts, P.; Ernst, D.; Wehenkel, L. Extremely randomized trees. *Mach. Learn.* **2006**, *63*, 3–42. [CrossRef]
18. Spirent. Testing Infinity: Bringing Realism into Your GNSS Testing with RF Record and Playback. 2022. Available online: https://www.spirent.com/ (accessed on 12 February 2025).
19. Cucchi, L.; Fortuny, J.; Baldini, G.; Fernandez-Hernandez, I.; Martinez, B.; Vecchione, G. A GNSS Jamming/Spoofing Test Suite for Smart Tachograph Applications. In Proceedings of the 33rd International Technical Meeting of the Satellite Division of The Institute of Navigation (ION GNSS+ 2020), Virtual, 22–25 September 2020; pp. 1490–1514. [CrossRef]
20. Kubo, N.; Kobayashi, K.; Furukawa, R. GNSS multipath detection using continuous time-series C/N0. *Sensors* **2020**, *20*, 4059. [CrossRef] [PubMed]

21. Yozevitch, R.; Moshe, B.B.; Weissman, A. A robust GNSS los/nlos signal classifier. *Navig. J. Inst. Navig.* **2016**, *63*, 429–442. [CrossRef]
22. Xu, H.; Angrisano, A.; Gaglione, S.; Hsu, L.T. Machine learning based LOS/NLOS classifier and robust estimator for GNSS shadow matching. *Satell. Navig.* **2020**, *1*, 15. [CrossRef]
23. Chen, J.; Wang, J.; Zheng, S.; Liu, Y.; Li, Z.; Xie, S.; Wang, Q. Improving NLOS/LOS Classification Accuracy in Urban Canyon Based on Channel-Independent Patch Transformer with Temporal Information. In Proceedings of the 2024 International Technical Meeting of the Institute of Navigation, Long Beach, CA, USA, 23–25 January 2024; pp. 869–882.

electronics

MDPI

Article

A Doppler Frequency-Offset Estimation Method Based on the Beam Pointing of LEO Satellites

Yanjun Song [1,2,*], Jun Xu [1], Chenhua Sun [2], Xudong Li [2] and Shaoyi An [2]

[1] School of Physics, University of Electronic Science and Technology of China, Chengdu 611731, China; xujun@uestc.edu.cn

[2] Scientific and Technical on Communication Network Laboratory, The 54th Research Institute of CETC, Shijiazhuang 050081, China; sun_chenhua@sina.com (C.S.); xu070416@163.com (X.L.); anshxy@foxmail.com (S.A.)

* Correspondence: 202212120508@std.uestc.edu.cn

Abstract: With the advancement of 5G-Advanced Non-Terrestrial Network (5G-A NTN) mobile communication technologies, direct satellite connectivity for mobile devices has been increasingly adopted. In the highly dynamic environment of low-Earth-orbit (LEO) satellite communications, the synchronization of satellite–ground signals remains a critical challenge. In this study, a Doppler frequency-shift estimation method applicable to high-mobility LEO scenarios is proposed, without reliance on the Global Navigation Satellite System (GNSS). Rapid access to satellite systems by mobile devices is enabled without the need for additional time–frequency synchronization infrastructure. The generation mechanism of satellite–ground Doppler frequency shifts is analyzed, and a relationship between satellite velocity and beam-pointing direction is established. Based on this relationship, a Doppler frequency-shift estimation method, referred to as DFS-BP (Doppler frequency-shift estimation using beam pointing), is developed. The effects of Earth's latitude and satellite orbital inclination are systematically investigated and optimized. Through simulation, the estimation performance under varying minimum satellite elevation angles and terminal geographic locations is evaluated. The algorithm may provide a novel solution for Doppler frequency-shift compensation in Non-Terrestrial Networks (NTNs).

Keywords: low-Earth-orbit (LEO) satellites; beam pointing; Doppler; frequency-offset estimation

1. Introduction

The relative motion between communication entities gives rise to the Doppler frequency effect, which induces carrier frequency shifts in communication systems. Due to their high orbital velocities, low-Earth-orbit (LEO) satellites experience significantly more pronounced Doppler effects, resulting in larger frequency deviations under fixed carrier frequency conditions [1,2]. Traditional satellite communication systems have predominantly adopted or are based on Digital Video Broadcasting (DVB) standards. DVB systems offer strong anti-interference capabilities and excellent resilience to Doppler-induced distortions. With the ongoing evolution of mobile communication technologies, terrestrial communication systems are increasingly being extended into space, driving the development of Non-Terrestrial Networks (NTNs). NTN systems typically employ orthogonal frequency-division multiplexing (OFDM), which offers robust multipath resistance through its subcarrier structure. However, the Doppler resilience of OFDM remains limited compared to that of DVB systems [3].

In satellite–ground communication, two primary approaches are generally used to address the carrier frequency offset (CFO) between the transmitter and the receiver.

The first involves enhancing signal processing capabilities to improve adaptability to CFO [4], such as developing demodulation algorithms that can tolerate larger offsets or designing signals in specialized formats.

The second involves applying frequency-offset compensation prior to signal processing, such as implementing CFO pre-compensation at the receiver side [5].

Both approaches face inherent limitations. Demodulation algorithms capable of handling large frequency offsets tend to increase signal processing complexity and implementation difficulty. Signals with special formats often lack flexibility and general applicability. Meanwhile, applying CFO compensation at the receiver adds to the system's overall complexity. As a result, these conventional techniques struggle to meet the performance requirements of high-speed LEO satellite communication and the broader application demands of NTN systems.

To address Doppler-induced frequency offsets in dynamic and diverse scenarios, the fundamental principles [6,7] of Doppler generation are revisited. The Doppler correction techniques used in navigation systems [8,9] are analyzed, and the method by which ground stations calculate Doppler offsets [10–12] for LEO satellites is studied. Based on these insights and extensive simulations, a novel approach for frequency-offset estimation and compensation is proposed. This method leverages beam-pointing information to infer Doppler frequency offsets. A satellite-centered coordinate system is established using satellite orbital altitudes, beam elevation angles, and azimuth angles as input parameters. The positions of the communicating terminals are mapped into this coordinate system. The frequency offset is then estimated based on variations in beam-pointing directions as the satellite moves according to the laws governing Doppler shifts.

This paper focuses on the modeling, optimization, and simulation analysis of Doppler frequency-offset estimations for LEO satellite signals based on beam-pointing perception. The remainder of this paper is organized as follows. In the Section 2, the Doppler frequency-offset estimation model based on the satellite coordinate system is established. The Section 3 optimizes the model according to the parameters affecting the model. In the Section 4, the model is simulated and analyzed and compared with the simulation results of the STK 11.6.0 software. The Section 5 is the discussion of the model, and the Section 6 is the conclusion of this article.

2. Model Establishment

In satellite communications, the orbital motion of a satellite is typically described within the Earth-centered inertial (ECI) coordinate system, where the satellite's position and velocity vectors are characterized by six orbital elements. To analyze the Doppler frequency effect and the resulting Doppler frequency offset, models of the relative position and relative velocity between the satellite and the ground terminal are commonly established. However, such models require real-time knowledge of the position, velocity, and other parameters of both communicating entities, and the associated computations are often complex and resource-intensive.

2.1. Satellite–Ground Doppler Frequency Offset

The Doppler frequency offset between satellite S and ground terminal M arises from their relative velocity. It can be expressed as follows:

$$F_d = F_c \frac{|V_{SM}|}{c} \tag{1}$$

As shown in Figure 1, V_{SM} is the relative speed between satellite S and ground terminal M, c is the speed of light, F_c is the working frequency of the communication system, and F_d is the Doppler frequency offset. Figure 1 is the satellite–ground Doppler frequency-offset model.

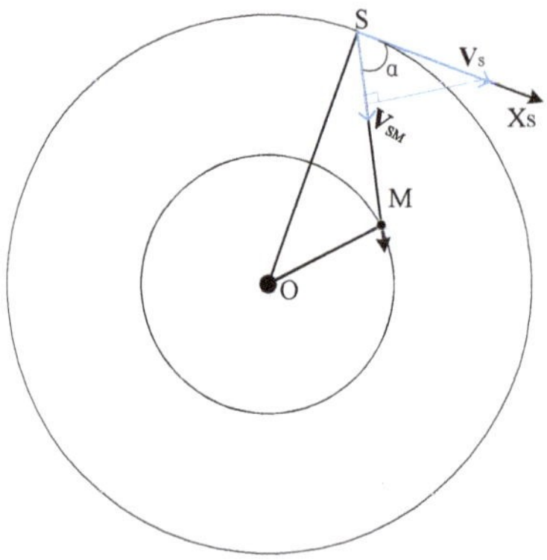

Figure 1. Simple satellite–ground Doppler frequency-offset model.

In Figure 1, V_S is the motion speed of the satellite, and α is the angle between the direction of the satellite's motion and the connection between the two sides of satellite–ground communication; we obtain the following:

$$|V_{SM}| = |V_S| \cos \alpha \tag{2}$$

Thus, Equation (1) yields F_d as follows:

$$F_d = F_c \frac{|V_S| \cos \alpha}{c} \tag{3}$$

In this model, F_c and c are constants, and the satellite speed V_S can be calculated according to the orbital altitude of the satellite.

However, the angle α is related to both the magnitude and direction of the satellite speed V_S, and it changes in real time, so the calculation process of $\cos \alpha$ is complex. Although Equation (3) is difficult to calculate, it provides an excellent approach for studying the Doppler frequency shift. We can use the projection of the satellite velocity onto the line connecting the satellite and the ground terminal (SM) to calculate the satellite-to-ground relative velocity, thus completing the estimation of the satellite-to-ground Doppler frequency shift. Based on the satellite velocity-based satellite-to-ground Doppler frequency-shift model, this paper establishes a simple Doppler frequency-shift estimation model based on satellite beam pointing. The model uses the beam-pointing information of the satellite's antenna to complete the velocity projection of the satellite's velocity onto the satellite-to-ground direction, which can simply estimate the satellite-to-ground Doppler frequency shift.

2.2. *Establishment of the Frequency-Offset Model*

Equation (3) indicates that the Doppler frequency offset F_d is dependent on the angle α. Within the satellite body's coordinate system, α can be expressed in terms of the satellite's azimuth angle θ, which can be expressed in terms of the satellite's elevation angle φ.

The satellite body's coordinate system is defined as follows:

- Origin (S): Located at the satellite's center of mass.
- Z-axis: Directed from the satellite's center of mass toward the center of the Earth.

- X-axis: Lies in the orbital plane, perpendicular to the Z-axis, and points in the direction of the satellite's velocity.
- Y-axis: Determined according to the right-hand rule with respect to the X-axis and Z-axis.

The satellite body's coordinate system and the Doppler frequency-offset model based on beam pointing (DFS-BP) are illustrated in Figure 2.

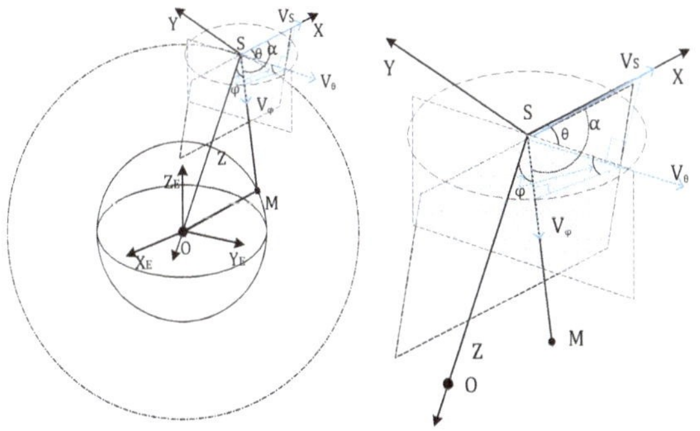

Figure 2. The satellite body coordinate system and the Doppler frequency-offset model based on satellite beam pointing.

In this paper, the satellite antenna's coordinate system is defined consistently with the satellite body's coordinate system; therefore, no coordinate transformation between the two systems is required.

According to the characteristics of the satellite body's coordinate system, we divide the satellite into three two-dimensional planes: plane XSY, plane XSZ, and plane YSZ. The satellite S, ground terminal M, and geocentric o form a plane SOM, and the satellite's Z-axis is in the plane SOM. Therefore, if solving the radial velocity between satellite S and ground terminal M is required, this can be divided into two steps:

- First, the satellite velocity V_S is projected onto the plane SOM. According to the included angle θ between V_S and the plane SOM, θ is the azimuth in the satellite antenna's coordinate system. The following can be obtained:

$$V_\theta = V_S \cos\theta \tag{4}$$

  where V_θ is on the plane SOM.
- Then, the velocity V_θ is projected onto the straight line SM. According to the included angle φ between V_θ and the straight line SM, φ is the elevation angle in the satellite antenna's coordinate system. The following can be obtained:

$$V_\varphi = V_\theta \cos\varphi \tag{5}$$

  where V_φ is the radial velocity of the satellite's velocity V_S along the straight line SM, and the following is obtained:

$$V_\varphi = V_S \cos\theta \cos\varphi \tag{6}$$

Since both V_φ and $|\boldsymbol{V}_{SM}|$ are the relative velocities between satellite S and ground terminal M, we obtain $|\boldsymbol{V}_{SM}| = V_\theta$.

Therefore, we substitute Equation (6) into Equation (1), and the Doppler frequency offset based on satellite beam pointing is as follows:

$$F_d = \frac{F_c}{c} V_S \cos\theta \cos\varphi \tag{7}$$

where F_c is the working frequency of the signal, c is the signal's propagation speed, V_S is the scalar value of V_S, θ is the azimuth of the satellite antenna, and φ is the elevation angle of the satellite antenna.

In Equation (7), the motion speed V_S of the satellite is as follows:

$$V_S = \sqrt{\frac{GM}{r}}, r = R + h \tag{8}$$

where G is the universal gravitation constant, M is the mass of the Earth, h is the orbit altitude of the satellite, r is the orbit radius of the satellite, and R is the radius of the Earth.

The Doppler frequency offset F_d of Equation (8) is obtained as follows:

$$F_d = \frac{F_c}{c}\sqrt{\frac{GM}{r}}\cos\theta\cos\varphi \tag{9}$$

In this model, F_c, c, G, M, and r are constants, and θ and φ are the pointing angle parameters of the satellite antenna, which can be calculated through the pointing of the satellite antenna to the ground terminal. In the case of satellite–ground communication, the satellite antenna's azimuth angle θ and elevation angle φ are the parameters that must be calculated by the satellite, so the angle θ and φ parameters can be directly used in the calculation of Doppler frequency offsets without recalculation.

3. Model Optimization

The Doppler frequency-offset model based on satellite beam pointing does not account for the effects of satellite orbital inclination, Earth's rotation, and other factors related to the frequency-offset estimation. In this section, the model is optimized and refined accordingly.

In an Earth-centered inertial coordinate system (ECI), the angular velocity is constant. However, due to the rotation of the Earth, the angular velocity of the satellite in an Earth-centered fixed coordinate system (ECF) varies with latitude. As shown in Figure 3, β is the geocentric latitude value. V_S represents the tangential velocity of the satellite in the ECI coordinate system, and $V_{E(\beta)}$ represents the velocity at latitude β due to the rotation of the Earth. i represents the inclination of the satellite orbit, i' represents the angle between V_S and $V_{E(\beta)}$, V_S represents the tangential velocity of the satellite in the ECI coordinate system, and $V_{E(\beta)}$ represents the velocity at latitude β due to the rotation of the Earth. r is the orbital radius of the satellite. The latitude β plane is parallel to the equatorial plane, and r' is the projection of r within the latitude β plane.

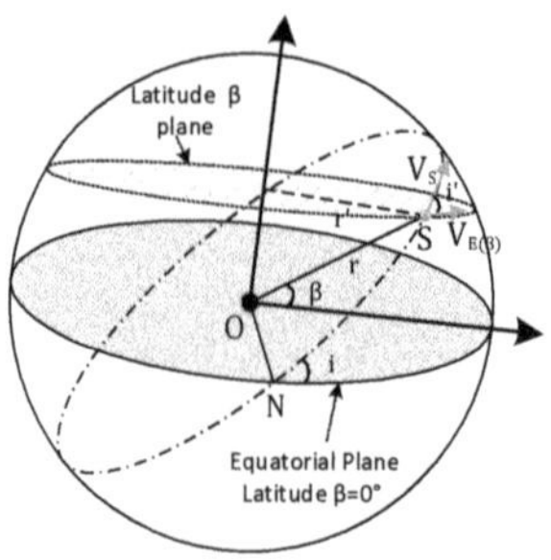

Figure 3. Relationship between Earth's rotation and the satellite's speed.

When the satellite orbital inclination is less than 90°, the direction of the Earth's rotation speed $V_{E(\beta)}$ is the same as that of the satellite's motion speed V_S. The Earth's rotation speed $V_{E(\beta)}$ will reduce the relative velocity between satellite S and ground terminal M.

Therefore, when the satellite is located at latitude β, its velocity in the ECF coordinate system is as follows:

$$\boldsymbol{V}_{\beta} - \boldsymbol{V}_{S} - \boldsymbol{V}_{E(\beta)} \tag{10}$$

Both sides of Equation (8) are squared, and we obtain the following:

$$\boldsymbol{V}_{\beta}{}^{2} = \left(\boldsymbol{V}_{S} - \boldsymbol{V}_{E(\beta)}\right)^{2} \tag{11}$$

Then, we obtain the following:

$$\left|\boldsymbol{V}_{\beta}\right|^{2} = |\boldsymbol{V}_{S}|^{2} + \left|\boldsymbol{V}_{E(\beta)}\right|^{2} - 2\boldsymbol{V}_{S}\cdot\boldsymbol{V}_{E(\beta)} \tag{12}$$

Since the angle between vector $\boldsymbol{V}_S$ and vector $\boldsymbol{V}_{E(\beta)}$ is i', it follows that

$$\boldsymbol{V}_{S}\cdot\boldsymbol{V}_{E(\beta)} = |\boldsymbol{V}_{S}||\boldsymbol{V}_{S}|\cos i' \tag{13}$$

Then, Equation (12) yields the following:

$$\left|\boldsymbol{V}_{\beta}\right|^{2} = |\boldsymbol{V}_{S}|^{2} + \left|\boldsymbol{V}_{E(\beta)}\right|^{2} - |\boldsymbol{V}_{S}|\left|\boldsymbol{V}_{E(\beta)}\right|\cos i' \tag{14}$$

Since $|\boldsymbol{V}_{\beta}| = |V_{\beta}|$, $\left|\boldsymbol{V}_{E(\beta)}\right| = \left|V_{E(\beta)}\right|$, Equation (14) yields

$$\left|V_{\beta}\right|^{2} = |V_{S}|^{2} + \left|V_{E(\beta)}\right|^{2} - 2|V_{S}|\left|V_{E(\beta)}\right|\cos i' \tag{15}$$

In the ECI coordinate system, ω_S is the angular velocity of the satellite; then, $|V_S|$ is

$$|V_{S}| = r\omega_{S} \tag{16}$$

Let r' be the radius of the latitude β plane and ω_E be the angular velocity of Earth's rotation. Then, Earth's rotational velocity at latitude β is

$$\left|V_{E(\beta)}\right| = r'\omega_{E} \tag{17}$$

Since r' is the projection of r within the latitude β plane, then r' is

$$r' = r\cos\beta \tag{18}$$

Thus, $\left|V_{E(\beta)}\right|$ is

$$\left|V_{E(\beta)}\right| = r\omega_{E}\cos\beta \tag{19}$$

Then, Equation (15) yields

$$\left|V_{\beta}\right|^{2} = r^{2}\omega_{S}^{2} + r^{2}\omega_{E}^{2}\cos\beta^{2} - 2r^{2}\omega_{S}\omega_{E}\cos\beta\cos i' \tag{20}$$

According to the three-dimensional cosine theorem in space, we obtain

$$\cos i' = \frac{\cos i}{\cos\beta} \tag{21}$$

Thus, $\left|V_{\beta}\right|^{2}$ is

$$\left|V_{\beta}\right|^{2} = r^{2}\omega_{S}^{2} + r^{2}\omega_{E}^{2}\cos\beta^{2} - 2r^{2}\omega_{S}\omega_{E}\cos i \tag{22}$$

As shown in Figure 3, **N** is the ascending node of satellite S. In the SON plane, according to the three-dimensional sine theorem in space, we obtain

$$\sin\beta = \sin i \sin \angle NOS \tag{23}$$

Since $0^\circ \le \angle NOS < 360^\circ$, then $-1 \le \sin\angle NOS \le 1$. When the satellite orbit is a prograde orbit, $\sin\beta$ is

$$-\sin i \le \sin\beta \le \sin i \tag{24}$$

The value range of latitude β is $-i \le \beta \le i$; then, using Equation (22), we obtain

$$|V_\beta|^2 = \begin{cases} r^2\omega_S^2 + r^2\omega_E^2 - 2r^2\omega_S\omega_E\cos i & \beta = 0^\circ \\ r^2\omega_S^2 + r^2\omega_E^2\cos i^2 - 2r^2\omega_S\omega_E\cos i & \beta = \pm i \\ r^2\omega_S^2 + r^2\omega_E^2\cos\beta^2 - 2r^2\omega_S\omega_E\cos i & \beta \text{ is others} \end{cases} \tag{25}$$

Simplifying Equation (24) obtains

$$|V_\beta| = \begin{cases} r\sqrt{\omega_S^2 + \omega_E^2 - 2\omega_S\omega_E\cos i} & \beta = 0^\circ \\ r\omega_S - r\,\omega_E\cos i & \beta = \pm i \\ r\sqrt{\omega_S^2 + \omega_E^2\cos\beta^2 - 2\omega_S\omega_E\cos i} & \beta \text{ is others} \end{cases} \tag{26}$$

Substituting Equation (26) into Equation (7) obtains

$$F_d = \begin{cases} \frac{F_c}{c} r\sqrt{\omega_S^2 + \omega_E^2 - 2\omega_S\omega_E\cos i}\cos\theta\cos\varphi & ,\beta = 0^\circ \\ \frac{F_c}{c} r(\omega_S - \omega_E\cos i)\cos\theta\cos\varphi & ,\beta = \pm i \\ \frac{F_c}{c} r\sqrt{\omega_S^2 + \omega_E^2\cos\beta^2 - 2\omega_S\omega_E\cos i}\;\cos\theta\cos\varphi & ,\beta \text{ is others} \end{cases} \tag{27}$$

When the eccentricity of the satellite orbit is approximately zero, ω_S is

$$\omega_S = \sqrt{\frac{GM}{r^3}} \tag{28}$$

where ω_E is

$$\omega_E = \frac{2\pi}{T} \tag{29}$$

T is the rotation period of the Earth, which is equal to the time of a stellar day. Then, Equation (27) is

$$F_d = \begin{cases} \frac{F_c}{c} r\sqrt{\frac{GM}{r^3} + \left(\frac{2\pi}{T}\right)^2 - \frac{4\pi}{T}\sqrt{\frac{GM}{r^3}}\cos i}\;\cos\theta\cos\varphi & ,\beta = 0^\circ \\ \frac{F_c}{c} r\left(\sqrt{\frac{GM}{r^3}} - \frac{2\pi}{T}\cos i\right)\cos\theta\cos\varphi & ,\beta = \pm i \\ \frac{F_c}{c} r\sqrt{\frac{GM}{r^3} + \left(\frac{2\pi}{T}\right)^2\cos\beta^2 - \frac{4\pi}{T}\sqrt{\frac{GM}{r^3}}\cos i}\;\cos\theta\cos\varphi & ,\beta \text{ is others} \end{cases} \tag{30}$$

F_c, c, G, M, r, i, and T are constants or approximate constants. When $\beta = 0^\circ$ or $\beta = \pm i$, the Doppler frequency shift F_d is only related to the azimuth angle θ and elevation angle φ of the satellite antenna's beam; then, $F_{d,\beta=0^\circ} > F_{d,\beta=\pm i}$. When $0^\circ < \beta < i$, the Doppler frequency shift F_d is related not only to the azimuth angle θ and elevation angle φ of the satellite antenna beam but also to β.

When the satellite orbital radius is 6878 km and the orbital inclination is 60°, $|V_{\beta=0^\circ}|$ = 7.3833 km/s, $|V_{\beta=20^\circ}|$ = 7.3799 km/s, $|V_{\beta=40^\circ}|$ = 7.3746 km/s, and $|V_{\beta=60^\circ}|$ = 7.3653 km/s. For satellites with low orbital altitudes, when β is different, the change in $|V_\beta|$'s value is less than 0.25%. Therefore, to simplify the calculation, β in Equation (30) can be approximated as the satellite's orbital inclination i, and thus, Equation (30) can be written as

$$F_d = \frac{F_c}{c} r \left(\sqrt{\frac{GM}{r^3}} - \frac{2\pi}{T} \cos i \right) \cos\theta \cos\varphi \quad (31)$$

4. Simulation Analysis

To analyze the frequency-offset estimation accuracy of the DFS-BP algorithm, this paper uses STK's SGP4 model and the DFS-BP algorithm to perform Doppler frequency-shift simulations for satellites in different scenarios, and it compares the frequency-offset estimation errors of the two methods.

This section mainly conducts simulation work in three aspects:

- Case 1: The DFS-BP algorithm is used to simulate the Doppler frequency offset with a minimum elevation angle of 0° and a terminal geographic latitude of 0°. The simulation results are compared against those obtained from the SGP4 model in STK. Under the condition of the same orbital altitude and orbital inclination, the influence of different latitude values on the satellite's beam-pointing frequency-offset estimation model is analyzed.
- Case 2: The DFS-BP algorithm is used to simulate the Doppler frequency offset with a minimum elevation angle of 0° and varying terminal geographic latitudes. The simulation results are compared with those from the SGP4 model in STK. The accuracy of the Doppler frequency-offset estimation by the DFS-BP algorithm is then analyzed. To analyze the accuracy of the Doppler frequency-offset estimation of the DFS-BP algorithm, the Root Mean Square Errors (RMSEs) of the Doppler frequency-offset estimation errors under different orbital altitudes and orbital inclinations are compared. The ±95% confidence intervals (CIs) are also analyzed.
- Case 3: The DFS-BP algorithm is employed to simulate the Doppler frequency offsets of various satellite minimum elevation angles. The simulation results are compared with those obtained from the SGP4 model in STK. The accuracy of the Doppler frequency-offset estimation by the DFS-BP algorithm is subsequently analyzed.

4.1. Simulation of Doppler Frequency Shift When the Minimum Elevation Angle Is 0° and the Geographical Location of the Terminal Is 0°

The Doppler frequency-shift simulation parameters for a minimum elevation angle of 0° are listed in Table 1.

Table 1. Doppler frequency-shift simulation parameter table for a minimum elevation angle of 0°.

Parameter Name	Parameter Value
Orbital altitude, h	480 km
Earth radius, R	6378 km
Eccentricity, e	0.0
Orbital inclination, i	50°
Satellite geocentric latitude, β	0°, 10°, 20°, 30°, 40°, 50°
Earth's rotation period, T	86,164 s
Signal propagation speed, c	2.998×10^8 m/s
Signal carrier frequency, Fc	2.0 GHz

The simulation process of Doppler frequency-offset estimations under this condition is as follows:

- According to the simulation parameters, the STK tool is utilized to simulate the Doppler frequency-offset variation of the LEO satellite during visible time, with a minimum elevation angle of 0° and a terminal geographic latitude of 0°. The simulation outputs include the Doppler frequency-offset variation over the visible period, as well as the satellite's azimuth and elevation angles.

- Using the satellite beam-pointing method, the Doppler frequency-offset variation is calculated according to the azimuth and elevation angles of LEO satellites.
- Compare the frequency-offset estimation error between the Doppler frequency-offset estimation based on STK simulations and the Doppler frequency-offset estimation based on the DFS-BP method.
- When the satellite geocentric latitudes are set to 0°, 10°, 20°, 30°, 40°, and 50°, compare the frequency-offset estimation error between the Doppler frequency-offset estimation based on STK simulations and the Doppler frequency-offset estimation based on the DFS-BP method. The differences among the three sub-equations in Equation (30) are analyzed through the RMSE of the frequency-offset estimation errors.

When the minimum elevation angle of the satellite is 0° and the geographical latitude of the terminal is 0°, the Doppler frequency-offset estimation of the satellite beam-pointing method is shown in Figure 4. Figure 4 compares the simulation results of the satellite beam-pointing method with STK simulation results. Within the visible time of the satellite, the range of frequency-offset estimation error is [−150 Hz, +150 Hz].

When the minimum satellite elevation angle is 0° and the terminal geographic latitude is 0°, the satellite's geocentric latitude β is set to 0°, 10°, 20°, 30°, 40°, and 50°. The error values of the Doppler frequency-offset estimation by the DFS-BP method are shown in Figure 5a. The RMSEs of the Doppler frequency-offset estimation errors at different satellite geocentric latitudes β are shown in Figure 5b. When the satellite's geocentric latitude β is set to different values, the RMSE of the Doppler frequency-offset estimation errors of the three sub-equations of the DFS-BP method are all less than 100 Hz, and the maximum difference in their RMSE values is 30 Hz.

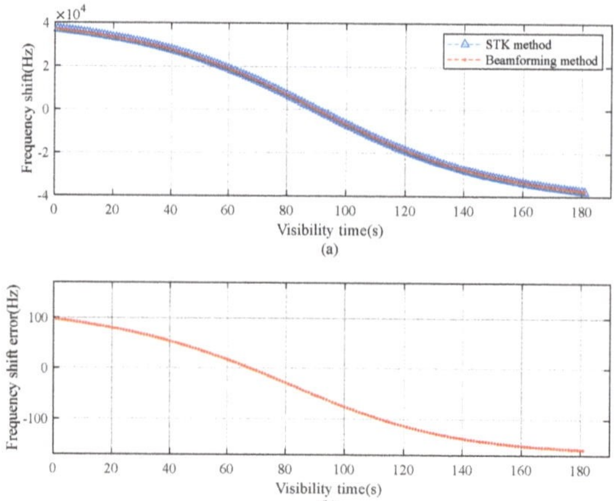

Figure 4. Simulation of Doppler frequency shifts when the minimum elevation angle is 0° and the geographical location of the terminal is 0°. (**a**) The Frequency shift; (**b**) The Frequency shift error.

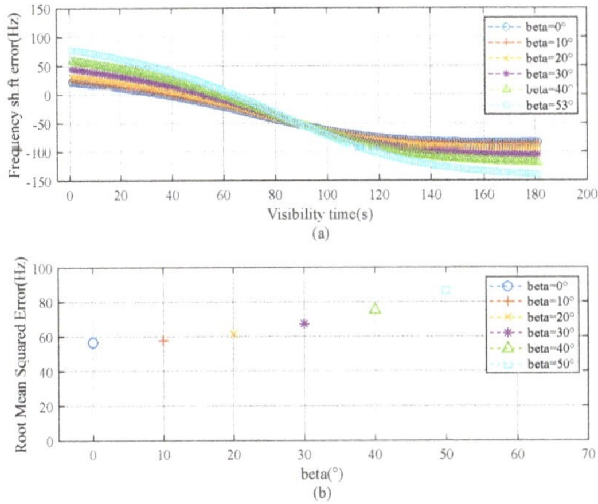

Figure 5. Doppler frequency-shift simulation and error statistics at different satellite geocentric latitudes. (**a**) The Frequency shift error; (**b**) The Root Mean Squared Error.

4.2. Simulation of Doppler Frequency Shift When the Minimum Elevation Angle Is 0° and the Geographical Location of the Terminal Is Adjustable

The Doppler frequency-shift simulation parameters of different terminal geographic positions and satellite orbits are listed in Table 2.

Table 2. Doppler frequency-shift simulation parameter table for different terminal geographic positions and satellite orbits.

Parameter Name	Parameter Value
Orbital altitude, h	480 km, 500 km, 600 km, 700 km, 800 km
Earth radius, R	6378 km
Eccentricity, e	0.0
Orbital inclination, i	40°, 50°, 60°, 70°, 80°, 90°
Geographical latitude of the terminal, Lat	0°, 15°, 30°, 45°
Earth's rotation period, T	86,164 s
Signal propagation speed, c	2.998×10^8 m/s
Signal carrier frequency, Fc	2.0 GHz

The simulation process of Doppler frequency-shift estimations is as follows:

- According to the simulation parameters, the STK tool is used to simulate the change value of the satellite–ground Doppler frequency offset under the condition that the minimum elevation angle is 0° and the geographic latitude of the terminal is set to 0°, 15°, 30°, and 45°. The simulation results of the Doppler frequency offset and azimuth and elevation angles of the satellite are outputted.
- Using the satellite beam-pointing method, the Doppler frequency offset is calculated according to the azimuth and elevation angles of the satellite. Compare the errors with the Doppler frequency-shift variations simulated by STK, analyze the RMSE of the Doppler frequency shift, and verify the adaptability of the DFS-BP method to different terminal geographic latitudes.
- According to the simulation parameters, the STK tool is used to simulate the change value of the satellite–ground Doppler frequency offsets under the condition that the minimum elevation angle is 0° and the satellite orbital altitude h is set to 480 km, 500 km, 600 km, 700 km, and 800 km. The simulation results of the Doppler frequency offset and azimuth and elevation angles of the satellite are outputted.

- Using the satellite beam-pointing method, the Doppler frequency offset is calculated according to the azimuth and elevation angles of the satellite. Compare the errors with the Doppler frequency-shift variations simulated by STK, analyze the RMSE of the Doppler frequency shift, and verify the adaptability of the DFS-BP method to different satellite orbital altitudes.
- According to the simulation parameters, the STK tool is used to simulate the change value of satellite–ground Doppler frequency offsets under the condition that the minimum elevation angle is 0° and the satellite orbital inclination i is set to 40°, 50°, 60°, 70°, 80°, and 90°. The simulation results of the Doppler frequency offset and azimuth and elevation angles of the satellite are output.
- Using the satellite beam-pointing method, the Doppler frequency offset is calculated according to the azimuth and elevation angles of the satellite. Compare the errors with the Doppler frequency-shift variations simulated by STK, analyze the RMSE of the Doppler frequency shift, and verify the adaptability of the DFS-BP method to different satellite orbital inclinations.
- Calculate the 95% CI of the Doppler frequency-offset estimation errors based on the RMSEs under different orbital altitudes and orbital inclinations. Due to the small sample size, the t-distribution model is selected for confidence interval analyses. The t-distribution model is $CI = \overline{x} \pm t_{0.025,n-1}\frac{s}{\sqrt{n}}$, where $\overline{x}$ is the average value of RMSE, $t_{0.025,n-1}$ is the distribution values at a 95% confidence level, s is the sample's standard deviation, and n is the sample size.

When the minimum elevation angle of the satellite is 0° and the geographical latitude of the terminal is set at 0°, 15°, 30°, and 45°, the Doppler frequency-offset estimation errors of the satellite beam-pointing method are shown in Figure 6. Figure 6a shows that the frequency-offset estimation error range of the two methods is between [−200 Hz, +200 Hz] in the whole visible period. Figure 6b shows that the RMSE of the frequency-offset estimation for ground terminals at different latitudes is less than 100 Hz. The geographical latitude of the terminal has little effect on the Doppler frequency-offset estimation error.

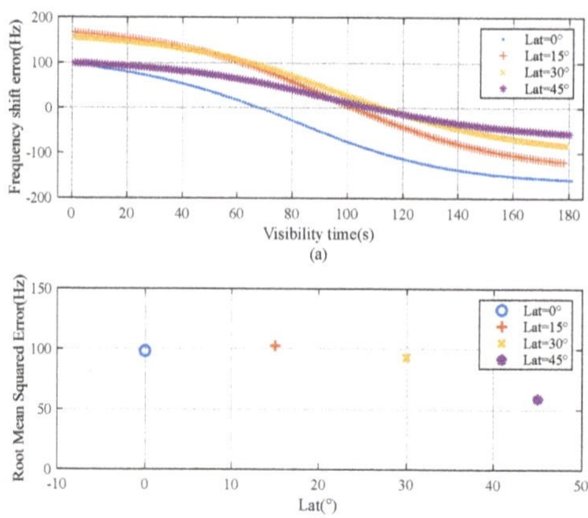

Figure 6. Simulation results of frequency-offset estimation error and RMSE for ground terminals at different latitudes. (**a**) The Frequency shift error; (**b**) The Root Mean Squared Error.

To analyze the adaptability of the DFS-BP algorithm to different satellite orbits, this paper designs simulations of the Doppler frequency-offset estimation and the RMSE of frequency-offset estimation errors for different orbital altitudes and orbital inclinations.

When the minimum satellite elevation angle is 0°, simulations of the Doppler frequency-offset estimation errors and the RMSE of the frequency-offset estimation errors are conducted for satellites with different orbital altitudes ranging from 480 km to 800 km. The simulation results are shown in Figure 7. The results indicate, based on the DFS-BP algorithm, that the satellite-to-ground frequency-offset estimation errors at different satellite orbital altitudes fall within the range of [−200 Hz, +200 Hz], and the RMSE of frequency-offset estimation errors is less than 120 Hz.

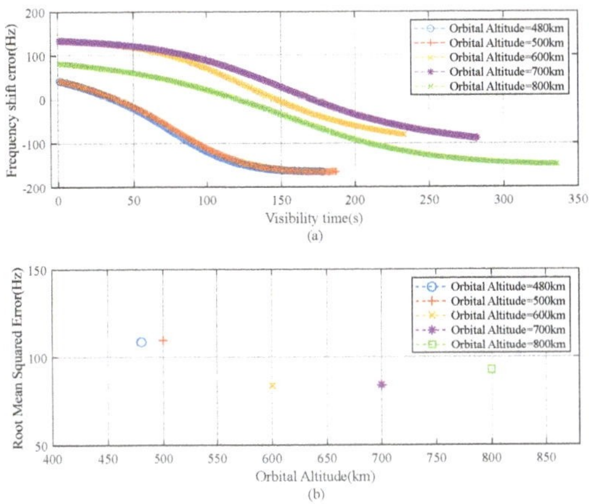

Figure 7. Simulation results of the RMSE of frequency-offset estimations for different orbital altitudes. (**a**) The Frequency shift error; (**b**) The Root Mean Squared Error.

When the minimum satellite elevation angle is 0°, simulations of the Doppler frequency-offset estimation errors and the RMSE of the frequency-offset estimation errors are conducted for satellites with different orbital inclinations ranging from 40° to 90°. The simulation results are shown in Figure 8. The results indicate, based on the DFS-BP algorithm, that the satellite-to-ground frequency-offset estimation errors at different satellite orbital inclinations fall within the range of [−240 Hz, +240 Hz], and the RMSE of frequency-offset estimation errors fall within the range of [50 Hz, 150 Hz].

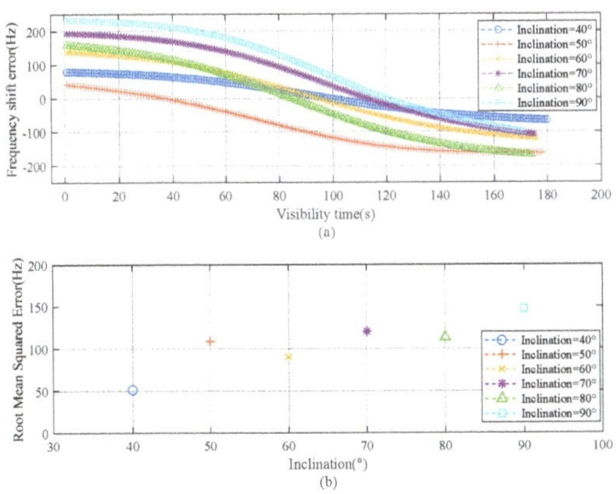

Figure 8. Simulation results of the RMSE of frequency-offset estimations for different orbital inclinations. (**a**) The Frequency shift error; (**b**) The Root Mean Squared Error.

Based on the RMSE results of the Doppler frequency-offset estimation errors for different orbital altitudes and orbital inclinations, the 95% CI analysis of the satellite-to-ground Doppler frequency-offset estimation errors of the DFS-BP algorithm is conducted. The RMSE results of the Doppler frequency-offset estimation errors for different orbital altitudes are shown in Table 3. The RMSE results of the Doppler frequency-offset estimation errors for different orbital inclinations are shown in Table 4.

Table 3. The RMSE results of Doppler frequency-offset estimation errors for different orbital altitudes.

Orbital Altitudes	480 km	500 km	600 km	700 km	800 km
RMSE	109 Hz	110 Hz	83 Hz	84 Hz	93 Hz

Table 4. The RMSE results of Doppler frequency-offset estimation errors for different orbital inclinations.

Orbital Inclinations	40°	50°	60°	70°	80°	90°
RMSE	52 Hz	108 Hz	90 Hz	120 Hz	113 Hz	147 Hz

Due to the small sample size for calculating the 95% CI, this paper selects the t-distribution model for confidence interval calculations. The calculated 95% CI of the Doppler frequency-offset estimation errors falls within the range of [80 Hz, 112 Hz].

4.3. Simulation of Doppler Frequency Shift When the Minimum Elevation Angle Parameter Is Adjustable

The Doppler frequency-shift simulation parameters for variable minimum satellite elevation angles are listed in Table 5.

Table 5. Table of Doppler frequency-shift simulation parameters for variable minimum satellite elevation angles.

Parameter Name	Parameter Value
Orbital altitude, h	480 km
Earth radius, R	6378 km
Eccentricity, e	0.0
Orbital inclination, i	50°
Maximum elevation angle, φ_{max}	0°, 10°, 20°, 30°, 40°, 50°
Earth's rotation period, T	86,164 s
Signal propagation speed, c	2.998×10^8 m/s
Signal carrier frequency, Fc	2.0 GHz

The simulation process of the Doppler frequency-offset estimation is as follows:

- According to the simulation parameters, when the minimum elevation angles of the satellite are set at 0°, 10°, 20°, 30°, 40°, and 50°, we use the STK tool to simulate the Doppler frequency-offset changes invisible time. The changes in the Doppler frequency offset and the azimuth and elevation of the satellite are simulated.
- Using the satellite beam-pointing method, the Doppler frequency-offset variation is calculated according to the azimuth and elevation angles of LEO satellites.
- Compare the error between the Doppler frequency-offset change value of the satellite beam-pointing method and the Doppler frequency-offset change value of the STK simulation.

φ_{max} represents the minimum elevation angle of the satellite. φ_{max} is set at 0°, 10°, 20°, 30°, 40°, and 50°. When the azimuth angle is less than 180°, the Doppler frequency-offset estimation error of the satellite beam-pointing method is shown in Figure 9. It can be seen from Figure 8 that the frequency-offset estimation error range of the two methods is between [−200 Hz, +2000 Hz] in the whole visible period. When the azimuth angle is less than 180°, the positive deviation error of the frequency-offset estimation increases with an increase in the minimum elevation angle of the satellite.

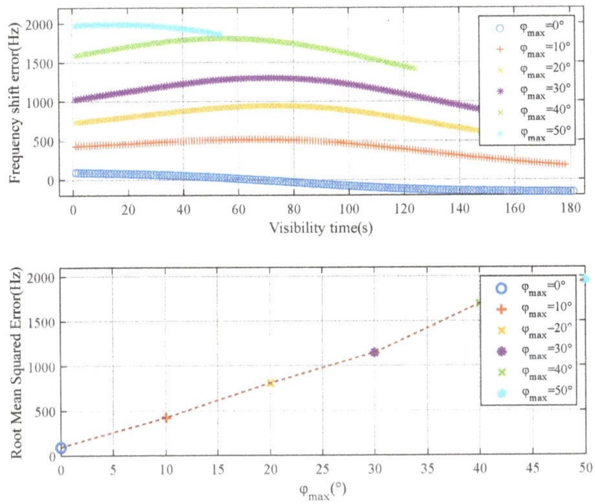

Figure 9. Simulation results of Doppler frequency shifts when the minimum elevation angle is adjustable (the ground terminal is on the right side of the subsatellite point).

We conducted simulations of the RMSE of the frequency-offset estimation for values of 0°, 10°, 20°, 30°, 40°, and 50°. The simulation results are shown in Figure 9. The results indicate that the RMSE of the frequency-offset estimation gradually increases. When φ_{max} = 0°, the RMSE of the frequency-offset estimation does not exceed 100 Hz, and when φ_{max} = 50°, the RMSE is approximately 1900 Hz.

The minimum elevation angles of the satellite are represented, and φ_{max} and φ_{max} are set to values such as 0°, 10°, 20°, 30°, 40°, and 50°. When the azimuth angle exceeds 180°, the Doppler frequency-offset estimation error of the satellite beam-pointing method is shown in Figure 10. During the entire visible period, the frequency-offset estimation error ranges of the two methods are between [−2000 Hz, +200 Hz]. The simulation results indicate that as the minimum elevation angle value of the satellite increases, the frequency-offset estimation error of the satellite beam-pointing method also increases.

We conducted simulations on the RMSE of the frequency-offset estimation for values of 0°, 10°, 20°, 30°, 40°, and 50°. The simulation results are shown in Figure 10. The results indicate that as φ_{max} increases, the RMSE of the frequency-offset estimation gradually increases. When φ_{max} = 0°, the RMSE of the frequency-offset estimation does not exceed 100 Hz, and when φ_{max} = 50°, the RMSE is approximately 2000 Hz.

The simulation results of the DFS-BP algorithm are compared with those obtained from STK's SGP4 model. The comparison shows that the terminal's geographic latitude has little effect on the Doppler frequency-offset estimation error. When the satellite's minimum elevation angle is 0°, the Doppler frequency-offset estimates from both methods are similar, with estimation errors ranging between −200 Hz and +200 Hz. The RMSE of the DFS-BP algorithm is less than 100 Hz. As the minimum elevation angle increases, the Doppler frequency-offset estimates from both methods also increase gradually, with estimation errors expanding to a range of approximately −2000 Hz to +2000 Hz. The RMSE of the DFS-BP algorithm is less than 2 kHz.

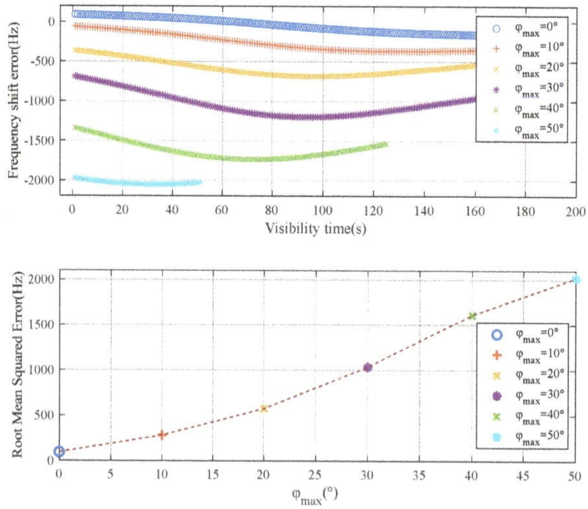

Figure 10. Simulation results of the Doppler frequency shift when the minimum elevation angle is adjustable (the ground terminal is on the left side of the subsatellite point).

Analyzing the frequency-offset estimation errors as the minimum elevation angle increases reveals that the difference between the frequency-offset estimates at the start and end of the satellite-to-ground communication remains under 500 Hz. Future work can focus on reducing frequency-offset estimation errors through more in-depth research. Depending on the specific satellite communication scenarios, techniques such as satellite beam-pointing awareness and fixed error compensation could be applied to optimize the frequency-offset estimation in the satellite beam-pointing method, thereby enhancing estimation accuracy.

This section simulates the estimation error accuracy of the DFS-BP algorithm in terms of the satellite's geocentric latitude, orbital altitude, orbital inclination, terminal position, etc., which uses statistical methods such as the RMSE and 95% CI to analyze the error range of the DFS-BP algorithm. This proposes follow-up improvement directions for the DFS-BP algorithm. However, this method does not consider the influence of clocks, equipment, and systems on the DFS-BP algorithm. This part of the influence will be discussed in Section 5.

5. Discussion

The accuracy of the DFS-BP algorithm proposed in this paper depends on the accuracy of the satellite antenna's azimuth and elevation angles, which are closely related to factors such as antenna pointing accuracy, satellite attitude control accuracy, clock stability, and radio frequency (RF) tolerance.

(1) Influence of Antenna Pointing Accuracy on the DFS-BP Algorithm

Traditional communication satellites typically have antenna pointing accuracies ranging from 0.1° to 0.5°. Military satellites and navigation satellites usually feature antenna pointing accuracies that are better than 0.01°. Low-Earth-orbit (LEO) satellites, due to the requirements of inter-satellite communication, often have antenna pointing accuracies that are superior to 0.01°.

Antenna pointing accuracy is primarily determined by two components: hardware errors and software algorithms. Hardware errors mainly arise from phase errors in phase shifters and radio frequency (RF) chain errors. Phase errors can be minimized by using digital phase shifters to increase the effective bits of the phase shifter and improve quantization accuracy. RF chain errors mainly stem from amplitude and phase errors between channels, which can be reduced by employing high-precision components and implementing amplitude predistortion adjustments. In terms of software, beam-pointing compensation algorithms can be used to reduce RF delays, avoid beam squint, and thus minimize antenna pointing errors.

(2) Influence of Satellite Attitude Control Accuracy on the DFS-BP Algorithm

Traditional communication satellites have attitude control accuracies of less than 0.1°, while remote sensing satellites achieve attitude accuracies that are better than 0.05°. Low-Earth-orbit (LEO) satellites for satellite constellation applications feature attitude control accuracies that are superior to 0.01°. For instance, the Chang'e-5 probe achieved an attitude control accuracy of $\leq$0.1° during lunar sampling, and the "Jilin-1" satellite can reach an attitude control accuracy of 0.02°.

The factors influencing satellite attitude control accuracy mainly include star sensors, gyroscopes, etc. The measurement error of star sensors directly affects attitude control accuracy, with high-precision star sensors achieving an accuracy of 0.0003°. The bias stability of fiber optic gyroscopes is 0.01°/h, while that of laser gyroscopes can be better than 0.001°/h. The combination of high-precision star sensors and laser gyroscopes enables high-precision satellite attitude control.

A satellite's attitude control accuracy and antenna pointing accuracy actually affect the antenna's pointing information. Based on the requirement that both the antenna pointing accuracy and satellite's attitude control accuracy are better than 0.5°, this paper adds random errors of 0.05° to 0.5° to the antenna's azimuth and elevation angles for simulation, and the simulation results are shown in Figure 11. The simulation results indicate that as the random errors added to the antenna's azimuth and elevation angles gradually increase, the Doppler frequency-shift estimation errors also increase progressively, with the estimation errors remaining within ±350 Hz. If the satellite's altitude control errors and antenna pointing errors reach a certain threshold, the accuracy of the DFS-BP algorithm will be affected.

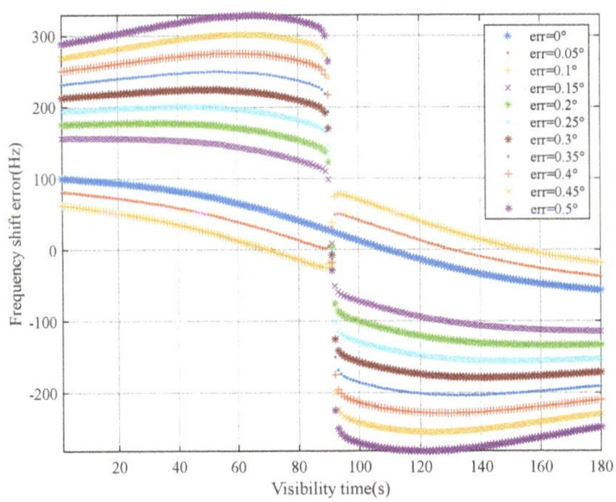

Figure 11. Influence of pointing errors in azimuth and elevation angles on the results of Doppler frequency-offset estimation.

(3) Influence of Clock Stability on the DFS-BP Algorithm

Clock stability is reflected in a signal's frequency stability. The stability of atomic clocks is superior to 10^{-11}, while that of ordinary oven-controlled crystal oscillators (OCXOs) is better than 10^{-8}. Most satellites use atomic clocks as clock sources. For a signal with an operating frequency of 2 GHz, the signal's frequency stability caused by either atomic clocks or ordinary OCXOs is less than 20 Hz. This value is much smaller than the satellite-to-ground Doppler frequency shift; thus, the influence of clock stability on the DFS-BP algorithm is negligible.

(4) Influence of Radio Frequency (RF) Tolerance on the DFS-BP Algorithm

Radio frequency (RF) tolerance generally refers to the permissible deviation range of characteristics such as frequency, power, phase, and amplitude–frequency when RF equipment is in operation. It is an important indicator for measuring the tolerance capability of RF components or systems relative to parameter fluctuations and their operational stability. Among RF characteristics like frequency, power, phase, and amplitude–frequency, frequency accuracy may impact the DFS-BP algorithm. Frequency accuracy is typically related to the precision grade of crystal oscillators or clocks, the operating temperature range, and manufacturing processes. For example, the frequency accuracy of atomic clocks is usually better than 10^{-10}, while that of ordinary oven-controlled crystal oscillators is typically better than 10^{-7}. For a 2 GHz signal, the maximum operating frequency error caused by the frequency's accuracy is 200 Hz, which translates to an estimation error of less than 1 Hz in the DFS-BP algorithm.

6. Conclusions

Building on the development of 5G-A NTN technology and the communication requirements of mobile terminals directly connected to satellites, we investigated a novel Doppler frequency-offset estimation method that operates without GNSS guidance and is tailored for the high dynamics of low-Earth-orbit (LEO) satellite communications. We analyzed the principles underlying satellite-to-ground Doppler frequency offsets, derived the relationship between satellite beam pointing and satellite velocities, and proposed a Doppler frequency-offset estimation method based on satellite beam pointing. Furthermore, we conducted a detailed study on how parameters such as the Earth's latitude and the satellite's orbital inclination affect this method. Under varying conditions of minimum satellite elevation angles and terminal geographical locations, we compared the Doppler frequency-offset estimation results using the satellite beam-pointing method to those from STK's SGP4 model. The simulation results show that when a satellite's minimum elevation angle is 0°, the frequency-offset estimation errors of both methods lie within ± 200 Hz, and the RMSE of the frequency-offset estimation does not exceed 100 Hz. When the minimum elevation angle is greater than 0°, the error range and the RMSE of the frequency-offset estimation increase to approximately ± 2000 Hz.

Author Contributions: Conceptualization, Y.S., J.X. and C.S.; Methodology, Y.S. and J.X.; Software, Y.S. and X.L.; Validation, Y.S., X.L. and S.A.; Resources, C.S.; Data curation, S.A.; Writing—review & editing, Y.S. All authors have read and agreed to the published version of the manuscript.

Funding: This research received no external funding.

Data Availability Statement: Data are contained within the article.

Conflicts of Interest: The authors declare no conflict of interest.

References

1. Ali, I.; Al-Dhahir, N.; Hershey, J.E. Doppler Characterization for LEO Satellites. Communications. *IEEE Trans. Commun.* **1998**, *46*, 309–313. [CrossRef]
2. You, B.H.; Jung, H.J.; Lee, I.H. Survey on Doppler Characterization and Compensation Schemes in LEO Satellite Communication Systems. In Proceedings of the 2022 27th Asia Pacific Conference on Communications (APCC), Jeju Island, Republic of Korea, 19–21 October 2022; pp. 637–638.
3. Nia, O.; Mark, B.L. Models for Frequency Doppler Shift Prediction for LEO Satellites at L-Band. In Proceedings of the 2022 IEEE Military Communications Conference (MILCOM), Rockville, MD, USA, 28 November–2 December 2022.
4. Lin, J.G.; Hou, Z.W.; Zhou, Y.Q.; Tian, L.; Shi, J.L. MAP Estimation Based on Doppler Characterization in Broadband Mobile LEO Satellite Communications. In Proceedings of the 2016 IEEE 83rd Vehicular Technology Conference (VTC Spring), Nanjing, China, 15–18 May 2016.
5. Li, Z.; Lin, W.H.; Deng, W.; Yang, J.F.; Zhu, Z.Y. Analysis of the Doppler Effect on Starlink Downlink Signals. *J. Commun.* **2024**, *45*, 131–141.
6. Khan, T.A.; Afshang, M. A Stochastic Geometry Approach to Doppler Characterization in a LEO Satellite Network. In Proceedings of the ICC 2020–2020 IEEE International Conference on Communications (ICC), Dublin, Ireland, 7–11 June 2020.
7. Wang, A.H.; Wang, X.Y.; Chen, L.T. A Geometric Algorithm of Doppler-Time Characterization for LEO Satellite. *Trans. Beijing Inst. Technol.* **2016**, *36*, 1294–1297.
8. Wang, D.L.; Fattouche, M.; Ghannouchi, F. Geometry-Based Doppler Analysis for GPS Receivers. *Wirel. Pers. Commun.* **2013**, *68*, 1–13. [CrossRef]
9. Peters, E.G.; Benson, C.R. A Doppler Correcting Software Defined Radio Receiver Design for Satellite Communications. *IEEE Aerosp. Electron. Syst. Mag.* **2020**, *35*, 38–48. [CrossRef]
10. Rouzegar, H.; Ghanbarisabagh, M. Estimation of Doppler Curve for LEO Satellites. *Wirel. Pers. Commun.* **2019**, *108*, 2195–2212. [CrossRef]
11. Amiri, S.; Mehdipour, M. Accurate Doppler Frequency Shift Estimation for any Satellite Orbit. In Proceedings of the 2007 3rd International Conference on Recent Advances in Space Technologies, Vols. 1 and 2, Istanbul, Turkey, 14–16 June 2007; pp. 602–607.
12. Han, L.; Wu, X.C.; Luo, W.X. Doppler shift characterization for Non-GEO satellite channels. *J. Beijing Inst. Technol.* **2005**, *25*, 143–146.

electronics

MDPI

Article

A Framework for Joint Beam Scheduling and Resource Allocation in Beam-Hopping-Based Satellite Systems

Jinfeng Zhang [1], Wei Li [1,2,*], Yong Li [1], Haomin Wang [1] and Shilin Li [2]

[1] School of Information and Communication Engineering, Beijing University of Posts and Telecommunications, Beijing 100876, China; zjf1@bupt.edu.cn (J.Z.); liyong@bupt.edu.cn (Y.L.); wanghaomin@bupt.edu.cn (H.W.)
[2] State Radio Monitoring Center, Beijing 100037, China; lishilin@srrc.org.cn
* Correspondence: liwei@srrc.org.cn

Abstract: With the rapid development of heterogeneous satellite networks integrating geostationary earth orbit (GEO) and low earth orbit (LEO) satellite systems, along with the significant growth in the number of satellite users, it is essential to consider frequency compatibility and coexistence between GEO and LEO systems, as well as to design effective system resource allocation strategies to achieve efficient utilization of system resources. However, existing beam-hopping (BH) resource allocation algorithms in LEO systems primarily focus on beam scheduling within a single time slot, lacking unified beam management across the entire BH cycle, resulting in low beam-resource utilization. Moreover, existing algorithms often employ iterative optimization across multiple resource dimensions, leading to high computational complexity and imposing stringent requirements on satellite on-board processing capabilities. In this paper, we propose a BH-based beam scheduling and resource allocation framework. The proposed framework first employs geographic isolation to protect the GEO system from the interference of the LEO system and subsequently optimizes beam partitioning over the entire BH cycle, time-slot beam scheduling, and frequency and power resource allocation for users within the LEO system. The proposed scheme achieves frequency coexistence between the GEO and LEO satellite systems and performs joint optimization of system resources across four dimensions—time, space, frequency, and power—with reduced complexity and a progressive optimization framework. Simulation results demonstrate that the proposed framework achieves effective suppression of both intra-system and inter-system interference via geographic isolation, while enabling globally efficient and dynamic beam scheduling across the entire BH cycle. Furthermore, by integrating the user-level frequency and power allocation algorithm, the scheme significantly enhances the total system throughput. The proposed progressive optimization framework offers a promising direction for achieving globally optimal and computationally tractable resource management in future satellite networks.

Keywords: beam-hopping; heterogeneous satellite networks; resource allocation; spectrum sharing

1. Introduction

1.1. Motivation

With the large-scale deployment of low earth orbit (LEO) satellite constellations, heterogeneous satellite networks are emerging as a key architectural paradigm for next-generation space–air communications. Reference [1] provides a detailed overview of their advantages, such as global coverage, low latency, and high capacity. It also highlights key technical challenges, including cross-orbit coordination, interference management, and efficient resource scheduling across LEO and geostationary earth orbit (GEO) satellite layers. These heterogeneous satellite networks combine LEO and GEO satellite systems to exploit their complementary strengths. LEO satellites enable high data rates and low communication latency, while GEO satellites deliver reliable coverage in a wide area [2]. However, this heterogeneous integration also introduces new technical challenges. In particular, the coexistence of LEO and GEO systems operating within overlapping frequency bands can lead to severe inter-system interference, making spectrum sharing a critical issue to be addressed [3]. In addition, to fully exploit the capacity of the system, satellite networks are required to allocate communication resources, such as frequency bands, transmission power, time slots, and beam directions in a highly adaptive and dynamic way. This requirement is particularly critical for LEO satellites, which have greater flexibility in resource scheduling [4]. However, interdependence among these resources across the temporal, spatial, spectral, and power domains results in strong multidimensional coupling, substantially increasing the complexity of joint resource optimization [5].

In conventional multi-beam satellite communication systems, each beam typically provides fixed coverage over a specific geographic region, lacking the flexibility to adapt resource allocation in response to dynamic user distribution. This often leads to resource shortages in high-traffic areas and resource underutilization in low-demand regions, resulting in inefficiencies and waste in the overall allocation of resources of the system [6]. Beam-hopping (BH) technology offers a promising solution to the resource allocation problem characterized by time–space–frequency coupling in heterogeneous satellite networks. By dividing operational time into discrete slots, the BH algorithm enables satellites to dynamically activate beams across different regions based on user demand and channel conditions, thus achieving multiplexed time-divided coverage and adapting to spatio-temporal variations in traffic load [7].

BH technology facilitates multi-dimensional resource allocation across the temporal, spatial, spectral, and power domains. Through this flexible and adaptive resource management paradigm, the system can dynamically schedule and optimize resources, ensuring rapid responsiveness to non-uniform traffic distributions and maintaining service quality guarantees [8]. Furthermore, by leveraging dynamic beam steering capabilities, BH technology enables precise control over spatial coverage, which not only increases the number of users that can be served simultaneously but also improves the efficiency of spatial resource utilization [9]. In addition, BH technology supports dynamic frequency and power allocation across beams, allowing the system to flexibly adapt to diverse user-traffic demands and service requirements [10].

1.2. Related Work

BH has emerged as a promising technique for enhancing the flexibility and efficiency of satellite communication systems. Its performance benefits have been extensively validated in the literature. Specifically, [11] quantifies the throughput improvement of BH over fixed multi-beam illumination at the link-level, while [12] further substantiates these gains at the system-level by explicitly accounting for inter-beam interference. Building upon these foundational studies, [13] characterizes BH-based resource management as a multidimensional optimization problem involving joint time-slot scheduling and frequency–power allocation.

In the domain of beam scheduling, with the growing density of satellite constellations, especially in LEO, the simultaneous coverage of overlapping ground areas by multiple satellites from different orbits has introduced new challenges for beam coordination and interference mitigation [14]. To support dynamic and uneven traffic demands while minimizing inter-beam interference within a satellite system, coordinated beam steering and resource scheduling have attracted growing research interest [15]. For instance, [16] proposes a two-stage approach that jointly minimizes the number of active beams and balances user load under a fixed power constraint, using graph-theoretic beam clustering and K-means-based user grouping. Similarly, [17] formulates beam placement as a multi-objective optimization problem and introduces a genetic algorithm to obtain Pareto-efficient trade-offs between beam count and spectrum usage. In [18], the authors further extend this line of work by jointly optimizing beam locations and transmitting power using both iterative and deep learning-based solutions, significantly improving computational efficiency.

In parallel, research on frequency and power allocation has progressed along a complementary track. The authors of [19] treat spectrum assignment and power control as separate optimization subproblems. In [20], the authors propose a greedy algorithm for frequency and power allocation in multi-beam LEO systems. The authors of [21] introduce an automatic power control (APC) mechanism to mitigate co-linear interference in the uplink and downlink between LEO and GEO systems, thereby effectively reducing inter-system interference. The authors of [2] presented a joint beam management and power allocation framework focused on maximizing GEO system throughput while ensuring continuity for LEO links. Furthermore, [22] presents a frequency and power allocation strategy tailored for multi-beam systems, designed to meet non-uniform user distribution and traffic demands under constrained power budgets, while exploring the trade-off between total system capacity and user fairness. In [23], the authors optimize subchannel and power allocation in a multi-beam satellite communication system to meet traffic requirements, while simultaneously minimizing power consumption and bandwidth overhead.

1.3. Contribution

Despite the advancements in the existing literature, two critical challenges remain inadequately addressed. Firstly, existing BH technology resource allocation methods primarily focus on beam assignment within a single time slot, lacking global optimization across the entire BH cycle. This limits the effective utilization of temporal resources within the system. Secondly, while these studies have provided important theoretical and algorithmic foundations, most either target a single resource dimension or rely on fully coupled optimization models that impose high computational complexity. Resource allocation in BH-based systems involves multiple coupled dimensions—beam-footprint partitioning, time-slot scheduling, and frequency–power allocation. The resulting joint optimization problem is high-dimensional and computationally complex. Therefore, it is essential to develop a scheduling framework that can effectively decouple spatio–tempora–spectral resources and adopt a hierarchical optimization strategy to achieve near-optimal system performance with reduced computational complexity.

To address the challenges of high-dimensional coupling and suboptimal temporal utilization in BH-based satellite systems, this paper investigates a BH-based beam scheduling and resource allocation framework. In contrast to conventional approaches that rely on joint iterative optimization, the proposed framework adopts a progressive, dimension-wise optimization strategy, in which each resource dimension is handled independently in a sequential manner, without requiring cross-iteration between modules. This enables efficient and scalable scheduling with significantly reduced computational complexity. Specifically, we develop a spatio–temporal–frequency resource allocation framework tailored for heterogeneous GEO–LEO satellite networks. The objective is to fully exploit the time, spatial, and spectral resources available to the LEO system, thereby enhancing the overall system throughput. Within the proposed framework, beam-footprint partitioning, time-slot assignment, beam-level power allocation, and user-specific frequency–power allocation are jointly addressed over the entire BH cycle. Each component is independently modeled and solved using modular algorithms, including clustering and integer linear programming (ILP) for spatial optimization, a hybrid greedy–simulated annealing approach for temporal scheduling, and a Hungarian–water-filling–convex optimization pipeline for spectral and power domains. Although the resource dimensions are optimized sequentially, cross-domain dependencies are minimized and there is no need for iterative feedback between stages, thereby enabling tractable and flexible resource coordination under realistic constraints.

The main contributions of this paper are summarized as follows:

- Beam-Footprint Partitioning: to improve the coverage efficiency of multi-beam LEO satellite systems, we develop a multi-phase beam partitioning strategy that integrates density-aware clustering with ILP. The initial beam layout is guided by user distribution, where regions of high user density are prioritized during clustering. Subsequently, ILP is applied to optimize the association between users and beams, aiming to adjust beam center positions so that users are located as close as possible to beam centers for enhanced channel gain. An iterative adjustment step eliminates low-efficiency beams and reallocates their users, thereby increasing beam utilization efficiency and enabling better adaptation to spatially non-uniform traffic patterns.
- Time-Slot Beam Scheduling: for scheduling beam positions across time slots within a BH cycle, we develop a hybrid approach, combining a greedy algorithm and simulated annealing. A feasible initial schedule is generated using a greedy filling strategy under constraints on minimum beam separation and per-slot beam limits. A simulated annealing algorithm is then applied to explore the solution space through probabilistic perturbations, enhancing the global optimality of beam allocation.
- Frequency–Power Joint Allocation: we propose a joint optimization method that integrates the Hungarian algorithm and water-filling theorem. The Hungarian algorithm is used to efficiently match users with sub-bands, while water-filling is applied to optimize intra-beam power distribution. Finally, we apply convex optimization to regulate inter-beam power allocation, leading to enhanced system throughput for the LEO network.

The remainder of this paper is structured as follows. Section 2 introduces the system model and formulates the joint resource allocation problem in the heterogeneous GEO–LEO satellite network, encompassing beam-footprint partitioning, time-slot beam scheduling, and frequency–power allocation with the goal of maximizing the overall system throughput. Section 3 details the proposed beam-hopping-based resource allocation framework, which employs a progressive, dimension-wise optimization strategy. Section 4 presents numerical simulations and analyzes the performance of the proposed algorithms in terms of throughput and computational efficiency. Finally, Section 5 summarizes the key findings and outlines potential directions for future research.

2. System Model

2.1. System Setup

This study investigates the resource scheduling problem within a single BH cycle, in line with the scheduling characteristics of BH systems. Based on the analysis in [24], the ground-track displacement of a LEO satellite within a single BH cycle is negligible compared to the beam coverage radius, and user positions are assumed to be static over this short duration. Thus, we adopt a snapshot-based model where satellite positions and user distribution remain unchanged within each BH cycle.

The two satellite systems operate over a shared frequency spectrum but serve distinct service objectives: the LEO network is designed for dynamic, high-density user access, whereas the GEO network is tasked with providing stable and continuous coverage. In the simulation scenario, a representative ground service area is selected, which is served by a heterogeneous satellite network consisting of N_{sat} LEO satellites and one GEO satellite as shown in Figure 1.

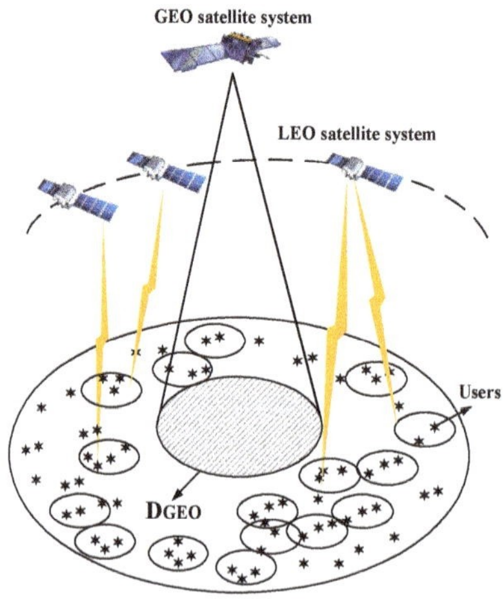

Figure 1. GEO–LEO scenario schematic diagram.

In the time domain, one BH cycle is divided into M discrete time slots. During each time slot, each satellite determines the positions of its activated beams. In the spatial domain, each LEO satellite is equipped with a multi-beam antenna array and can simultaneously illuminate up to N_{beam} beams, with each beam covering a circular area of radius R. Within a single BH time slot, the total number of active beams across all N_{sat} satellites is constrained by $N = N_{\text{sat}} \cdot N_{\text{beam}}$. To mitigate intra-system interference, the illuminated beams in each time slot must maintain a minimum inter-beam distance $d_{\min} \geq 4R$ as suggested in [25]. To mitigate inter-system interference, a protection region D_{GEO} is established around the GEO users, within which the activation of LEO satellite beams is prohibited. In the remainder of this study, all LEO users are assumed to be located outside of the GEO protection region to ensure inter-system interference is avoided during resource scheduling [26]. In the frequency domain, each LEO satellite is allocated the entire system bandwidth B_s, and each beam can serve up to L users, with the bandwidth equally distributed among them. The total transmit power of each satellite is limited to P_s and the sum of the power allocated to all beams must not exceed this limit.

By partitioning the time, space, and frequency domains, a three-dimensional spatio–temporal–frequency resource grid is constructed. Within a single BH cycle, the N_{sat} LEO satellites can schedule up to MNL resource units, where each user is limited to occupying at most, one unit of this grid. To represent user-resource assignment, we define a binary variable $\alpha^{i}_{t,b,f} \in \{0,1\}$, which indicates whether the i-th user is assigned to the t-th time slot, the b-th beam, and the f-th sub-band. For a given user i, the received signal power can be expressed as follows:

$$S^{i}_{t,b,f} = P^{i}_{t,b,f} \cdot G^{i}_{b} \cdot PL^{i}_{b,f} \tag{1}$$

where $P^{i}_{t,b,f}$ denotes the transmit power allocated to the i-th user on the f-th sub-band by the b-th beam in the t-th time slot. G^{i}_{b} represents the beamforming gain from the b-th beam to the i-th user, and $PL^{i}_{b,f}$ denotes the path loss between the b-th beam and the i-th user.

The receiver antenna at the user terminal is assumed to be omnidirectional, and thus its gain is omitted.

The interference received by the i-th user is defined as follows:

$$I^i_{t,b,f} = \sum_{b' \neq b} P^i_{t,b',f} \cdot G^i_{b'} \cdot PL^i_{b',f} \tag{2}$$

where $b' \neq b$ represents the aggregated interference from non-serving beams associated with irrelevant links.

Accordingly, the signal-to-interference-plus-noise ratio (SINR) γ^i of the i-th user can be expressed as follows:

$$\gamma^i_{t,b,f} = \frac{P^i_{t,b,f} \cdot G^i_b \cdot PL^i_{b,f}}{\sum_{b' \neq b} P^i_{t,b',f} \cdot G^i_{b'} \cdot PL^i_{b',f} + N_i} \tag{3}$$

where N_i denotes the thermal noise power associated with the sub-band allocated to the i-th user, which is determined by the sub-band bandwidth and receiver characteristics.

The achievable data rate $R^i_{t,b,f}$ for the i-th user can be calculated based on the Shannon capacity formula:

$$R^i_{t,b,f} = B^i_{t,b,f} \log_2(1 + \gamma^i) \tag{4}$$

where $B^i_{t,b,f}$ is the bandwidth of the sub-band assigned to the i-th user via the b-th beam.

2.2. Channel Model

Considering realistic satellite-to-ground propagation conditions, the path loss of the i-th user when served by the b-th beam on the f-th sub-band is composed of three main components: free-space attenuation, atmospheric losses, and Rician fading. The overall path loss is expressed as follows [27]:

$$PL^i_{b,f} = \left(\frac{c}{4\pi d f}\right)^2 A^{-1}(d)\rho_1 \tag{5}$$

where ρ_1 is the Rician fading factor, d denotes the distance between the b-th beam and the i-th user, and $A(d)$ represents atmospheric attenuation, which can be further expressed as follows: $A(d) = 10^{\frac{d(4.343\rho_2+\rho_3)}{10H_S}}$, where ρ_2 and ρ_3 are the cloud and rain attenuation coefficients, respectively. H_S is the orbital altitude of the satellite.

2.3. Problem Formulation

Within a single BH cycle, the N_{sat} LEO satellites can schedule up to MNL resource units and each user is allowed to occupy at most, one such unit. The overall optimization objective is to maximize the utilization of these multi-dimensional resources while jointly optimizing the allocation of frequency and power for each satellite's beams and associated users, so as to maximize the total system throughput. However, due to the strong coupling across the spatio–temporal–frequency-power dimensions, direct optimization of the global objective incurs prohibitively high computational complexity. To address this, we decompose the original problem into two sub-objectives, enabling a stepwise optimization approach. This hierarchical strategy allows for the design of low-complexity algorithms capable of achieving near-optimal solutions with significantly reduced computational overhead.

The first optimization objective aims to maximize the number of scheduled system resources, which can be formulated as the following objective function, P1:

$$\text{P1: } \max \sum_{t=1}^{M} \sum_{b=1}^{N} \sum_{f=1}^{L} \alpha_{t,b,f}^{i} \tag{6a}$$

$$\text{s.t. } \sum_{i} \alpha_{t,b,f}^{i} \leq 1, \forall \text{triplet}\{t,b,f\} \tag{6b}$$

$$\sum_{t} \sum_{b} \sum_{f} \alpha_{t,b,f}^{i} \leq 1, \forall i \tag{6c}$$

$$\alpha_{t,b,f}^{i} \in \{0,1\}. \tag{6d}$$

In objective function P1, constraint (6b) ensures that each spatio–temporal–frequency resource unit can be assigned to at most, one user within a single BH cycle. Constraint (6c) guarantees that each user can be allocated no more than one such resource unit per BH cycle. Constraint (6d) specifies that $\alpha_{t,b,f}^{i}$ is a binary allocation variable.

The second optimization objective aims to allocate satellite beam power, intra-beam user power, and user frequency resources in order to maximize the total throughput of users served by the satellite system. This objective can be formulated as the following optimization problem, P2:

$$\text{P2: } \max \sum_{t=1}^{M} \sum_{b=1}^{N} \sum_{i=1}^{U_{t,b}} R_{t,b,f_i}^{i} \tag{7a}$$

$$\text{s.t. } \sum_{i=1}^{U_{t,b}} P_t^i \leq P_t^b, \forall t, b \tag{7b}$$

$$P_t^i \geq 0, \forall i, t \tag{7c}$$

$$P_t^b \leq P_{\max}, \forall t, b \tag{7d}$$

$$\sum_{b=1}^{N} P_t^b \leq P_s, \forall t \tag{7e}$$

$$\sum_{i=1}^{U_{t,b}} f_{t,b}^i = B_s, \forall t, b \tag{7f}$$

$$f_{t,b}^i \geq 0, \forall i, t \tag{7g}$$

$$i \notin D_{\text{GEO}}. \tag{7h}$$

$U_{t,b}$ denotes the number of users served by the b-th beam during the t-th time slot and $\left\{u_{t,b}^1, u_{t,b}^2, \ldots, u_{t,b}^{U_{t,b}}\right\}$ represents the corresponding user set.

In objective function P2, the associated constraints are interpreted as follows:

- Constraint (7b) specifies that the total power allocated to users under each beam must not exceed the power assigned to that beam.
- Constraint (7c) specifies that the power allocated to each user must be non-negative.
- Constraint (7d) specifies that the power allocated to each beam must not exceed the predefined beam-level power limit.
- Constraint (7e) specifies that the total power allocated across all beams must not exceed the satellite's onboard power budget.
- Constraint (7f) specifies that the total bandwidth allocated to users under each beam must equal the total bandwidth available to the satellite.
- Constraint (7g) specifies that the bandwidth allocated to each user must be non-negative.
- Constraint (7h) specifies that scheduled LEO users must be located outside the GEO protection zones in order to eliminate potential interference to GEO system operations.

Based on the two objective functions defined above, we perform dimension-wise optimization for spatio–temporal–frequency–power resource allocation in the LEO system.

3. A Beam-Hopping-Based Global Beam Scheduling and User Resource Allocation Algorithm

In this section, we propose a beam-hopping-based joint beam scheduling and resource allocation framework for a resource-constrained heterogeneous GEO–LEO satellite network. The proposed framework consists of three core components: beam-footprint partitioning, time-slot-based beam scheduling, and joint frequency–power allocation. The first objective function P1 is addressed through a two-step process involving beam partitioning and beam scheduling, with the goal of maximizing the number of scheduled spatio–temporal frequency resources. The second objective function P2 is solved via frequency and power allocation, aiming to maximize the total system throughput.

Section 3.1 presents the optimization of beam-footprint partitioning using an ILP approach. Section 3.2, we address the scheduling of beam activations across time slots by proposing a global beam–slot allocation strategy that accounts for spatio-temporal dynamics. Section 3.3 introduces a joint optimization of beam-level power allocation and user-level frequency and power assignment.

3.1. *Algorithm A: Beam-Footprint Partitioning Algorithm*

In this section, we propose an ILP-based beam-footprint partitioning algorithm. The Algorithm A predefines the set of beams that can be scheduled for the entire BH cycle of the LEO system, based on the spatial distribution of LEO users located outside the GEO protection zones. The objective is to maximize the number of users served per beam, ensuring that each beam is utilized as fully as possible.

3.1.1. Initial Beam Candidate Generation

In the initial beam candidate generation phase, a density-weighted clustering strategy is applied to account for the spatial distribution of users. For each user, the number of neighboring users within a radius R is computed as a local density indicator. Users in high-density regions are selected and replicated to form a virtual user set, thereby guiding the K-means algorithm to place more cluster centers in areas with concentrated user demand. Clustering this density-enhanced user set yields a preliminary set of beam center coordinates, denoted as $\{\mu_1, \mu_2, \ldots, \mu_{MN}\}$, by minimizing the sum of squared distances between users and their assigned cluster centers:

$$\min \sum_{j=1}^{MN} \sum_{x_i \in C_j} ||x_i - \mu_j||^2 \tag{8}$$

where x_i denotes the coordinate of i-th user, and C_j represents the set of user coordinates covered by the j-th beam.

3.1.2. Beam–User Matching

After obtaining the initial beam center coordinates, an optimization model is constructed to maximize the number of users served by each beam. A binary decision matrix of size $U \times MN$ is defined, where U denotes the number of users and MN denotes the number of candidate beams. Each element in this matrix is a binary decision variable denoted as x_{ij}, where $i \in \{1,2,\ldots,U\}$ and $j \in \{1,2,\ldots,MN\}$. If $x_{ij} = 1$, it indicates that the i-th user is scheduled by the j-th beam ; conversely, if $x_{ij} = 0$, the i-th user is not served by the j-th beam.The total number of successfully scheduled users can be computed by summing all elements across both user and beam dimensions of the decision matrix. Accordingly, the optimization objective can be expressed as follows:

$$\max \sum_{i=1}^{U} \sum_{j=1}^{MN} x_{ij} \tag{9a}$$

$$\text{s.t. } x_{ij} \cdot d_{ij} \leq R, \forall i, j \tag{9b}$$

$$\sum_{i=1}^{U} x_{ij} \leq F, \forall j \tag{9c}$$

$$\sum_{j=1}^{MN} x_{ij} \leq 1, \forall i. \tag{9d}$$

where d_{ij} denotes the distance between the i-th user and the center of the j-th beam. The constraints are interpreted as follows: constraint (9b) represents the coverage constraint, where a user is eligible for scheduling by a beam only if it is located within the beam's coverage radius R. If $d_{ij} > R$, then x_{ij} must be 0. Constraint (9c) is a capacity constraint, which states that each beam can serve no more than F users. Constraint (9d) is also a capacity constraint, ensuring that each user can be scheduled by at most, one beam.

3.1.3. Iterative Optimization

Due to the possibility that the algorithms in step 1 and step 2 may generate beams with insufficient user coverage, an iterative refinement mechanism is introduced to improve overall beam utilization. Specifically, we define a beam as inefficient if it serves fewer than F users. After the initial beam–user assignment is completed, all beams are examined sequentially. The users assigned to inefficient beams are released and merged with the pool of unscheduled users. A new round of clustering and beam–user assignment is then performed on the updated user set. The procedure is repeated until either the user capacity requirement is met for all beams, or a maximum number of iterations is reached.

3.2. *Algorithm B: Globally Optimized Beam–Slot Assignment Algorithm*

In this section, we propose Algorithm B, a globally optimized beam–slot assignment, which allocates the beam footprints obtained from Algorithm A to specific time slots within a BH cycle. The objective is to allocate MN beams across M time slots, where each slot is constrained to activate no more than N beams. To mitigate intra-slot interference and ensure reliable system performance, a minimum spatial separation $d_{\min} = 4R$ must be maintained between any two simultaneously active beams.

The globally optimized beam–slot assignment algorithm combines a greedy initialization with a simulated annealing strategy to maximize the total number of scheduled beams, while satisfying the minimum inter-beam distance constraint and the per-slot capacity limit. The optimization problem is formulated as follows:

$$\max \sum_{t=1}^{M} |S_t| \tag{10a}$$

$$\text{s.t. } |S_t| \leq N, \forall t \tag{10b}$$

$$\sqrt{(x_i - x_j)^2 + (y_i - y_j)^2} \geq d_{\min}, \quad \forall t,\ \forall p_i, p_j \in S_t,\ i \neq j. \tag{10c}$$

where S_t denotes the set of beams assigned to the t-th time slot, and N represents the maximum number of beams that can be activated per time slot. Each beam is associated with a center coordinate $p_i = (x_i, y_i), i \in \{1, 2, \ldots, MN\}$. Constraint (10b) ensures that the number of beams assigned to any time slot does not exceed N, constraint (10c) enforces that the distance between any two active beams within the same time slot must be greater than a predefined minimum separation $d_{\min}$. We define a state variable $state(b) = t$, $b = \{1, 2, \ldots, MN\}, t = \{1, 2, \ldots, M\}$ to represent that the b-th beam is assigned to the t-th time slot , where $state(b) = -1$ indicates that the b-th beam is not scheduled. The principal parameter settings of the simulated annealing procedure are summarized in Table 1.

Table 1. Simulated annealing optimization simulation parameter settings.

Parameter	Value
Initialize temperature T	500
Minimum temperature $T_{\min}$	10^{-3}
Cooling rate α	0.95
Max iterations $N_{\max}$	500
Acceptance rule	Metropolis criterion $P = \mathrm{e}^{-\Delta E/T}$

The pseudocode of Algorithm 1 is as follows:

Algorithm 1 Global Optimization Beam-Time Slot Allocation Algorithm

Phase 1: Initial Solution Construction Based on Greedy Algorithm Principles
Initialize: $S_t = \varnothing$, $state(b) = -1, b = \{1, 2, \ldots, MN\}, t = \{1, 2, \ldots, M\}$;
for each time slot t **do**
  for each beam b **do**
    if $state(b) = -1$ and $\forall j \in S_t, \sqrt{(x_i - x_j)^2 + (y_i - y_j)^2} \geq d_{\min}$ **then**
      Update the beam allocation set $S_t \leftarrow b$;
      Update the state variables, $state(b) = t$;
    end if
    if $|S_t| = N$ **then**
      break
    end if
  end for
end for
Phase 2: Refinement Based on Simulated Annealing Optimization
Initialize temperature T, minimum temperature $T_{\min}$, cooling rate α, max iterations $N_{\max}$;
Set the greedy algorithm result as the initial solution S_{current};
while $T > T_{\min}$ **do**
  for 1 to $N_{\max}$ **do**
    Randomly select the b-th beam and remove it from the current allocation S_t;
    Update beam allocation S_t and state variables $state(b) = -1$;
    for each time slot t **do**
      if $|S_t| < N_s$ and $\forall j \in S_t, \sqrt{(x_i - x_j)^2 + (y_i - y_j)^2} \geq d_{\min}$ **then**
        Update the beam set $S'_t \leftarrow b$;
        Update the state variables, $state(b) = t$;
        break
      end if
    end for
    if $\sum_{t=1}^{M} |S'_t| > \sum_{t=1}^{M} |S_t|$ **then**
      Accept the new solution immediately;
    else
      Accept the new solution with certain probability;
    end if
  end for
  Update temperature $T \leftarrow \alpha T$.
end while

3.3. Algorithm C: Joint Frequency–Power Allocation Algorithm

This section presents Algorithm C, which performs the joint optimization of frequency and power resources based on the overall objective defined in objective function P2. For each beam, the total bandwidth is equally divided into $U_{t,b}$ sub-bands, where $U_{t,b}$ is set as equal to the number of users $N_{t,b}$ served by the b-th beam in the t-th time slot, resulting in a sub-band set $\{B_f^{sub}|f = 1, 2, \ldots, U_{t,b}\}$. For each user $u_{t,b,f}^i$, the achievable data rate $R_{t,b,f}^i$ across all sub-bands B_f^{sub} is computed and used to construct a reward matrix. The Hungarian algorithm is then applied to solve the resulting assignment problem, aiming to maximize the total system throughput. The frequency allocation problem follows the objective in objective function P2 and is subject to constraints (7f) and (7g).

For power allocation, we first consider intra-beam power distribution among users. Given the beam-specific channel parameters and the total power P_t^b allocated to the b-th beam in the t-th time slot, the intra-beam power allocation can be computed using the classical water-filling algorithm. The objective is to maximize the sum throughput of users within each beam, subject to constraints (7b) and (7c). The channel parameter is denoted as $\lambda_{b,f}^i = \frac{H_{b,f}^i}{N_i B_{b,f}^i}$, where $H_{b,f}^i$ represents the channel gain, including both the transmit gain and path loss. Let μ_b denote the water-filling level corresponding to each beam and $S_{t,b} = \sum_{i=1}^{U_{t,b}} \frac{1}{\lambda_b^i}$. Then, the intra-beam user power allocation result can be obtained as follows:

$$P_{t,b}^i = \mu_b - \frac{1}{\lambda_{b,f}^i} \tag{11}$$

$$\mu_b = \frac{1}{N}(P_t^b + S_{t,b}) \tag{12}$$

The power allocation for users under a single beam of a satellite is given by the following:

$$P_{t,b}^i = \frac{1}{N}(P_t^b + S_{t,b}) - \frac{1}{\lambda_b^i} \tag{13}$$

The inter-beam power allocation is performed based on the results of intra-beam power distribution. At this stage, the throughput of an individual user can be expressed as a function of the beam-level transmit power P_t^b as the following:

$$R_{t,b,f}^i = B_{b,f}^i \log_2\left[1 + \lambda_{b,f}^i P_{t,b}^i\right] = \log_2\left[\frac{\lambda_{b,f}^i}{N}\left(P_t^b + S_{t,b}\right)\right] \tag{14}$$

The total throughput of all users served by a single beam can then be expressed as follows:

$$\begin{aligned} R_t^b &= \sum_i^{U_{t,b}} B_{b,f}^i R_{t,b}^i = \sum_i^{U_{t,b}} B_{b,f}^i \log_2\left[\frac{\lambda_{b,f}^i}{N}\left(P_t^b + S_{t,b}\right)\right] \\ &= \sum_i^{U_{t,b}} B_{b,f}^i\left[\log_2\lambda_b^i + \log_2(P_t^b + S_{t,b}) - \log_2 U_{t,b}\right] \end{aligned} \tag{15}$$

The objective function can then be expressed as follows:

$$\max \sum_t^M \sum_b^N \frac{B_s}{U_{t,b}}\left[U_{t,b}\log_2(P_t^b + S_{t,b}) + \sum_i^{U_{t,b}} \log_2\lambda_{b,f}^i - U_{t,b}\log_2(U_{t,b})\right] \tag{16}$$

The objective function in (16) aims to maximize the total system throughput and is formulated as a weighted summation of multiple logarithmic terms, where P_t^b is the optimization variable, $S_{t,b}$ is a fixed constant obtained from the preceding water-filling algorithm, and $\log_2\lambda_{b,f}^i$ and $U_{t,b}\log_2(U_{t,b})$ are constant terms independent of the optimization variable. Since $\log_2(P_t^b + S_{t,b})$ is a strictly concave function over its domain and both $P_t^b > 0$ and $S_{t,b} > 0$ hold, the objective remains concave. Moreover, the constraints of the optimization problem (7d) and (7e) are linear and do not affect the concavity of the objective. The form of this problem is consistent with the power allocation formulation in [23], and can therefore be efficiently solved using standard convex optimization methods.

4. Simulation Results and Analysis

4.1. Simulation Parameters

In the simulation scenario considered in this paper, a circular ground area with a radius of 700 km is selected as the target service region. This area is jointly covered by four LEO satellites and one GEO satellite, with each LEO satellite operating in a multi-beam mode and capable of activating up to four beams simultaneously. Over six time slots, the LEO satellite system can form up to 96 beam positions that can be scheduled under this configuration. In each time slot, the GEO satellite activates one beam and establishes a protection zone with a radius of 150 km around GEO user. Beam coverage from the LEO satellites is prohibited within this protection zone to avoid interference.The simulation parameters are summarized in Table 2.

Table 2. Simulation parameter settings.

Parameter	Value
LEO Satellite Orbital Altitude	500 km/600 km
Number of LEO Satellites	4
GEO Satellite Position	40° E, 100° N
GEO Earth Station Location	40° E, 100° N
Radius of Target Ground Area	700 km
LEO Satellite Beam Radius	50 km
Peak Antenna Gain of LEO Satellite	35 dB
3 dB Beamwidth	1.65°
Number of Beams per LEO Satellite	4
Number of Users to Be Served	750
Number of Time Slots	6
System Bandwidth	400 MHz
Carrier Frequency	20 GHz
Noise Temperature	150 K
Maximum LEO Satellite Transmit Power	800 W
Rician Fading Factor	0.95
Cloud/Rain Attenuation Coefficient	0.1/0.05
Maximum Beam Power	250 W

The antenna gain used in the simulation is calculated according to the following formula [28,29]:

$$G_b^i = G_0\left[\frac{J_1(\mu)}{2\mu} + 36\frac{J_3(\mu)}{\mu^3}\right]^2 \tag{17}$$

where G_0 denotes the peak transmit antenna gain of the b-th beam , which is calculated as follows: $G_0 = \eta\frac{4\pi D}{(c/f)^2}$, where η is the antenna aperture efficiency, D is the antenna aperture diameter, f is the carrier frequency, and c is the speed of light. Then $\mu = 2.07123\frac{\sin\Theta_b}{\sin\Theta_{3dB}}$, where Θ_{3dB} denotes the 3 dB beamwidth of the b-th beam, and Θ_b represents the off-axis angle between the i-th user and the pointing direction of the b-th beam. $J_1(\cdot)$ and $J_3(\cdot)$ are the first- and third-order Bessel functions of the first kind, respectively.

4.2. Performance of the Beam-Footprint Partitioning Algorithm

In this section, we evaluate the performance of the beam-footprint partitioning algorithm. The beam positions generated by Algorithm A are assigned to time slots using Algorithm 1. If Algorithm A produces only MN beam positions, the assignment performance of Algorithm 1 may degrade in certain scenarios due to the spatial separation constraints among beams within each time slot. This can lead to a reduced number of active beams scheduled within a BH cycle, thereby lowering overall resource utilization. To address this issue, an adjustment factor ∂ is introduced. Specifically, Algorithm A generates ∂MN candidate beams, and Algorithm 1 selects MN beams from this larger pool for actual scheduling. In the simulation, $\partial = 1.5$. Each LEO satellite can activate up to 4 beams simultaneously, resulting in a maximum of 24 beam positions that can be scheduled over six time slots. Based on this configuration, a total of 144 candidate beams are generated across four LEO satellites. Theoretically, the maximum number of users that can be scheduled is 432.

Without the iterative optimization mechanism, a significant number of beams serve fewer than F users, resulting in a total of 408 users being scheduled, with a resource utilization rate of 94.4%, as illustrated in Figure 2a. After introducing iterative refinement, the number of inefficient beams is significantly reduced. Consequently, the number of scheduled users increases to 430, and the resource utilization rate improves to 99.5%, as shown in Figure 2b.

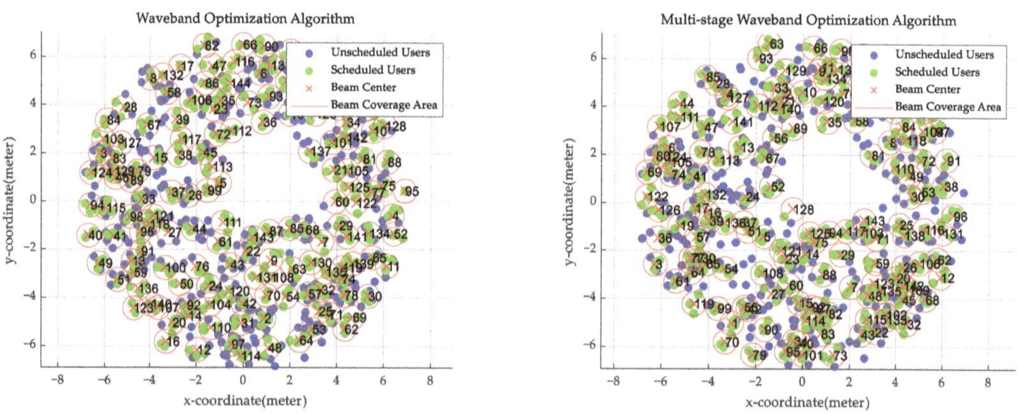

(a) (b)

Figure 2. Performance of beam-footprint partitioning algorithm. (**a**) Beam–slot assignment without iterative optimization; (**b**) beam–slot assignment with iterative optimization.

4.3. Performance of the Globally Optimized Beam–Slot Assignment Algorithm

In this section, we evaluate the performance of the globally optimized beam–slot assignment algorithm by comparing the beam distributions across time slots before and after applying simulated annealing. Under the simulation settings described in Section 4.1, the maximum number of beams that can be activated over six time slots is 96. Without simulated annealing, the actual number of activated beams is 90, as shown in Figure 3a. After applying the annealing-based iterative optimization, the number increases to 96, as illustrated in Figure 3b.

The comparison reveals that, without simulated annealing, beam allocation in the first five time slots is based purely on a greedy strategy without global optimization. As a result, the sixth time slot suffers from a lack of available beam positions, significantly reducing the total number of active beams during the BH cycle and degrading overall system performance. In contrast, the inclusion of simulated annealing enables global optimization across all time slots. By introducing random perturbations, the algorithm escapes local optima that may arise from greedy allocation strategies, thereby increasing the number of activated beams from 90 to 96, which corresponds to the theoretical upper bound of system performance.

The proposed hybrid scheduling algorithm, combining greedy initialization with simulated annealing, exhibits strong scalability. Its computational complexity grows polynomially with the number of candidate beams MN and is largely insensitive to variations in user density. Experimental results under a large-scale scenario (10 satellites, 240 beams, and 2500 users) show that the algorithm completes in 0.015 s while activating all 240 beams, achieving near-global optimality and effectively avoiding the local optima commonly encountered by purely greedy strategies. Moreover, the simulated annealing phase can be further accelerated through multi-threaded perturbation, enabling the proposed method to scale efficiently to even larger LEO constellation systems.

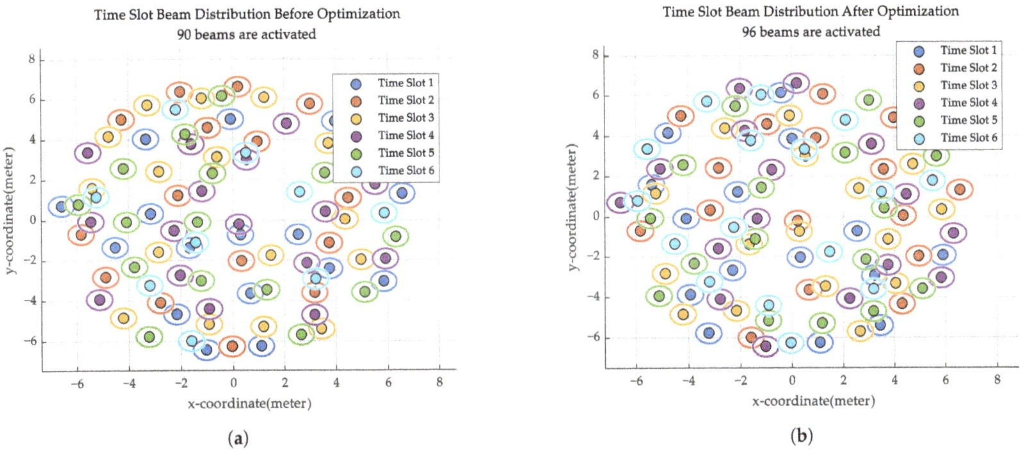

Figure 3. Performance of globally optimized beam–slot assignment algorithm. (**a**) Beam–slot distribution before optimization; (**b**) beam–slot distribution after optimization.

4.4. Performance of the Joint Frequency–Power Allocation Algorithm

In this section, we evaluate the performance of the proposed joint frequency–power optimization algorithm through simulation. For power allocation, three baseline schemes are used for comparison: (1) uniform power allocation across both beams and users, (2) uniform power allocation across beams only, and (3) uniform power allocation across users only. The simulation results comparing all four schemes are presented in Figure 4.

The simulation results indicate that optimizing power allocation either across beams or within beams can significantly enhance system throughput. Among the four evaluated schemes, Algorithm C, which performs joint power allocation for both satellite beams and users, achieves the highest performance gain. It improves the total system throughput by approximately 7.2% to 20% compared to the other uniform power allocation strategies.

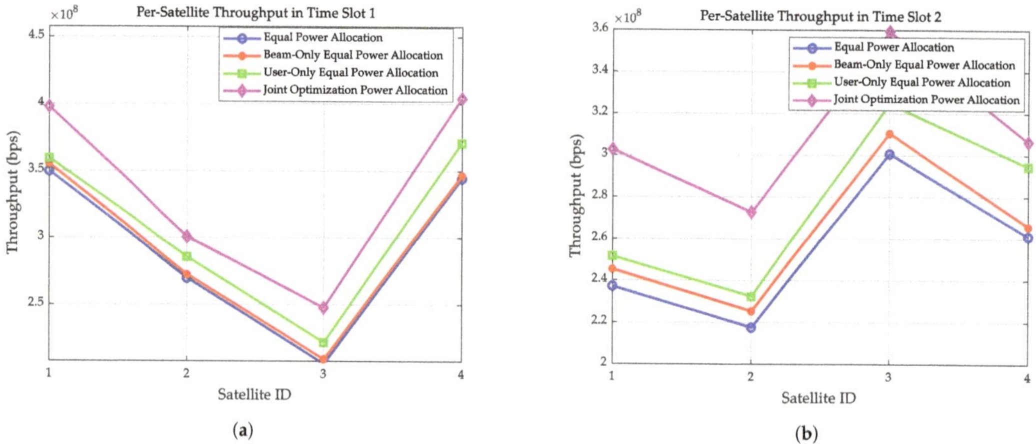

Figure 4. Performance of joint frequency–power allocation algorithm. (**a**) Throughput comparison of different power allocation algorithms for each satellite in time slot 1; (**b**) throughput comparison of different power allocation algorithms for each satellite in time slot 2.

4.5. Computational Complexity and Runtime Evaluation

We analyze the computational complexity of each sub-algorithm as follows.

For Algorithm A, in the beam generation stage, K-means clustering is applied. Each iteration requires computing the distances between U users and MN beam centers. If the number of iterations is denoted by N_k, the complexity of this process is $O(N_k UMN)$. In beam–user matching, an ILP formulation is adopted. Based on empirical results from commercial solvers, the complexity of the ILP grows approximately with the number of variables to the power of 1.3 to 1.6. Therefore, the complexity can be estimated as $O((UMN)^{1.5})$. Hence, the total complexity of Algorithm A is $O(N_k UMN + (UMN)^{1.5})$. This complexity scales polynomially with both the number of users U and the number of candidate beams MN, making it sensitive to large user populations. However, the modular structure ensures that beam generation and assignment are handled efficiently and converge quickly in practice. Therefore, even for large-scale LEO constellations, Algorithm A remains computationally feasible.

For Algorithm 1, during greedy initialization, each candidate beam must be evaluated for placement within a time slot by checking the minimum distance constraint with up to N existing beams. This results in a complexity of $O(MN^2)$. In the simulated annealing phase, assume the total number of random perturbations is N_{SA}. Each perturbation involves randomly selecting a beam and checking distances against up to N other beams in the new slot, yielding a complexity of $O(N_{SA}N)$. Therefore, the overall complexity of Algorithm 1 is $O(MN^2) + O(N_{SA}N) = O(MN^2)$. Since Algorithm 1 only depends on the number of beams and time slots, but not on the number of users U, it exhibits excellent scalability in scenarios with dense user deployments. Its complexity increases quadratically with the number of beams per slot but remains tractable for typical satellite system parameters. This makes it suitable for real-time or near-real-time scheduling in large constellations.

For Algorithm C, this algorithm involves sub-band assignment, intra-beam power allocation, and inter-beam power coordination. Sub-band assignment is performed using the Hungarian algorithm. For each beam, a reward matrix of dimension $L \times L$ is constructed, where L is the maximum number of users per beam. The complexity of Hungarian matching per beam is $O(L^3)$, and the total complexity is $O(MNL^3)$. Intra-beam power allocation using the water-filling method requires ordering the channel gain values. The per-beam complexity is $O(L \log L)$, thus a total complexity of $O(MNL \log L)$. Inter-beam power control is formulated as a convex optimization problem, resulting in a total of $O(MN)$. Therefore, the total complexity of Algorithm C is $O(MNL^3) + O(MNL \log L) + O(MN)$. The complexity of Algorithm C grows with the number of beams MN and the number of users per beam L, but remains polynomial overall. This ensures tractability even in high-throughput systems where each beam serves many users. The use of efficient algorithms like Hungarian matching and water-filling further improves practicality, making this approach well-suited for dynamic resource management in large-scale satellite systems.

The experiments were conducted under a hardware environment consisting of a Windows 11 operating system, an Intel i9-13900H processor (Intel Corporation, Santa Clara, CA, USA), and 16 GB of RAM. Based on the simulation settings described in Section 4.1, the average runtime of each algorithm is as follows: Algorithm A requires approximately 1.76 s, Algorithm 1 requires approximately 0.003 s, and Algorithm C requires approximately 5 s.

5. Conclusions

This paper proposes a beam-hopping-based joint beam scheduling and resource allocation framework, which leverages beam hopping technology to perform global scheduling of beam positions across time slots, enabling efficient multi-dimensional optimization in time, space, frequency, and power domains. The overall resource allocation problem is decomposed into three subproblems: beam-footprint partitioning, beam-to-slot assignment, and joint frequency–power allocation.

First, in the beam partitioning stage, a multi-stage optimization algorithm is introduced to maximize the number of users served by each beam through ILP-based modeling and iterative refinement. Second, for beam-to-slot assignment, a combination of greedy and simulated annealing algorithms is employed to optimize beam allocation across time slots, aiming to reduce inter-beam interference and improve resource utilization efficiency. Finally, in the frequency and power allocation stage, the Hungarian matching algorithm and water-filling method are used to efficiently allocate sub-bands and beam power, with the objective of maximizing user throughput.

Simulation results demonstrate that the proposed dimension-wise optimization approach significantly improves resource utilization and throughput with relatively low computational complexity. Compared to traditional joint iterative optimization methods, the proposed method avoids the heavy computational burden caused by multi-dimensional iterative coupling. Under the simulation settings used in this study, the system achieves a resource utilization rate of 99.5%, with a total user throughput improvement ranging from 7.2% to 20%.

The proposed framework provides a flexible and scalable solution for dynamic resource management in LEO satellite systems. By decoupling the complex joint optimization into well-structured subproblems and integrating both exact and heuristic algorithms, it significantly enhances system performance in terms of throughput and resource efficiency. Importantly, this framework also offers a promising direction for addressing the coexistence challenges in heterogeneous satellite networks, such as LEO–GEO integration, by enabling coordinated and interference-aware scheduling across multiple dimensions. As such, it lays a solid foundation for the development of intelligent, adaptive, and high-capacity satellite systems in future space–air–ground integrated networks.

However, the current scheduling framework is primarily based on offline computation and assumes static system states within each BH cycle. In highly dynamic orbital environments or under burst-like traffic demands, the responsiveness and timeliness of scheduling decisions become critical. Future research could incorporate real-time scheduling mechanisms that address computational latency and rescheduling strategies in response to dynamic user access and satellite mobility. For example, task priority and link-state awareness could be integrated to design fast-response dynamic scheduling strategies suited for more complex operational scenarios.

While this paper has demonstrated the effectiveness of the proposed method in medium-scale systems, larger-scale satellite constellations and higher user densities will significantly increase resource coupling and optimization dimensions. To this end, future work may consider developing distributed or hierarchical scheduling frameworks tailored for ultra-large-scale constellations—for instance, by enabling inter-layer coordination across different orbits to achieve globally efficient resource allocation. In addition, incorporating intelligent methods such as reinforcement learning or graph neural networks to enable context-aware scheduling strategies presents a promising direction for further exploration.

Author Contributions: Conceptualization, J.Z., W.L. and Y.L.; Methodology, J.Z., W.L. and S.L.; Software, J.Z. and H.W.; Writing—original draft, J.Z.; Writing—review and editing, H.W., Y.L. and J.Z.; Funding acquisition, W.L. and S.L. All authors have read and agreed to the published version of the manuscript.

Funding: This work was supported in part by the Beijing Natural Science Foundation under grant L222003, and in part by the National Key R&D Program of China under grant 2024YFB2907600.

Data Availability Statement: Data is contained within the article.

Acknowledgments: The authors would like to thank all reviewers for their helpful comments and suggestions regarding this paper.

Conflicts of Interest: The authors declare no conflicts of interest.

References

1. Gu, P.; Li, R.; Hua, C.; Tafazolli, R. Dynamic cooperative spectrum sharing in a multi-beam LEO-GEO co-existing satellite system. *IEEE Trans. Wire. Commun.* **2022**, *21*, 1170–1182. [CrossRef]
2. Ding, X.; Ren, Z.; Lu, H.; Zhang, G. Improving SINR via joint beam and power management for GEO and LEO spectrum-sharing satellite communication systems. *China Commun.* **2022**, *19*, 25–36. [CrossRef]
3. Li, B.; Park, J.; Al-Hourani, A.; Pokhrel, S.R.; Choi, J. A novel frequency reuse model for co-existing LEO and GEO satellites. *IEEE Wire. Commun. Lett.* **2024**, *13*, 1024–1028. [CrossRef]
4. Lin, Z.; Ni, Z.; Kuang, L.; Jiang, C.; Huang, Z. NGSO satellites beam hopping strategy based on load balancing and interference avoidance for coexistence with GSO systems. *IEEE Commun. Lett.* **2023**, *27*, 278–282. [CrossRef]
5. Feng, W.; Lin, Y.; Wang, Y.; Wang, J.; Chen, Y.; Ge, N.; Zhu, H. Radio map-based cognitive satellite-UAV networks towards 6G on-demand coverage. *IEEE Trans. Cognitive Commun. Netw.* **2024**, *10*, 1075–1089. [CrossRef]
6. He, M.; Cui, G.; Wu, M.; Wang, W. Collaborative interference avoidance technology in GEO-LEO co-existing satellite system. *Int. J. Satellite Commun. Netw.* **2024**, *42*, 257–272. [CrossRef]
7. Chan, S.; Lee, H.; Kim, S.; Oh, D. Intelligent low complexity resource allocation method for integrated satellite-terrestrial systems. *IEEE Wire. Commun. Lett.* **2022**, *11*, 1087–1091. [CrossRef]
8. Sun, H.; Jing, W.; Lu, Z.; Wen, X.; Li, W. An energy-efficient beam hoping strategy in interference coexistence satellite systems. In Proceedings of the 2024 IEEE/CIC International Conference on Communications in China (ICCC Workshops), Hangzhou, China, 7–9 August 2024; pp. 711–716. [CrossRef]
9. Anzalchi, J.; Couchman, A.; Gabellini, P.; Gallinaro, G.; D'agristina, L.; Alagha, N.; Angeletti, P. Beam hoping in multi-beam broadband satellite systems: System simulation and performance comparison with non-hopped systems. In Proceedings of the 2010 5th Advanced Satellite Multimedia Systems Conference and the 11th Signal Processing for Space Communications Workshop (ASMS), Cagliari, Italy, 13–15 September 2010; pp. 248–255. [CrossRef]
10. Wang, A.; Lei, L.; Lagunas, E.; Pérez-Neira, A.I.; Chatzinotas, S.; Ottersten, B. Joint optimization of beam-hopping design and NOMA-assisted transmission for flexible satellite systems. *IEEE Wire. Commun. Lett.* **2022**, *21*, 8846–8858. [CrossRef]
11. Wang, L.; Zhang, C.; Qu, D.; Zhang, G. Resource allocation for beam-hoping user downlinks in multi-beam satellite system. In Proceedings of the 2019 15th International Wireless Communications Mobile Computing Conference (IWCMC), Tangier, Morocco, 24–28 June 2019; pp. 925–929. [CrossRef]

12. Du, X.; Hu, X.; Wang, Y.; Wang, W. Dynamic resource allocation for beam hopping satellites communication system: An exploration. In Proceedings of the 2022 IEEE International Conference on Trust, Security and Privacy in Computing and Communications (TrustCom), Wuhan, China, 9–11 December 2022; pp. 1296–1301. [CrossRef]
13. Han, Y.; Zhang, C.; Zhang, G. Dynamic beam hopping resource allocation algorithm based on deep reinforcement learning in multi-beam satellite systems. In 2021 3rd International Academic Exchange Conference on Science and Technology In-novation (IAECST), Guangzhou, China, 10–12 December 2021; pp. 68–73. [CrossRef]
14. Su, Y.; Liu, Y.; Zhou, Y.; Yuan, J.; Cao, H.; Shi, J. Broadband LEO satellite communications: Architectures and key technologies. *IEEE Wire. Commun.* **2019**, *26*, 55–61. [CrossRef]
15. Xv, H.; Sun, Y.; Zhao, Y.; Peng, M.; Zhang, S. Joint beam scheduling and beamforming design for cooperative positioning in multi-beam LEO satellite networks. *IEEE Trans. Veh. Technol.* **2023**, *73*, 5276–5287. [CrossRef]
16. Bui, V.P.; Van Chien, T.; Lagunas, E.; Grotz, J.; Chatzinotas, S.; Ottersten, B. Joint beam placement and load balancing optimization for non-geostationary satellite systems. In Proceedings of the 2022 IEEE International Mediterranean Conference on Communications and Networking (MeditCom), Athens, Greece, 5–8 September 2022; pp. 316–321. [CrossRef]
17. Pachler, N.; Crawley, E.F.; Cameron, B.G. A genetic algorithm for beam placement in high-throughput satellite constellations. In Proceedings of the 2021 IEEE Cognitive Communications for Aerospace Applications Workshop (CCAAW), Cleveland, OH, USA, 21–23 June 2021; pp. 1–6. [CrossRef]
18. Choi, H.H.; Park, G.; Heo, K.; Lee, K. Joint optimization of beam placement and transmit power for multibeam LEO satellite communication systems. *IEEE Internet Things J.* **2024**, *11*, 14804–14813. [CrossRef]
19. Zhang, S.; Cui, G.; Long, Y.; Wang, W. Joint computing and communication resource allocation for satellite communication networks with edge computing. *China Commun.* **2021**, *18*, 236–252. [CrossRef]
20. Tian, F.; Huang, L.; Liang, G.; Jiang, X.; Sun, S.; Ma, J. An efficient resource allocation mechanism for beam-hopping based LEO satellite communication system. In Proceedings of the 2019 IEEE International Symposium on Broadband Multimedia Systems and Broadcasting (BMSB), Jeju, Republic of Korea, 5–7 June 2019; pp. 1–5. [CrossRef]
21. Sharma, S.K.; Chatzinotas, S.; Ottersten, B. In-line interference mitigation techniques for spectral coexistence of GEO and NGEO satellites. *Int. J. Satellite Commun. Netw.* **2014**, *34*, 11–39. [CrossRef]
22. Choi, J.P.; Chan, V.W. Optimum power and beam allocation based on traffic demands and channel conditions over satellite downlinks. *IEEE Trans. Wire. Commun.* **2005**, *4*, 2983–2993. [CrossRef]
23. Abdu, T.S.; Kisseleff, S.; Lagunas, E.; Chatzinotas, S. Flexible resource optimization for GEO multibeam satellite communication system. *IEEE Trans. Wire. Commun.* **2021**, *20*, 7888–7902. [CrossRef]
24. Yuan, S.; Sun, Y.; Peng, M.; Yuan, R. Joint beam direction control and radio resource allocation in dynamic multi-beam LEO satellite networks. *IEEE Trans. Veh. Technol.* **2024**, *73*, 8222–8237. [CrossRef]
25. Lin, Z.; Ni, Z.; Kuang, L.; Jiang, C.; Huang, Z. Multi-satellite beam hopping based on load balancing and interference avoidance for NGSO satellite communication systems. *IEEE Trans. Commun.* **2022**, *71*, 282–295. [CrossRef]
26. Kibria, M.G.; Lagunas, E.; Maturo, N.; Spano, D.; Chatzinotas, S. Precoded cluster hopping in multi-beam high throughput satellite systems. In Proceedings of the 2019 IEEE Global Communications Conference (GLOBECOM), Waikoloa, HI, USA, 9–13 December 2019; pp. 1–6. [CrossRef]
27. Yuan, Y.; Lei, L.; Vu, T.X.; Chang, Z.; Chatzinotas, S.; Sun, S. Adapting to dynamic LEO-B5G systems: Meta-critic learning based efficient resource scheduling. *IEEE Trans. Wire. Commun.* **2022**, *21*, 9582–9595. [CrossRef]
28. Khan, W.U.; Ali, Z., Lagunas, E.; Mahmood, A.; Asif, M.; Ihsan, A.; Dobre, O.A. Rate splitting multiple access for next generation cognitive radio enabled LEO satellite networks. *IEEE Trans. Wire. Commun.* **2023**, *22*, 8423–8435. [CrossRef]
29. Huang, Q.; Lin, M.; Wang, J.B.; Tsiftsis, T.A.; Wang, J. Energy efficient beamforming schemes for satellite-aerial-terrestrial networks. *IEEE Trans. Wire. Commun.* **2020**, *68*, 3863–3875. [CrossRef]

electronics

MDPI

Article

SCNOC-Agentic: A Network Operation and Control Agentic for Satellite Communication Systems

Wenyu Sun [1], Chenhua Sun [2,*], Yasheng Zhang [1], Zhan Yin [1] and Zhifeng Kang [1]

[1] Department of Satellite Communications, Academy for Network & Communications of CETC, Shijiazhuang 050050, China; swy755565055@gmail.com (W.S.); zys163@163.com (Y.Z.); yinzhancetc54@163.com (Z.Y.); charles_1613@163.com (Z.K.)

[2] National Key Laboratory of Advanced Communication Network Technology, Academy for Network & Communications of CETC, Shijiazhuang 050050, China

* Correspondence: sun_chenhua@sina.com

Abstract: Large language models (LLMs) have demonstrated powerful capability to solve practical problems through complex step-by-step reasoning. Specifically designed LLMs have begun to be integrated into terrestrial communication networks. However, relevant research in the field of satellite communications remains exceedingly rare. To address this gap, we introduce SCNOC-Agentic, a novel architecture especially designed to integrate the management and control of satellite communication systems in LLMs. SCNOC-Agentic incorporates four components tailored to the characteristics of satellite communications: intent refinement, multi-agent workflow, personalized long-term memory, and graph-based retrieval. Furthermore, we define four typical real-world scenarios that can be effectively addressed by integrating with LLMs: network task planning, carrier and cell optimization, fault analysis of satellites, and satellite management and control. Utilizing the SCNOC-Agentic framework, a series of open-source LLMs have achieved outstanding performance on the four tasks under various baselines, including zero-shot CoT, CoT-5, and self-consistency. For example, qwen2.5-70B with SCNOC-Agentic significantly improves the parameter generation accuracy in the network task planning task from 15.6% to 32.2%, while llama-3.3-70B increases from 16.2% to 29.0%. In addition, ablation studies were conducted to validate the importance of each proposed component within the SCNOC-Agentic framework.

Keywords: satellite communication; large language model (LLM); network operation and control; network task planning

1. Introduction

AI algorithms, exemplified by large language models (LLMs), are effectively applied across various vertical domains, particularly in communication network applications of space–air–ground–sea integrated networks (SAGSINs). In terrestrial communications, research on communication LLMs [1–4] has explored their integration with 5G networks [5,6], which could facilitate the development of LLMs for satellite-terrestrial networks. Benefiting from the coverage capability of a single communication satellite far exceeding that of ground base stations, satellite communications have emerged as an effective means to meet the growing demands of 6G and Internet of Everything (IoE) applications [7,8]. However, research on LLMs aimed at supporting ubiquitous satellite communication network services remains extremely limited.

In telecommunications, LLMs (e.g., GPT series [9,10], Qwen [11,12] series) have been extensively introduced to address various challenges within network systems. For instance, by designing traffic characterization and adjusting network output structures, these models facilitate a deeper understanding of network traffic [13–15]. Network operation and maintenance are also facilitated by the use of prompt engineering, network design, and copilot services [16–18]. Moreover, communication network security is enhanced by identifying network attacks and generating response plans [19]. Compared with traditional task-specific models, "LLM-powered networks" offer additional capabilities to communication systems. These include achieving intent understanding, altering the interaction methods between networks and operators, and enabling automated network operations. In addition, compared to terrestrial communications, satellite communications rely more heavily on the intentions of operations and maintenance personnel for network management, making applications based on LLMs in satellite communications more focused on network management and control aspects.

Despite the demonstrated capability of LLMs in executing complex step-by-step reasoning to solve problems, recent research in satellite communications has not kept pace with advances in model inference capabilities and agent functionalities. Specifically, there is no established definition or framework on how LLMs can be effectively integrated into satellite communication networks to address practical challenges within this domain, such as satellite networking planning [20–22], network management [23,24], traffic scheduling [25,26], and resource allocation [27–30]. Currently, there is also an absence of evaluation regarding whether LLMs can resolve management and control issues within satellite communication networks through step-by-step reasoning to derive solutions, alongside generating reliable parameters via agent invocation. Taking network construction in satellite communications as an example, when communication tasks for a specific region arise, it becomes necessary to quickly establish a satellite communication network tailored to these requirements, ensuring communication network coverage for that area. However, current practices in the operational management of satellite networks still rely on manually entering task parameters. This approach suffers from two main disadvantages: (i) Firstly, it requires high levels of expertise and reliability from personnel, which cannot guarantee the optimization of the parameters. (ii) Secondly, the networking process is time-consuming and requires several minutes to complete all procedures. Given the potential urgency of satellite communication tasks, such minute-level networking operations could result in missed communication opportunities.

To address these challenges, we propose a specially designed agentic framework for satellite communication operations and management, SCNOC-Agentic, as shown in Figure 1. This framework is designed for the application of LLMs in satellite communication systems, encompassing four critical components and making full use of the capabilities of planning, memory, and tool usage within the agentic method. In the SCNOC-Agentic framework, leveraging the intent parsing capabilities of LLMs, we have achieved a complete closed-loop process from communication requirements to the generation and distribution of network configuration parameters, realizing the goal of rapid and agile response. Looking ahead, we believe that the SCNOC-Agentic architecture can serve as the control center in satellite communication systems, capable of achieving full-process management and control of satellites, terminals, and ground stations.

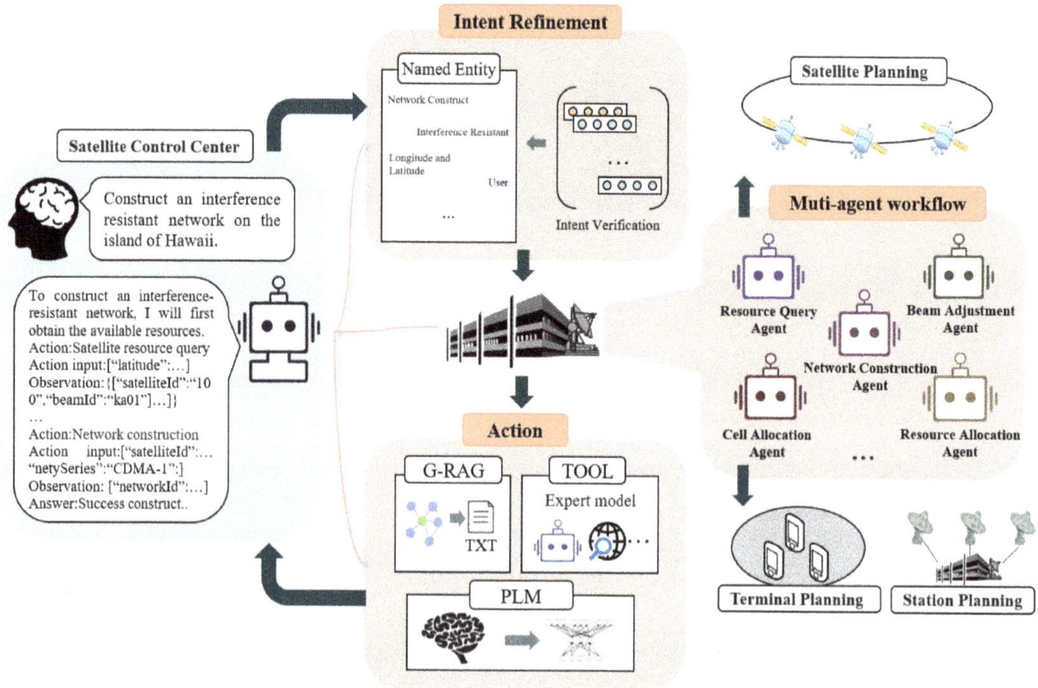

Figure 1. Overview of the proposed SCNOC-Agentic framework for satellite communication operations and management.

The contributions of this work can be summarized as follows:

1. We propose the SCNOC-Agentic framework for the operation and control of satellite communications, firstly marking the implementation of the LLMs agent architecture into this domain, which enhances the capability for rapid and agile response.
2. In SCNOC-Agentic, we introduce four components specifically designed for the characteristics of the satellite communication field: intent refinement, multi-agent workflow, personalized long-term memory, and graph-based retrieval. The performance influence of each component is evaluated through ablation experiments.
3. We design four typical scenarios for applying LLMs in satellite communications: network task planning, carrier and cell optimization, fault analysis of satellites, and satellite management and control. The integration of the SCNOC-Agentic framework within these scenarios is elaborated. Comparative experiments against current state-of-the-art general models and agents show that SCNOC-Agentic achieves superior performance across these four scenarios.

The remainder of this paper is organized as follows. Section 2 describes the related works. Section 3 details the proposed SCNOC-Agentic framework, which comprises four key components. Subsequently, Section 4 defines the scenarios for four typical applications. Section 5 describes the experimental results of SCNOC-Agentic in these four typical scenarios. Finally, Section 6 concludes the paper.

2. Related Work

2.1. AI-Agentic

LLMs serving as foundation models (FM), such as GPT, Qwen, Llama, and others, have demonstrated a powerful capability to perform complex step-by-step reasoning to solve practical problems [31–34]. This advancement is attributed to the in-depth research into AI agentic work. To enable LLMs to achieve more accurate and reliable responses when tackling complex real-world issues, it is generally necessary to integrate these models within specifically designed agentic systems that can access external tools such as database queries, proprietary model invocations, search engines, etc. An agentic system typically comprises three modules: planning [35,36], memory [37], and tool use [38,39], where the LLM plays a crucial role in decision-making and reasoning processes [40]. Furthermore, pioneering efforts to explore agents for general tasks, including AutoGPT and AutoGen [41], are designed to address a broad spectrum of general tasks by dissecting user inquiries into actionable components. These advancements signify a significant leap towards achieving versatile task automation facilitated by sophisticated AI agents.

In addition, researchers have developed a variety of building blocks and design patterns for different applications. This enables LLMs to solve complex problems by generating a series of intermediate steps, where each step addresses a simpler sub-problem, thereby progressively advancing towards the final solution [42]. For example, in [32], the Chain-of-Thought (CoT) prompting method was proposed. In [43], CoT-SC was introduced, which aggregates multiple parallel responses from CoT to produce more accurate responses. The self-refining method proposed in [44] allows for iterative self-reflection aimed at correcting errors made in previous attempts. Furthermore, LLM-Debate, introduced in [45], enables different LLMs to debate among themselves, leveraging diverse perspectives to arrive at superior solutions. In [46], Intelligent Go-Explore was presented, which facilitated the generation and integration of various responses to better explore potential solutions. Additionally, other crucial building blocks for agentic systems include prompting techniques [47,48], reasoning methodologies [34], reflection mechanisms [49], developing new skills for embodied agents [50], external memory and Retrieval-Augmented Generation (RAG) [51], assigning distinct roles to FM modules within the agentic system and enabling their collaboration [52–54], among others. These elements collectively contribute to enhancing the capability and versatility of AI agents to tackle complex tasks in various domains.

The design of LLM-based agent systems needs to be guided by industry-specific problems. Currently, such systems have shown significant potential in applications in domains such as finance [55], education [56], and medical diagnostics [54,57]. However, the field of satellite communications has not kept pace with the advancements in the reasoning capabilities of LLMs and the latest progress in agents' abilities. Current research on applying LLMs to satellite communication tasks remains limited, with insufficient exploration of practical integration frameworks between satellite systems and agentic architectures.

2.2. LLM for Communication

Current research in the communication field has explored applications of LLMs. For instance, in [58,59], potential application directions of LLMs in communications are discussed. Specifically, in [60], Maatouk et al. introduced TeleQnA, a benchmark dataset to evaluate LLMs on telecom knowledge. In [61], Karim et al. proposed SPEC5G, a knowledge-based network protocol Q&A system. In [62], Miao et al. presented NetEval, which focuses on training LLMs for network operations. In [16], Wei et al. introduced an intention-based framework using LLM agents to translate network configurations and ensure network operation. In [63,64], approaches to training specialized LLMs for communications with small, curated text datasets were proposed. In [65], Wang et al. introduced NetAssistant, an LLM-based method for network fault diagnosis that selects appropriate troubleshooting processes based on user intent. In [66], Chen et al. focused on storing historical fault analysis records in a vector database, achieving cloud fault root cause analysis RCACopilot via LLM-based retrieval and analysis. In [67], Yuan et al. proposed UniST, a user mobility modeling method based on network-LLM. Similarly to the use of general LLMs, research on LLMs in the communication field focuses primarily on knowledge-based dialogue and Q&A to provide copilot services for daily maintenance.

Furthermore, research on multi-modal LLMs has also been conducted in the field of communications. For example, the performance of 6G [68], networks [17], and specific issues [14] in the communication domain has been enhanced by incorporating visual information. Simultaneously, multi-modality in the field of communications is not limited to images and text; network traffic also constitutes a significant modality. For example, in [19], Wang et al. proposed ShieldGPT, a feature engineering-based network LLM that utilizes text representations of binary traffic data, which is capable of interpreting slowbody attack behaviors and generating diverse defense instructions. In [69], Wang et al. adopted a unidirectional Mamba architecture to propose an efficient pre-trained traffic classification model, NetMamba. In [17], Wu et al. introduced NetLLM, which designs a multimodal encoder, removes the default language model head, and adds a trainable network head tailored for network tasks. In [8], Sun et al. proposed satellite orchestration in 6G via AI-native frameworks and multi-agent DRL to enhance edge computing efficiency and explained the role of AI agents in achieving cross-domain collaboration. In [70], Rong et al. explored LLMs for intelligent control in 6G TN-NTN to address resource management, interference cancellation, and handover challenges. In [71], Hao et al. proposed ToolkenGPT, which represents each network tool as a new "toolken" and allows the LLM to learn the embedding representations of these toolkens, thus directly generating toolkens to trigger tool invocations.

3. Overview of SCNOC-Agentic

This section introduces the proposed SCNOC-Agentic framework designed for satellite communication operation and control, as shown in Figure 1, elaborating on its four key components.

3.1. Intent Refinement

LLMs inherently possess the ability to understand and parse input intent [16], enabling a more effective interaction with human intentions. However, the intent parsed by LLMs is not always precise. To address this, we introduce an intent refinement (IR) component in SCNOC-Agentic, inspired by the Intent-Driven Networks (IDN) approach [72]. The IR component validates and refines the user input requirements. As shown in Figure 2, using a satellite beam adjustment task as an example, when the input is "Adjust the pointing beam of Ku1 of the TT03 satellite to the Hawaii region," the IR component corrects the parsed intent parameters to their accurate values. This significantly improves the accuracy of intent translation and simplifies traditional intent management and network operation processes.

The IR component encompasses several steps, including intent translation, template matching, and conflict detection. Among these, the intent template matching step facilitates the alignment of parameter templates. It ensures the correctness of both the number of parameters and their keywords by aligning the parameter names generated by the model with the actual parameters issued through intent template matching. The intent translation step converts the intent into parameters that can be understood by satellite communication agents. For instance, "Query the beam coverage of the Hawaii region" would translate "Hawaii region" into specific latitude and longitude coordinates [20 N,

−157 W]. The conflict detection step verifies resource conflicts, such as checking whether a specific satellite beam frequency resource is already occupied when applying for it, thereby preventing conflicts.

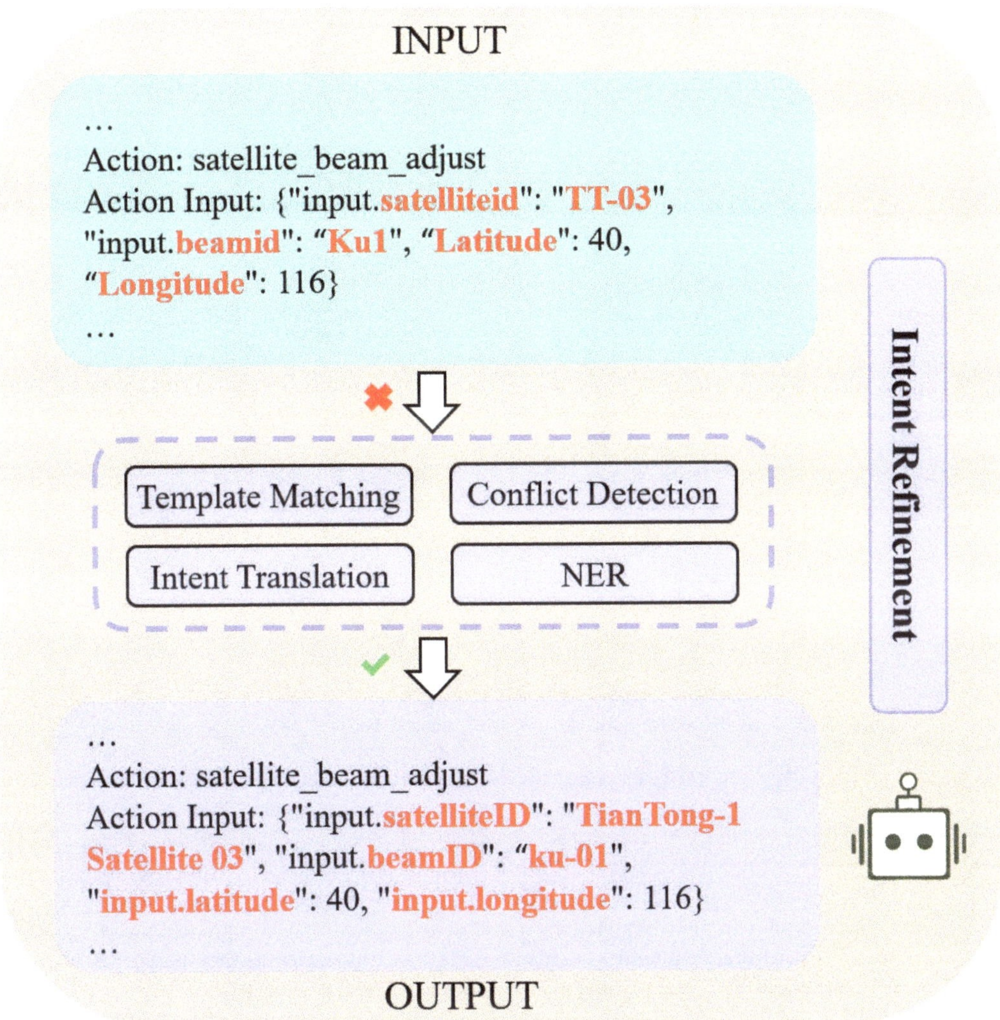

Figure 2. Illustration of the intent refinement (IR) component. For satellite beam adjustment tasks, the IR component proceeds to revise the parsed intent parameters to the correct values.

Notably, we have integrated Named Entity Recognition (NER) [73] into the IR component to handle cases where users employ abbreviated terms instead of full names, such as using "TT03" for "TianTong-2 Satellite 03". By incorporating Named Entity Recognition for matching and correction, the actual system's robustness and usability are significantly enhanced.

3.2. Multi-Agent Workflow

Multi-agent collaboration [52–54] is a crucial design for LLMs to address complex satellite communication tasks. In real-world satellite communication scenarios, LLMs generally need to invoke different agents through multiple steps to resolve various issues separately. For example, when fulfilling the requirements of network task planning, resource querying and allocation are typically performed first, followed by network construction. Therefore, we have introduced a multi-agent workflow (MaW) component, as shown in Figure 3, to establish a fixed workflow for handling complex network management and control tasks.

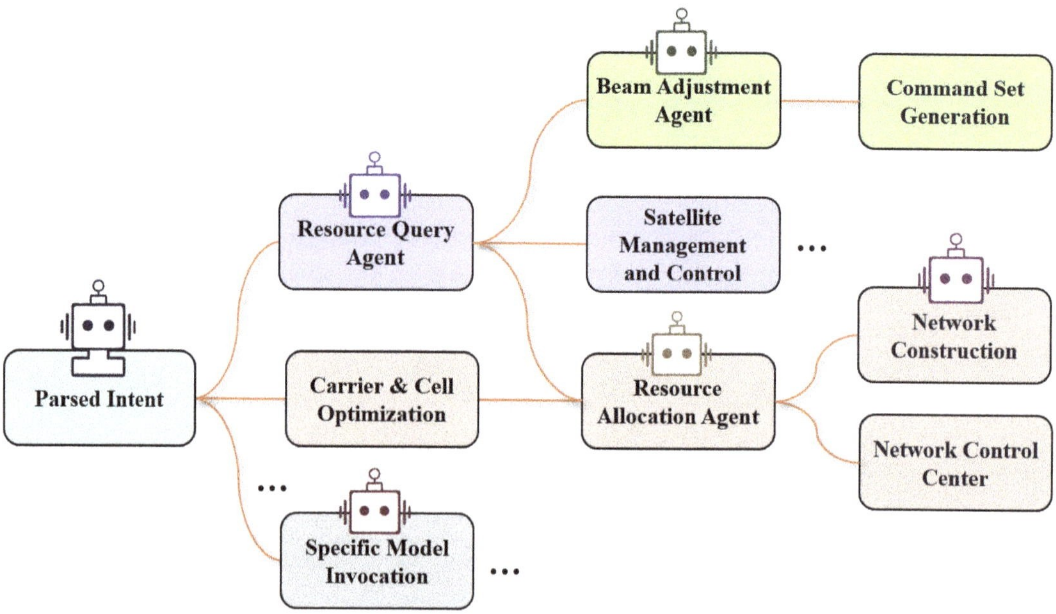

Figure 3. Illustration of the multi-agent workflow (MaW) component in the SCNOC-Agentic framework.

The MaW component is designed to handle functional requests (as opposed to question–answering) in satellite communications. It operates by sequentially executing a series of nodes, making it suitable for automated satellite operations, network formation, and resource optimization. At its core, the MaW component consists of nodes, each of which is an independent agent with a specific function. These agents perform critical operations such as task analysis of satellite communications, resource management, and network control. By connecting the outputs of different agents to subsequent inputs, multiple agents can be linked together, enabling plug-in integration and workflow orchestration.

The multi-agents are called through predefined connection relationships. For example, a beam adjustment agent is restricted from invoking a traffic scheduling model but is allowed to invoke a resource query agent. When predefined tasks arise, they are decomposed into workflows, thereby enhancing the handling of intricate satellite communication operations and control challenges. Furthermore, the MaW incorporates the invocation of specific small-scale models. All agents can leverage specialized models within satellite communications, such as traffic prediction models, resource allocation models, and anti-jamming models, to facilitate decision-making support. The distinction between the MaW and CoT lies in the fact that the CoT method merely provides templates for LLM agent invocations, whereas the MaW imposes stringent constraints on system-level agent invocations. The MaW component significantly improves the accuracy of multistep agent invocations for complex tasks within the SCNOC-Agentic framework.

3.3. Personalized Long-Term Memory

The uniqueness of satellite communications lies in its diverse user base, where different users, such as maritime, vehicular, and airborne users, exhibit distinct behavioral patterns. To address this, we propose a personalized long-term memory (PLM) component that personalizes the recording of usage habits and behaviors for each type of user. During model inference, this component provides context-specific prompts to the model, leveraging advances in long-context LLMs. The PLM component consists of two parts: preference and memory. Preference focuses on the rules in the field of satellite communications, and memory focuses on historical user data.

As shown in Figure 4, the PLM component sets different preferences for different users and stores the memory of previous usages for each user. This design can have notably beneficial effects in the field of satellite communications. For example, maritime users typically rely on a fixed number of satellites and prefer to use the Ku/Ka band. Given that various types of satellite users have relatively stable behaviors and preferences, incorporating and customizing the PLM component allows LLMs to distinguish between different types of users. Consequently, this customization facilitates the provision of valuable insights to assist LLMs in analysis and decision making.

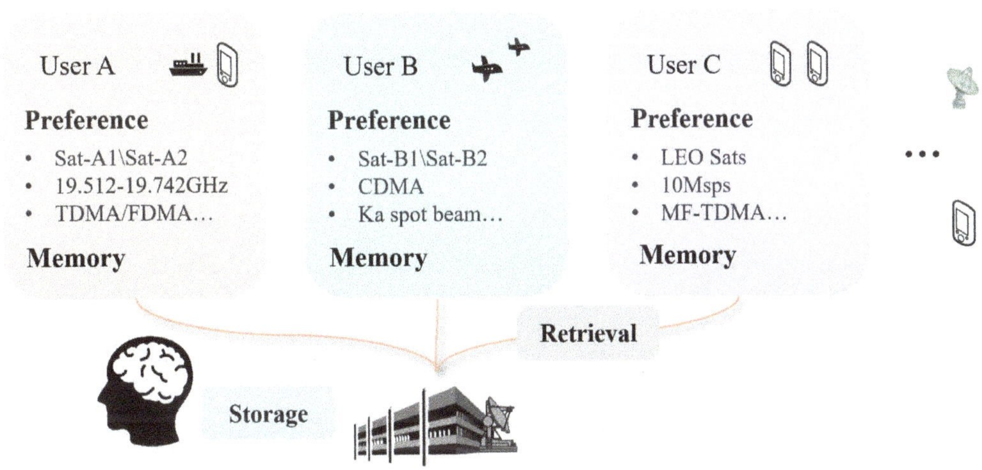

Figure 4. Illustration of the personalized long-term memory (PLM) component in the SCNOC-Agentic framework.

During the reasoning process, the storage and retrieval process of the PLM component operate as follows: (1) Storage: The system automatically identifies and records personalized information provided by users in conversations, such as mobile users or ship-based users. Long-term memory in satellite communications should be mutually isolated among users. (2) Retrieval: Long-term memory can be retrieved either through user prompts or at designated steps within the MaW component. When users request retrieval of relevant long-term memory, SCNOC-Agentic generates the final response by summarizing the personalized information as prior knowledge.

3.4. Graph-Based Retrieval

In light of the complexity of satellite communication systems, to enhance the accuracy of invocation more reliably, we have constructed the graph-based retrieval (GBR) component utilizing the Graph-RAG approach [73]. The abundance of research on Graph-RAG underscores its aptitude for encoding heterogeneous and relational information intrinsic to satellite communication systems, making it highly suitable for practical applications. By integrating real-time data retrieval capabilities, the Graph-RAG system can effectively enhance the functionalities of SCNOC-Agentic.

Furthermore, recognizing the privacy concerns associated with certain satellite communication systems, where satellites, beams, and terminal names are often absent from the pre-training datasets of LLMs, the introduction of the GBR component enables SCNOC-Agentic to identify named entities not encountered during the LLMs' pre-training phase. Specifically, our GBR component is built upon an accumulated corpus of satellite communication documents, including satellite system design documents, user manuals, educational materials on satellite communications, and other relevant texts. During the graph construction phase, LLMs extract entities and relationships from structured texts, visually representing these connections in a graph format. This allows for joint modeling of entities and their relationships during the retrieval process. In the inference phase, the vectors generated based on user queries facilitate the retrieval of related entities and corresponding relationships, indexing knowledge blocks pertinent to queried entities, thereby enhancing the performance of the LLMs.

4. Scene and Problem Formulation

In this section, we provide detailed definitions of agents for four typical scenarios in which SCNOC-Agentic is applied and describe how LLMs can be integrated with these satellite communication scenarios.

4.1. Network Task Planning

With the development of satellite internet, the scale of space-based networks is further expanding, necessitating a higher level of abstraction for traditional satellite resource allocation and network construction methods. In response to this, our project proposes the integration of LLMs based on the SCNOC-Agentic architecture with intention-driven networks (IDN) [72], as illustrated in Figure 5. This integration aims to facilitate intent-based task planning in satellite networks.

The network task planning task for satellite communications based on the SCNOC-Agentic framework is defined as follows: understand the communication intent requirements, then complete the allocation of satellite, earth station, and terminal resources, and finally complete the network construction. For example, if the user inputs 2025.2.10 to ensure a 30M TDMA communication network in Hawaii, SCNOC-Agentic can understand and parse the intent input and perform accurate agent selection and parameter generation step by step. Specifically, under the given conditions of a satellite communication system, SCNOC-Agentic utilizes the proposed PLM component to prompt on the basis of the user's previous task requirements. Currently, it retrieves relevant knowledge bases using the proposed GBR component to generate prompts. Intent parsing and validation are performed using general LLMs, as outlined in Section 3.1. The proposed MaW component formulates the requirements for satellite communication tasks into a multistep execution workflow. In each step, different agents are invoked by the LLM to execute corresponding functions.

As shown in Figure 5, the workflow of the network task planning task is divided into three steps, with each step invoking the corresponding agent for processing: (1) Network resource query: This initial step involves querying the available resources at the specified networking location (e.g., Hawaii region) for network task planning. SCNOC-Agentic generates the necessary parameters and invokes the appropriate agent. The interface definitions utilize standard MCP and Function Call protocols. The input format is defined as *Action input:["input.networktype ": 2, "input.latitude": 40, "input.longitude": 116, "input.time": datetime.now()]* . The query results include available satellites, the area of beams cover, ground stations, and terminals within the region, formatted as *{[satelliteId1, satelliteId2], [beamId1, beamId2], [terminalId1, terminalId2, terminalId3]}*. (2) Network resource allocation: Following the resource query, this step involves selecting resources based on the query results. It consists of two parts: selecting satellites, beams, and transponders and choosing available ground stations and terminals. Subsequently, frequency resources from the selected satellites and beams are allocated (e.g., dividing 30M from a 100M transponder resource while ensuring no time conflicts), similar to satellite communication resource allocation studies [27,28]. (3) Network construction: In the final step, SCNOC-Agentic generates the parameters for network construction and sends them to the network management system to complete the construction of the satellite communication network involving satellites, beams, and user terminals. This structured approach ensures efficient and effective management and control of satellite communication networks, tailored to meet specific user requirements.

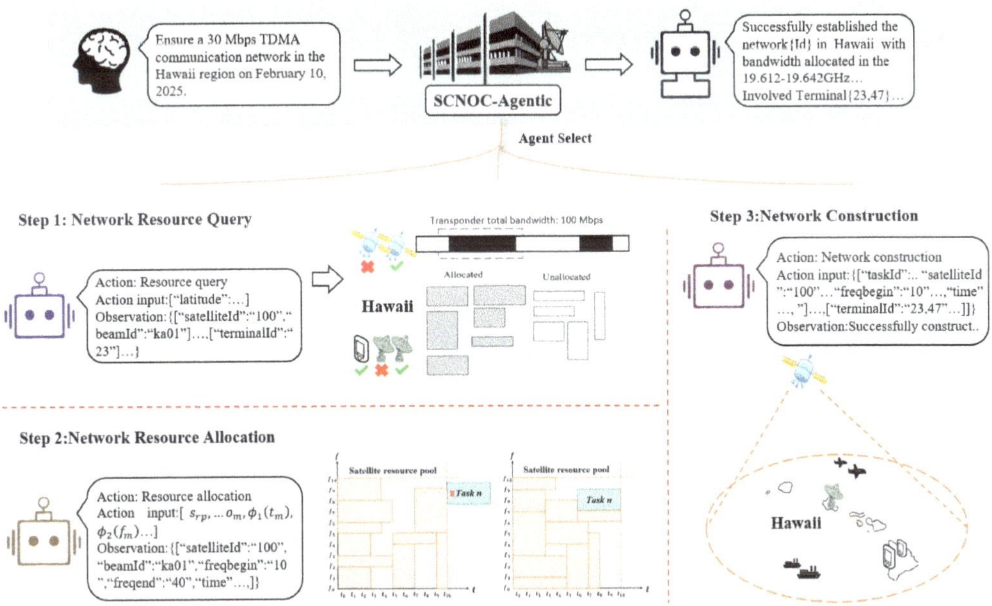

Figure 5. Illustration of the proposed SCNOC-Agentic framework for the network task planning. Upon receiving task requirements from communication assurance users, SCNOC-Agentic will sequentially construct the satellite communication network involving satellites, beams, and user terminals.

Compared to traditional satellite communication network task planning, SCNOC-Agentic-based network task planning offers additional gains by facilitating an end-to-end process from user intent input to network construction. Traditional methods [74] exclusively encompass resource planning in terms of frequency and time slots (Step 2 in Figure 5). In contrast, SCNOC-Agentic network task planning includes satellite and beam selection, along with the generation of networking parameters.

4.2. Carrier and Cell Optimization

The next generation of broadband communication satellites and satellite internet node satellites generally adopt onboard processing systems, equipped with phased array antennas. Optimization of the carrier and cell of phased array antennas is a critical technology for achieving low-latency satellite communications for the ubiquitous Internet of Everything (IoE). Phased array antennas significantly enhance the spatial diversity of the system, enabling time division coverage of multiple dispersed areas on the Earth's surface within certain periods, thereby covering more regions with fewer beam-hopping events. Furthermore, methods such as [29,30] provide communication capabilities that match service demands by adjusting the dwell time of beams in different areas within a period.

In the carrier and cell optimization task, SCNOC-Agentic compensates for link disparities to ensure communication quality by adjusting the Modulation and Coding Scheme (MCS) instead of slot allocation, as shown in Figure 6. This approach fundamentally addresses limited-response problems. By predefining M sets of carrier groups, each representing an MCS, SCNOC-Agentic selects among these to reconcile transmission capability differences caused by varying antenna aperture sizes and output powers of ground-user stations. Different carriers are configured with distinct MCSs to accommodate various types of stations. By adjusting the MCSs, SCNOC-Agentic aims to guarantee access for most station types while maximizing the bandwidth utilization. Moreover, it optimizes link impacts from phased array antennas operating in high-frequency bands such as Ku/Ka and Q/V (e.g., mitigating rain fade). SCNOC-Agentic achieves carrier and cell optimization by generating parameters for MCS adjustments and terminal priority settings, but it cannot calculate the optimal End-to-End Latency (E2EL). Hence, three evaluation metrics are employed: Agent Select Accuracy (AGA), Parameter Generation Accuracy (PGA), and Average End-to-End Latency (E2EL), with E2EL reflecting the average delay of all terminals under specific carrier templates and priority settings. Given that carrier and cell optimization is NP-hard, making the calculation of optimal E2EL unfeasible, comparisons are made using a fixed number N of terminals to evaluate the relative performance.

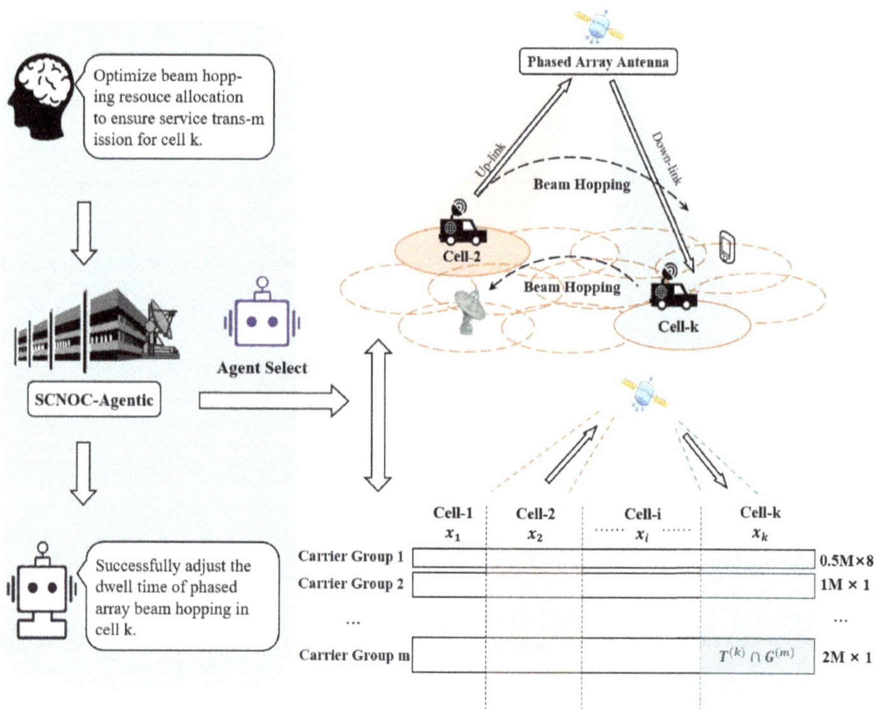

Figure 6. Illustration of the proposed SCNOC-Agentic framework for the carrier & cell optimization task.

4.3. Fault Analysis of Satellite

The fault analysis of a satellite via SCNOC-Agentic essentially constitutes a multi-classification problem, thereby enabling the formulation of tasks as multiple-choice questions. Utilizing SCNOC-Agentic not only facilitates the identification of single-point equipment failures, such as satellite platform and payload malfunctions but also extends to the analysis of network-level and link failures, including inter-satellite link and feeder link failures. This capability is attributed to the SCNOC-Agentic's foundation on LLMs, which enables simultaneous analysis of onboard telemetry data, log data, and parameter data from various devices—capabilities beyond the reach of traditional single-fault analysis models.

We employed a self-developed LEO satellite simulation center to construct experimental scenarios by receiving telemetry data from payload unit boards. This ground-based simulation system is designed as a payload controller for LEO satellite constellations, featuring integrated processing simulations for two satellites. It emulates the operational status of a satellite communication network, comprising two satellites in 500 km nearly polar orbits, and supports both payload remote control and telemetry functions.

Artificially setting the payload fault parameters, we simulated four types of payload anomalies: control board failure (main control unit payload), satellite–ground exchange unit failure, modem board unit failure, and inter-satellite exchange unit failure. A normal payload state was also configured for comparison. For each state, 90 min of telemetry data were collected to establish the experimental dataset, as shown in Table 1. For the fault analysis of the satellite task, the evaluation primarily adopted the top-k recall rate (Hit@k, where k = 1, 3) as the key metric, with the Hit@k measure accuracy by verifying whether the correct fault analysis appeared within the top k predictions.

Table 1. Telemetry information of satellite payload fault status.

Telemetry Status	Fault Payload	Data Size
Normal	-	12,124
Fault state 1	main control unit	5312
Fault state 2	satellite-ground switching unit	5769
Fault state 3	modem board card unit	4793
Fault state 4	inter-satellite switching unit	4902

4.4. Satellite Management and Control

The implementation of satellite management and control tasks based on the SCNOC-Agentic framework involves controlling the satellite platform and its payloads. As shown in Figure 7, first, SCNOC-Agentic interprets the intentions of the control commands from maintenance personnel and performs the correct agent selection. Subsequently, the selected agent translates these intentions into a set of instructions and uploads them to the satellite. Finally, adjustments to the satellite are verified through telemetry data. Typical control command intentions include: "Adjust Ka beam pointing to Hawaii", "Activate inter-satellite laser payload", "Switch X satellite phased array mode to stare", and "Switch X satellite to secure communication mode". Maintenance personnel customize common command intents based on the SCNOC-Agentic architecture. The PLM module provides prompts tailored to different individuals, enabling various methods of satellite operation and control, significantly enhancing the efficiency of management. This task is evaluated through ASA and PGA.

In satellite and network management, smaller scales allow focus on detailed parameter configurations of individual satellites. However, as the satellite internet expands, along with the broadening operational dimensions, managing all parameter details becomes increasingly impractical. Consequently, SCNOC-Agentic represents a higher level abstraction of traditional single-satellite parameter-based management and control, emphasizing intent understanding for management. Furthermore, the proposed SCNOC-Agentic framework is designed to adapt to more complex network configurations in the future, with the continuous development of LLMs enhancing intent-driven capabilities.

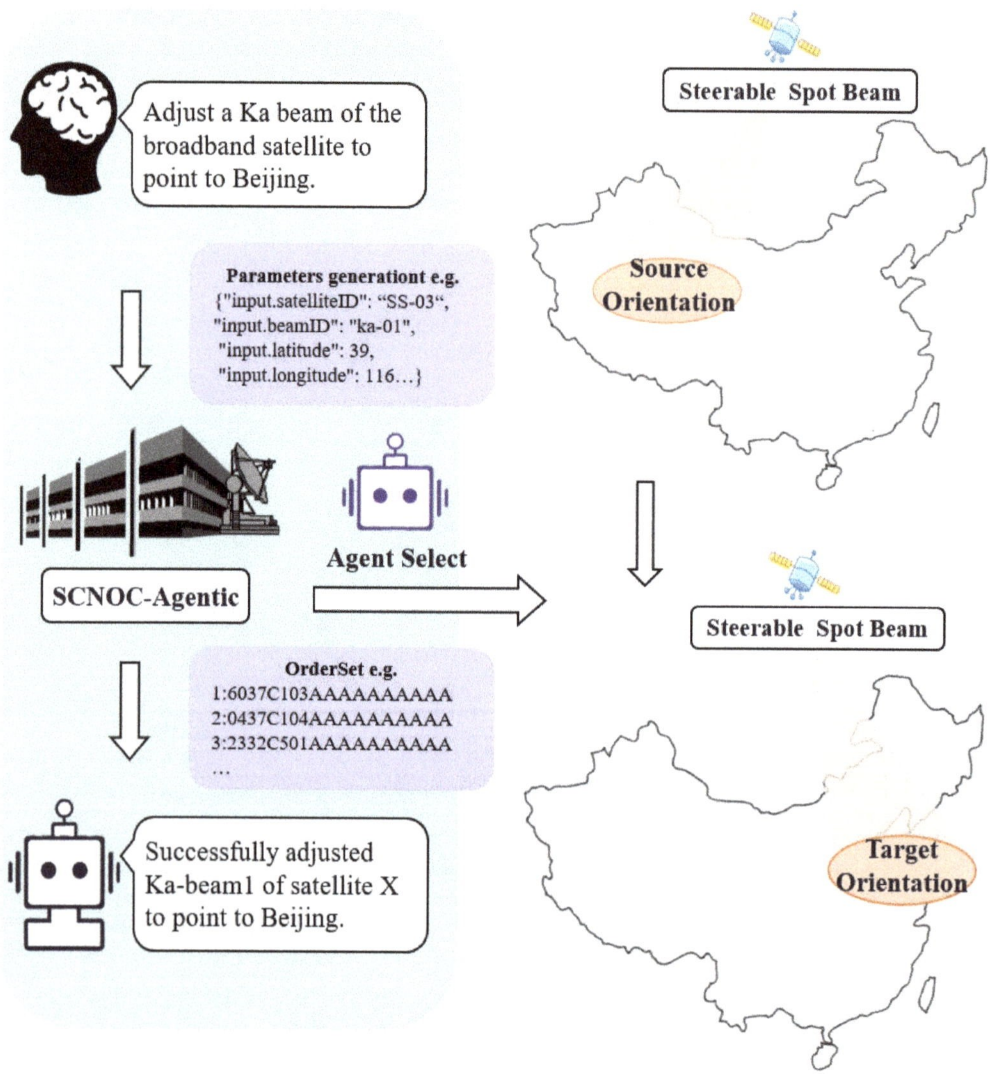

Figure 7. Illustration of the proposed SCNOC-Agentic framework for the satellite management and control task: adjust spot beam pointing according to operators' intent, with parameters automatically generated by SCNOC-Agentic and translated into command sets uploaded to satellites.

5. Experiments

In this section, the proposed SCNOC-Agentic framework is validated on various general-purpose LLMs, including the latest versions of Llama3.3-70B, qwen2.5-70B, GPT-4o, and GPT-3.5-turbo-0125. All models were assessed using zero-shot CoT [32], by default, with the temperature parameter set to 0. Various baseline approaches were used, including zero-shot CoT, which uses the prompt "let us think step by step," and the CoT prompting technique with five examples (CoT5) [33]. Regarding self-consistency (SC) [43], the accuracy was determined based on the majority decision of 20 generations. All baselines completed experiments on four typical satellite communication operation and control application scenarios presented in Section 4, along with ablation studies on components. Moreover, on the basis of the experimental results, we discuss how LLMs can be practically applied to satellite communication operation and control systems.

5.1. Main Results

We evaluated the performance of the SCNOC-Agentic framework under various baselines. Table 2 presents the comparative results of the proposed SCNOC-Agentic framework with different general LLMs and prompting methods on network task planning and carrier and cell optimization. The best two results are highlighted, with the best result bolded, underlined, and marked in red, while the second-best is underlined and marked in blue.

Table 2. Comparative experiments on network task planning and carrier and cell optimization tasks.

Model	Network Task Planning			Carrier & Cell Optimization		
	ASA	PGA	AoP	ASA	PGA	E2EL (ms)
General LLMs (default with zero-shot CoT)						
GPT-4o	0.6736	0.2046	0.835	0.7511	0.5146	302
GPT-4o+CoT5	0.7321	0.2339	0.863	0.8164	0.5943	286
GPT-4o+SC	0.8101	0.3898	0.903	0.8449	0.7177	290
GPT3.5	0.6171	0.1754	0.835	0.7968	0.4577	343
GPT3.5+CoT5	0.6853	0.1968	0.846	0.7970	0.4962	318
GPT3.5+SC	0.7594	0.3294	0.887	0.8132	0.6037	294
Llama-3.3-70B-Instruct (default with zero-shot CoT)						
RawAgent	0.6035	0.1559	0.812	0.6697	0.4608	346
RawAgent+CoT5	0.7029	0.2202	0.840	0.7784	0.5120	280
RawAgent+SC	0.7769	0.3664	0.863	0.7943	0.5753	285
SCNOC-Agentic	0.7029	0.3216	0.887	0.8101	0.6575	312
SCNOC+CoT5	0.7867	0.3762	0.927	0.8354	0.7335	**<u>260</u>**
SCNOC+SC	**<u>0.8491</u>**	**<u>0.4619</u>**	**<u>0.945</u>**	0.8670	<u>0.7683</u>	277
Qwen-2.5-70B-Instruct (default with zero-shot CoT)						
RawAgent	0.6346	0.1617	0.829	0.6840	0.4703	314
RawAgent+CoT5	0.7243	0.2358	0.843	0.7689	0.5151	280
RawAgent+SC	0.7638	0.3586	0.872	0.8006	0.6006	274
SCNOC-Agentic	0.7185	0.2904	0.854	0.8575	0.6797	298
SCNOC+CoT5	0.7828	0.3411	0.921	<u>0.8829</u>	0.7145	273
SCNOC+SC	<u>0.8276</u>	<u>0.4249</u>	<u>0.939</u>	**<u>0.9082</u>**	**<u>0.7841</u>**	<u>264</u>

As observed in Table 2, the SCNOC-Agentic-based framework achieved optimal and suboptimal results across various tasks. In the network task planning task, Llama-3.3-

70B+SCNOC-Agentic+SC demonstrated the best performance in all metrics. The utilization of the SCNOC-Agentic architecture resulted in improvements of 7.2% and 9.5% in the AGA (0.8491 vs. 0.7769) and PGA (0.4619 vs. 0.3664) metrics, respectively. Compared to the baseline using CoT by default, the SCNOC-Agentic architecture still yielded enhancements of 9.9% and 16.7% in the AGA (0.7029 vs. 0.6035) and PGA (0.3216 vs. 0.1559) metrics, respectively. Similarly, in the carrier and cell optimization task, the SCNOC-Agentic architecture outperformed the Llama and Qwen models under RawAgent, delivering superior AGA and PGA performance, as well as reduced channel latency. These results collectively substantiate the efficacy of the SCNOC-Agentic framework.

Meanwhile, although we observe that the proposed SCNOC-Agentic has achieved significant improvements over state-of-the-art general LLMs and the GPT-4o model, we believe that the current agents framework still falls short of practical application in satellite communication systems, particularly in network task planning and carrier and cell optimization tasks. For practical use, it is essential to ensure that the performance of AGA and PGA is as high as possible, with parameter generation being as reliable as possible. Therefore, more robust foundational models and engineering tricks or, alternatively, the design of a superior agent architecture are required.

Similarly, as shown in Table 3, we conducted comparisons using the same baselines and LLMs on the tasks of fault analysis of satellites and satellite management and control to fully evaluate the proposed SCNOC-Agentic framework. It can be observed that the proposed framework achieved the best and second-best results in both tasks. Particularly in the satellite management and control task, Qwen2.5-70B+SCNOC-Agentic+SC demonstrated exceptional performance with an ASA metric of 0.9865 and a PGA metric of 0.8667. This indicates that LLMs based on SCNOC-Agentic are well-suited for relatively simpler satellite management and control tasks. It is evident that SCNOC-Agentic outperforms the state-of-the-art GPT-4o and RawAgent in the same zero-shot CoT, CoT5, and SC settings, with significant improvements in both ASA and PGA metrics. This enhancement is attributed to the intent resolution capabilities provided by various components within SCNOC-Agentic. The MaW component improves the agent's ability to accurately invoke functions, while the PLM and IR components significantly enhance parameter generation capabilities. The ablation experiments for each component will be discussed in Section 5.2.

In addition, we observed that the performance of Hit@1 in the fault analysis of the satellite task was merely 25.9%, which is far from satisfactory for practical applications. However, if the Hit@5 accuracy metric is adopted, the optimal result achieved by SCNOC-Agentic reached 85.2%. Therefore, for the deployment of the fault analysis of satellite task in real-world satellite communication systems, it is strongly recommended to utilize the best multiple recommendations provided by LLMs, with final confirmation by the operation and maintenance personnel. Finally, from a qualitative perspective, applying the agent capabilities of LLMs to the fault analysis of satellite tasks proved to be less effective compared to satellite management and control tasks. This discrepancy arises because fault analysis involves model-based reasoning and decision-making, whereas satellite management and control primarily concern the generation of interface parameters. Our findings, as outlined in Tables 2 and 3, indicate that although current SCNOC-Agentic systems can effectively replace satellite management and control systems in practical applications, substantial advancements in LLM agent frameworks are still required for tasks involving computational reasoning in satellite communications before these systems can be confidently implemented in operational environments.

Table 3. Comparative experiments on fault analysis of satellite and satellite management and control tasks.

Model	Fault Analysis of Satellite			Satellite Management and Control	
	Hit@1	Hit@3	Hit@5	ASA	PGA
General LLMs (default with zero-shot CoT)					
GPT-4o	0.190	0.315	0.725	0.8908	0.5285
GPT-4o+CoT5	0.214	0.360	0.796	0.9430	0.5761
GPT-4o+SC	0.246	0.404	0.818	0.9791	0.6238
GPT3.5	0.164	0.287	0.683	0.7952	0.4809
GPT3.5+CoT5	0.182	0.373	0.742	0.8821	0.5285
GPT3.5+SC	0.227	0.388	0.794	0.9256	0.5761
Llama-3.3-70B-Instruct (default with zero-shot CoT)					
RawAgent	0.171	0.269	0.693	0.8121	0.4285
RawAgent+CoT5	0.208	0.349	0.760	0.9169	0.5761
RawAgent+SC	0.203	0.370	0.781	0.9343	0.6714
SCNOC-Agentic	0.185	0.341	0.755	0.9430	0.7238
SCNOC+CoT5	0.224	0.404	0.796	0.9517	0.8714
SCNOC+SC	0.253	**0.417**	0.827	0.9691	**0.9142**
Qwen-2.5-70B-Instruct (default with zero-shot CoT)					
RawAgent	0.177	0.272	0.684	0.8213	0.4285
RawAgent+CoT5	0.190	0.343	0.770	0.9082	0.5238
RawAgent+SC	0.216	0.362	0.783	0.9430	0.6714
SCNOC-Agentic	0.198	0.354	0.742	0.9704	0.7714
SCNOC+CoT5	0.238	0.394	0.805	0.9778	0.8190
SCNOC+SC	0.259	0.410	0.852	0.9865	0.8667

In this section, we conducted experiments to evaluate the importance of each component of SCNOC-Agentic under various scenarios, with the results presented in Tables 4 and 5. The experiments were conducted under the default setting of zero-shot CoT, where 'w/o' is an abbreviation for 'without', such as 'w/o PLM' indicating the exclusion of the PLM component within the SCNOC-Agentic framework while still utilizing the other three components. Arrows represent the performance gain(↑) or drop(↓) from component removal compared to baseline, as well as the percentage change compared to the baseline. The results of the ablation study demonstrate that all components contribute positively, highlighting the utility of each component in the design of SCNOC-Agentic. We observed that the removal of the IR component leads to a decline in PGA performance but has almost no impact on ASA. This is in line with our expectations. The reason is that the IR component enhances the accuracy of parameter selection by performing intent correction using named entity recognition. The role of the IR component is merely to correct parameters, thus it does not affect the choices made by the agent.

Table 4. Ablation study on the importance of components on network task planning and carrier and cell optimization tasks.

Model	Network Task Planning			Carrier & Cell Optimization		
	ASA	PGA	Ro	ASA	PGA	E2EL (ms)
Llama-3.3-70B-Instruct (default with zero-shot CoT)						
w/o IR	-	0.2317	0.834	-	0.5082(↓)	330(↓)
w/o WAG	0.6629	0.2472	0.841	0.7991	0.5746	328
w/o PLM	0.6450	0.2191	0.825	0.7825(↓)	0.5284	322
w/o GBR	0.6912	0.2904	0.879	0.8029	0.6386	293
SCNOC-Agentic	0.7029	0.3216	0.887	0.8101	0.6575	321
Qwen-2.5-70B-Instruct (default with zero-shot CoT)						
w/o IR	-	0.2241	0.849	-	0.5664	293
w/o WAG	0.6891	0.2598	0.831	0.7943	0.5341(↓)	346(↓)
w/o PLM	0.6639	0.2205	0.847	0.7879(↓)	0.5468	304
w/o GBR	0.7084	0.2745	0.862	0.8354	0.6802(↑)	284
SCNOC-Agentic	0.7185	0.2904	0.854	0.8575	0.6797	298

Table 5. Ablation study on the importance of components on fault analysis in satellite and satellite management and control tasks.

Model	Fault Analysis of Satellite			Satellite Management and Control	
	Hit@1	Hit@3	Hit@5	ASA	PGA
Llama-3.3-70B-Instruct (default with zero-shot CoT)					
w/o IR	0.179	0.303	0.693	-	0.5809(**↓19.7%**)
w/o WAG	0.184	0.315	0.719	0.8721(**↓7.5%**)	0.6455(↓10.8%)
w/o PLM	0.177	0.297	0.707	0.8847(↓6.4%)	0.6109(↓15.6%)
w/o GBR	0.190	0.338	0.736	0.9343(↓0.9%)	0.6761(↓6.6%)
SCNOC-Agentic	0.185	0.341	0.755	0.9430	0.7238
Qwen-2.5-70B-Instruct (default with zero-shot CoT)					
w/o IR	0.180	0.301	0.710	-	0.6092(**↓21.0%**)
w/o WAG	0.187	0.318	0.706	0.9095(↓6.2%)	0.6761(↓12.4%)
w/o PLM	0.183	0.306	0.692	0.8908(**↓8.2%**)	0.6285(↓18.5%)
w/o GBR	0.195	0.320	0.714	0.9617(↓0.8%)	0.7238(↓6.2%)
SCNOC-Agentic	0.198	0.354	0.742	0.9704	0.7714

5.2. Components Ablation Study

Additionally, as shown in Figure 8, we calculated the performance drop in each component across four tasks. The removal of WAG or PLM components was observed to lead to a significant performance drop, indicating that incorporating WAG and PLM components is crucial for practical tasks. However, it was noted that eliminating the GBR component results in only a minor performance drop, suggesting that the introduction of satellite communication-related knowledge does not significantly enhance the performance of the LLM agents. However, the inclusion of the GBR component remains meaningful; for instance, when operators ask question-and-answer type questions (e.g., how many satellites are there in the TianTong constellation?), the GBR component can assist the LLM in generating more definitive responses by retrieving information from the knowledge base.

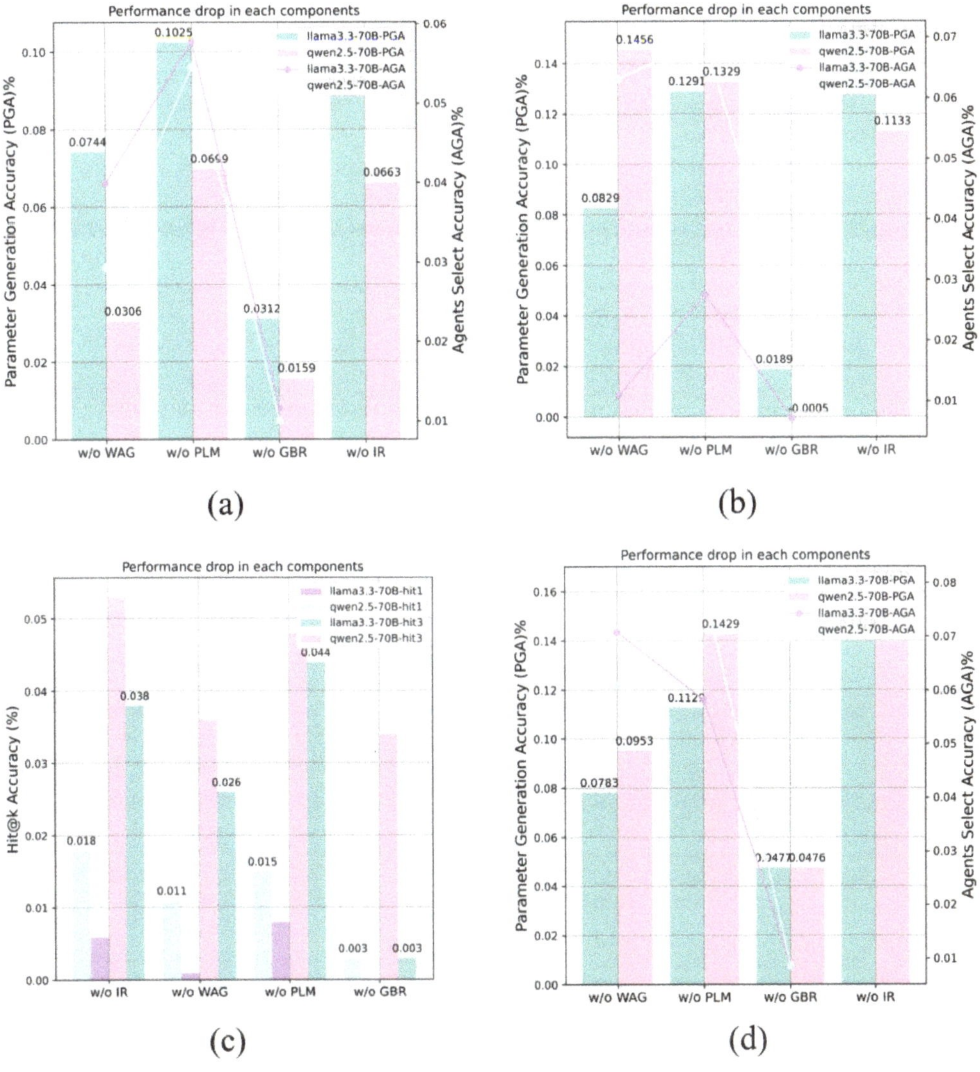

Figure 8. Ablation Study of performance drop of four scenarios in each component. (**a**) Network task planning; (**b**) Carrier & cell optimization; (**c**) Fault analysis of satellite; (**d**) Satellite management and control.

5.3. Discussion and Analysis

Compared with traditional algorithms: Compared to traditional feature learning models, the introduction of LLMs in the field of satellite communications brings not only performance improvements but also additional application scenarios, achieving objectives that small models cannot achieve. Meanwhile, the integration of LLMs enables comprehensive state management across the entire satellite network system, whereas smaller models are typically optimized for specific domains or tasks without considering the overall system comprehensively. Unlike small task-specific models, LLMs possess the capability to orchestrate tasks, allowing an end-to-end process from initial communication requirements to decision-making outcomes, thus significantly enhancing the efficiency of satellite communication operations and control. Furthermore, from the perspective of generalization, versatile LLMs can better adapt and transfer across various tasks and scenarios in satellite communications, a feat unachievable by traditional models.

Additionally, we propose that the architecture "LLMs + specific small models" represents the optimal approach for integrating LLMs with existing satellite communication systems and algorithms. This combination brings new benefits, rather than replacing traditional small models outright. Given that LLMs have clear limitations and may not be as precise or efficient in specific areas or tasks (such as traffic scheduling and resource allocation), they can invoke smaller models to handle these detailed tasks. For instance, during communication network formation, LLMs can first call upon smaller models for resource allocation, then proceed to issue network formation tasks once idle resources are identified. This synergy maximizes the strengths of both LLMs and small specialized models.

Benefits of SCNOC-Agentic: We posit that incorporation of SCNOC-Agentic into satellite communications operations and maintenance control yields significant enhancements, specifically: (1) SCNOC-Agentic advances typical communication scenario matching and intelligent parsing of communication intents based on intent-based methodologies. This elevates traditional manual conversion of communication requirements or the human designation of network parameters to an intelligent automated task analysis driven by LLMs and intent comprehension, transitioning from a "human-in-the-loop" to a "human-on-the-side, fully automated" operational paradigm. (2) After integration, SCNOC-Agentic streamlines information access for communication maintenance personnel, with LLMs acting as the "next-generation operating system" central to inputting requirements. Maintenance personnel interact solely with the Copilots system, issuing communication intents via LLMs to orchestrate comprehensive satellite communication system responses, such as querying available bandwidth over Hawaii, adjusting beam directions, or tallying beams covering the region. (3) Functioning as the control nucleus, LLMs adapt to the evolving state of satellite communication networks, coordinating subsystems and specialized "mini-models" for end-to-end management. This realizes a dynamic resource adjustment approach driven by service demand and optimized for user experience, effectively balancing resource allocation across the entire domain and optimizing resource utilization to enhance both satellite resource efficiency and service assurance capabilities. With the continuous advancement of LLMs, it holds the potential to become a critical technological pathway towards achieving autonomy in satellite networks.

In addition, each of the four application scenarios mentioned above is complex and deserves separate study. For example, constructing a satellite network involves various constellations, satellite types, antennas, and transponders. This paper only presents the simplest possible examples of how the LLMs' agentic architecture might be applied in the field of satellite communications. Subsequent research needs to be developed based on different satellite communication systems, such as GEO and LEO, spot beam and multi-beam, etc. The details of satellite networks vary across different systems, making this a complex issue in itself. It is hoped that future research will explore these applications in greater depth.

6. Conclusions

In this paper, we introduced the SCNOC-Agentic framework, a large language model architecture specifically designed for the management and control of satellite communication systems. Within SCNOC-Agentic, we designed four components tailored to the characteristics of satellite communications: intent refinement, multi-agent workflow, personalized long-term memory, and graph-based retrieval. Subsequently, we conducted ablation experiments to evaluate the performance impact of each individual component. Furthermore, we defined four typical scenarios for the application of LLMs within satellite communications: network task planning, carrier and cell optimization, fault analysis of satellites, and satellite management and control. We provided detailed explanations of how SCNOC-Agentic could be applied in these scenarios, demonstrating superior performance compared to current state-of-the-art general models and agents.

It should be noted that the definable agent scenarios within the satellite communication domain extended beyond the four outlined in this study. Each of the four application scenarios mentioned above is complex and deserves a separate study. For example, constructing a satellite network involves various constellations, satellite types, antennas, and transponders. This paper presents only the simplest possible examples of how the LLMs agentic architecture might be applied in the field of satellite communications. Potential research areas within satellite communication management and control include user service assurance, fault analysis, traffic analysis, and situation generation, among others. For example, future work may focus on how to enable LLMs to understand communication protocols and assist in network management. In addition, the proposed SCNOC-agentic system is still limited by the workflow and prompting methods. Therefore, autonomous planning of multi-agents is a worthy research direction in the future.

In the future, subsequent research should continue to be conducted according to different satellite communication architectures, such as GEO and LEO, spot beam and multi-beam, etc. The details of satellite networks vary across different systems, making this a complex issue in itself. It is hoped that future research will explore these applications in greater depth.

Author Contributions: Conceptualization, W.S. and C.S.; methodology, W.S. and Y.Z.; software, W.S. and Z.Y.; formal analysis, W.S.; writing—original draft preparation, W.S.; writing—review and editing, C.S. and Z.K.; project administration, C.S.; funding acquisition, C.S. All authors have read and agreed to the published version of the manuscript.

Funding: This research received no external funding.

Data Availability Statement: Access to the experimental data presented in this article can be obtained by contacting the corresponding author.

Conflicts of Interest: The authors declare no conflicts of interest.

References

1. Gao, H.; Cao, R.; Xu, W.; Yuan, C.; Chen, H.H. Space-Air-Ground Integrated Networks with Task-Driven Connected Intelligence. *IEEE Wirel. Commun.* **2024**, *32*, 254–261. [CrossRef]
2. Zhang, R.; Du, H.; Niyato, D.; Kang, J.; Xiong, Z.; Jamalipour, A.; Zhang, P.; Kim, D.I. Generative AI for Space-Air-Ground Integrated Networks. *IEEE Wirel. Commun.* **2024**, *31*, 10–20. [CrossRef]
3. Tang, J.; Tang, F.; Long, S.; Zhao, M.; Kato, N. Utilizing Large Language Models for Advanced Optimization and Intelligent Management in Space-Air-Ground Integrated Networks. *IEEE Netw.* **2024**, 1. [CrossRef]
4. Cui, H.; Zhang, J.; Geng, Y.; Xiao, Z.; Sun, T.; Zhang, N.; Liu, J.; Wu, Q.; Cao, X. Space-air-ground integrated network (SAGIN) for 6G: Requirements, architecture and challenges. *China Commun.* **2022**, *19*, 90–108. [CrossRef]
5. Kan, K.B.; Mun, H.; Cao, G.; Lee, Y. Mobile-LLaMA: Instruction Fine-Tuning Open-Source LLM for Network Analysis in 5G Networks. *IEEE Netw.* **2024**, *38*, 76–83. [CrossRef]
6. Zhou, H.; Hu, C.; Yuan, Y.; Cui, Y.; Jin, Y.; Chen, C.; Wu, H.; Yuan, D.; Jiang, L.; Wu, D.; et al. Large language model (llm) for telecommunications: A comprehensive survey on principles, key techniques, and opportunities. *arXiv* **2024**, arXiv:2405.10825. [CrossRef]
7. Miraz, M.H.; Ali, M.; Excell, P.S.; Picking, R. A review on Internet of Things (IoT), Internet of everything (IoE) and Internet of nano things (IoNT). In Proceedings of the 2015 Internet Technologies and Applications (ITA), Wrexham, UK, 8–11 September 2015; pp. 219–224. [CrossRef]

8. Sun, G.; Ayepah-Mensah, D.; Maale, G.T.; Omer, M.B.; Kuadey, N.A.; Kwantwi, T.; Liu, Y.; Liu, G. Toward AI-Native Task Orchestration for Collaborative Computing in SAGSINs. *IEEE Commun. Mag.* **2025**, 1–7. [CrossRef]
9. Achiam, J.; Adler, S.; Agarwal, S.; Ahmad, L.; Akkaya, I.; Aleman, F.L.; Almeida, D.; Altenschmidt, J.; Altman, S.; Anadkat, S.; et al. Gpt-4 technical report. *arXiv* **2023**, arXiv:2303.08774. [CrossRef]
10. Floridi, L.; Chiriatti, M. GPT-3: Its nature, scope, limits, and consequences. *Minds Mach.* **2020**, *30*, 681–694. [CrossRef]
11. Bai, J.; Bai, S.; Chu, Y.; Cui, Z.; Dang, K.; Deng, X.; Fan, Y.; Ge, W.; Han, Y.; Huang, F.; et al. Qwen technical report. *arXiv* **2023**, arXiv:2309.16609. [CrossRef]
12. Yang, A.; Yang, B.; Zhang, B.; Hui, B.; Zheng, B.; Yu, B.; Li, C.; Liu, D.; Huang, F.; Wei, H.; et al. Qwen2.5 technical report. *arXiv* **2024**, arXiv:2412.15115. [CrossRef]
13. Zhang, S.; Fu, D.; Liang, W.; Zhang, Z.; Yu, B.; Cai, P.; Yao, B. Trafficgpt: Viewing, processing and interacting with traffic foundation models. *Transp. Policy* **2024**, *150*, 95–105. [CrossRef]
14. de Zarzà, I.; de Curtò, J.; Roig, G.; Calafate, C.T. LLM multimodal traffic accident forecasting. *Sensors* **2023**, *23*, 9225. [CrossRef]
15. Liu, C.; Yang, S.; Xu, Q.; Li, Z.; Long, C.; Li, Z.; Zhao, R. Spatial-temporal large language model for traffic prediction. *arXiv* **2024**, arXiv:2401.10134. [CrossRef]
16. Wei, Y.; Xie, X.; Zuo, Y.; Hu, T.; Chen, X.; Chi, K.; Cui, Y. Leveraging LLM Agents for Translating Network Configurations. *arXiv* **2025**, arXiv:2501.08760. [CrossRef]
17. Wu, D.; Wang, X.; Qiao, Y.; Wang, Z.; Jiang, J.; Cui, S.; Wang, F. Netllm: Adapting large language models for networking. In Proceedings of the ACM SIGCOMM 2024 Conference, Sydney, Australia, 4–8 August 2024; pp. 661–678. [CrossRef]
18. He, Z.; Gottipati, A.; Qiu, L.; Yan, F.Y.; Luo, X.; Xu, K.; Yang, Y. LLM-ABR: Designing Adaptive Bitrate Algorithms via Large Language Models. *arXiv* **2024**, arXiv:2404.01617. [CrossRef]
19. Wang, T.; Xie, X.; Zhang, L.; Wang, C.; Zhang, L.; Cui, Y. Shieldgpt: An llm-based framework for ddos mitigation. In Proceedings of the 8th Asia-Pacific Workshop on Networking, Sydney, Australia, 3–4 August 2024; pp. 108–114. [CrossRef]
20. Bonnet, J.; Gleizes, M.P.; Kaddoum, E.; Rainjonneau, S.; Flandin, G. Multi-satellite mission planning using a self-adaptive multi-agent system. In Proceedings of the 2015 IEEE 9th International Conference on Self-adaptive and Self-organizing Systems, IEEE, Cambridge, MA, USA, 21–25 September 2015; pp. 11–20. [CrossRef]
21. Hilton, S.; Thangavel, K.; Gardi, A.; Sabatini, R. Intelligent mission planning for autonomous distributed satellite systems. *Acta Astronaut.* **2024**, *225*, 857–869. [CrossRef]
22. Luis, J.J.G.; Crawley, E.; Cameron, B. Applicability and challenges of deep reinforcement learning for satellite frequency plan design. In Proceedings of the 2021 IEEE Aerospace Conference (50100), IEEE, Virtual, 6–13 March 2021; pp. 1–11. [CrossRef]
23. Ma, T.; Qian, B.; Qin, X.; Liu, X.; Zhou, H.; Zhao, L. Satellite-terrestrial integrated 6G: An ultra-dense LEO networking management architecture. *IEEE Wirel. Commun.* **2022**, *31*, 62–69. [CrossRef]
24. Chen, L.; Tang, F.; Li, X.; Liu, J.; Zhu, Y.; Yu, J. Adaptive Network Management Service Based on Control Relation Graph for Software-Defined LEO Satellite Networks in 6G. *IEEE Trans. Serv. Comput.* **2024**, *17*, 3122–3139. [CrossRef]
25. Huang, Y.; Jiang, X.; Chen, S.; Yang, F.; Yang, J. Pheromone incentivized intelligent multipath traffic scheduling approach for LEO satellite networks. *IEEE Trans. Wirel. Commun.* **2022**, *21*, 5889–5902. [CrossRef]
26. Tao, B.; Masood, M.; Gupta, I.; Vasisht, D. Transmitting, fast and slow: Scheduling satellite traffic through space and time. In Proceedings of the 29th Annual International Conference on Mobile Computing and Networking, Madrid, Spain, 2–6 October 2023; pp. 1–15. [CrossRef]
27. Fu, S.; Gao, J.; Zhao, L. Collaborative multi-resource allocation in terrestrial-satellite network towards 6G. *IEEE Trans. Wirel. Commun.* **2021**, *20*, 7057–7071. [CrossRef]
28. Guerster, M.; Luis, J.J.G.; Crawley, E.; Cameron, B. Problem representation of dynamic resource allocation for flexible high throughput satellities. In Proceedings of the 2019 IEEE Aerospace Conference, Big Sky, MT, USA, 2–9 March 2019; IEEE: Piscataway, NJ, USA, 2019; pp. 1–8. [CrossRef]
29. Sun, G.; Ayepah-Mensah, D.; Boateng, G.O.; Kuadey, N.A.; Omer, M.B.; Liu, G. Holistic roadmap of trends in radio access network slicing: A survey. *IEEE Commun. Mag.* **2023**, *61*, 118–124. [CrossRef]
30. Wang, E.; Li, H.; Zhang, S. Load balancing based on cache resource allocation in satellite networks. *IEEE Access* **2019**, *7*, 56864–56879. [CrossRef]
31. Zhang, Z.; Zhang, A.; Li, M.; Smola, A. Automatic chain of thought prompting in large language models. *arXiv* **2022**, arXiv:2210.03493. [CrossRef]
32. Wei, J.; Wang, X.; Schuurmans, D.; Bosma, M.; ichter, b.; Xia, F.; Chi, E.; Le, Q.V.; Zhou, D. Chain-of-Thought Prompting Elicits Reasoning in Large Language Models. In Proceedings of the Advances in Neural Information Processing Systems, New Orleans, LA, USA, 28 November–9 December 2022; Koyejo, S., Mohamed, S., Agarwal, A., Belgrave, D., Cho, K., Oh, A., Eds.; Curran Associates, Inc.: New York, NY, USA, 2022; Volume 35, pp. 24824–24837.
33. Chu, Z.; Chen, J.; Chen, Q.; Yu, W.; He, T.; Wang, H.; Peng, W.; Liu, M.; Qin, B.; Liu, T. A survey of chain of thought reasoning: Advances, frontiers and future. *arXiv* **2023**, arXiv:2309.15402. [CrossRef]

34. Yuan, L.; Cui, G.; Wang, H.; Ding, N.; Wang, X.; Deng, J.; Shan, B.; Chen, H.; Xie, R.; Lin, Y.; et al. Advancing llm reasoning generalists with preference trees. *arXiv* **2024**, arXiv:2404.02078. [CrossRef]
35. Yao, S.; Zhao, J.; Yu, D.; Du, N.; Shafran, I.; Narasimhan, K.; Cao, Y. React: Synergizing reasoning and acting in language models. *arXiv* **2022**, arXiv:2210.03629.
36. Shinn, N.; Cassano, F.; Gopinath, A.; Narasimhan, K.; Yao, S. Reflexion: Language agents with verbal reinforcement learning. In Proceedings of the Advances in Neural Information Processing Systems, New Orleans, LA, USA, 10–16 December 2024; Volume 36.
37. Zhong, W.; Guo, L.; Gao, Q.; Ye, H.; Wang, Y. Memorybank: Enhancing large language models with long-term memory. In Proceedings of the AAAI Conference on Artificial Intelligence, Vancouver, BC, Canada, 20–27 February 2024; Volume 38, pp. 19724–19731. [CrossRef]
38. Schick, T.; Dwivedi-Yu, J.; Dessì, R.; Raileanu, R.; Lomeli, M.; Hambro, E.; Zettlemoyer, L.; Cancedda, N.; Scialom, T. Toolformer: Language models can teach themselves to use tools. *Adv. Neural Inf. Process. Syst.* **2023**, *36*, 68539–68551.
39. Qu, C.; Dai, S.; Wei, X.; Cai, H.; Wang, S.; Yin, D.; Xu, J.; Wen, J.R. Tool learning with large language models: A survey. *Front. Comput. Sci.* **2025**, *19*, 198343. [CrossRef]
40. Liu, X.; Yu, H.; Zhang, H.; Xu, Y.; Lei, X.; Lai, H.; Gu, Y.; Ding, H.; Men, K.; Yang, K.; et al. Agentbench: Evaluating llms as agents. *arXiv* **2023**, arXiv:2308.03688. [CrossRef]
41. Wu, Q.; Bansal, G.; Zhang, J.; Wu, Y.; Zhang, S.; Zhu, E.; Li, B.; Jiang, L.; Zhang, X.; Wang, C. Autogen: Enabling next-gen llm applications via multi-agent conversation framework. *arXiv* **2023**, arXiv:2308.08155.
42. Park, J.S.; O'Brien, J.; Cai, C.J.; Morris, M.R.; Liang, P.; Bernstein, M.S. Generative agents: Interactive simulacra of human behavior. In Proceedings of the 36th Annual ACM Symposium on User Interface Software and Technology, San Francisco, CA, USA, 29 October–1 November 2023; pp. 1–22. [CrossRef]
43. Wang, X.; Wei, J.; Schuurmans, D.; Le, Q.; Chi, E.; Narang, S.; Chowdhery, A.; Zhou, D. Self-consistency improves chain of thought reasoning in language models. *arXiv* **2022**, arXiv:2203.11171. [CrossRef]
44. Madaan, A.; Tandon, N.; Gupta, P.; Hallinan, S.; Gao, L.; Wiegreffe, S.; Alon, U.; Dziri, N.; Prabhumoye, S.; Yang, Y.; et al. Self-refine: Iterative refinement with self-feedback. In Proceedings of the Advances in Neural Information Processing Systems, New Orleans, LA, USA, 10–16 December 2024; Volume 36.
45. Du, Y.; Li, S.; Torralba, A.; Tenenbaum, J.B.; Mordatch, I. Improving factuality and reasoning in language models through multiagent debate. *arXiv* **2023**, arXiv:2305.14325. [CrossRef]
46. Lu, C.; Hu, S.; Clune, J. Intelligent Go-Explore: Standing on the Shoulders of Giant Foundation Models. *arXiv* **2024**, arXiv:2405.15143. [CrossRef]
47. Zamfirescu-Pereira, J.; Wong, R.Y.; Hartmann, B.; Yang, Q. Why Johnny can't prompt: How non-AI experts try (and fail) to design LLM prompts. In Proceedings of the 2023 CHI Conference on Human Factors in Computing Systems, Hamburg, Germany, 23–28 April 2023; pp. 1–21. [CrossRef]
48. Li, L.; Zhang, Y.; Chen, L. Prompt distillation for efficient llm-based recommendation. In Proceedings of the 32nd ACM International Conference on Information and Knowledge Management, Eastside Rooms, UK, 21–25 October 2023; pp. 1348–1357. [CrossRef]
49. Zhao, A.; Huang, D.; Xu, Q.; Lin, M.; Liu, Y.J.; Huang, G. Expel: Llm agents are experiential learners. In Proceedings of the AAAI Conference on Artificial Intelligence, Vancouver, BC, Canada, 20–27 February 2024; Volume 38, pp. 19632–19642. [CrossRef]
50. Song, C.H.; Wu, J.; Washington, C.; Sadler, B.M.; Chao, W.L.; Su, Y. Llm-planner: Few-shot grounded planning for embodied agents with large language models. In Proceedings of the IEEE/CVF International Conference on Computer Vision, Paris, France, 2–3 October 2023; pp. 2998–3009.
51. Edge, D.; Trinh, H.; Cheng, N.; Bradley, J.; Chao, A.; Mody, A.; Truitt, S.; Larson, J. From local to global: A graph rag approach to query-focused summarization. *arXiv* **2024**, arXiv:2404.16130. [CrossRef]
52. Hong, S.; Zheng, X.; Chen, J.; Cheng, Y.; Wang, J.; Zhang, C.; Wang, Z.; Yau, S.K.S.; Lin, Z.; Zhou, L.; et al. Metagpt: Meta programming for multi-agent collaborative framework. *arXiv* **2023**, arXiv:2308.00352.
53. Qian, C.; Xie, Z.; Wang, Y.; Liu, W.; Dang, Y.; Du, Z.; Chen, W.; Yang, C.; Liu, Z.; Sun, M. Scaling Large-Language-Model-based Multi-Agent Collaboration. *arXiv* **2024**, arXiv:2406.07155. [CrossRef]
54. Tang, X.; Zou, A.; Zhang, Z.; Li, Z.; Zhao, Y.; Zhang, X.; Cohan, A.; Gerstein, M. Medagents: Large language models as collaborators for zero-shot medical reasoning. *arXiv* **2023**, arXiv:2311.10537. [CrossRef]
55. Li, Y.; Wang, S.; Ding, H.; Chen, H. Large language models in finance: A survey. In Proceedings of the Fourth ACM International Conference on AI in Finance, Brooklyn, NY, USA, 27–29 November 2023; pp. 374–382. [CrossRef]
56. Zhang, Z.; Zhang-Li, D.; Yu, J.; Gong, L.; Zhou, J.; Liu, Z.; Hou, L.; Li, J. Simulating classroom education with llm-empowered agents. *arXiv* **2024**, arXiv:2406.19226. [CrossRef]
57. Chen, X.; Jin, Y.; Mao, X.; Wang, L.; Zhang, S.; Chen, T. RareAgents: Autonomous Multi-disciplinary Team for Rare Disease Diagnosis and Treatment. *arXiv* **2024**, arXiv:2412.12475. [CrossRef]

58. Bariah, L.; Zhao, Q.; Zou, H.; Tian, Y.; Bader, F.; Debbah, M. Large generative ai models for telecom: The next big thing? *IEEE Commun. Mag.* **2024**, *62*, 84–90. [CrossRef]
59. Soman, S.; Ranjani, H.G. Observations on LLMs for telecom domain: Capabilities and limitations. In Proceedings of the Third International Conference on AI-ML Systems, Bangalore, India, 25–28 October 2023; pp. 1–5. [CrossRef]
60. Maatouk, A.; Ayed, F.; Piovesan, N.; De Domenico, A.; Debbah, M.; Luo, Z.Q. Teleqna: A benchmark dataset to assess large language models telecommunications knowledge. *arXiv* **2023**, arXiv:2310.15051. [CrossRef]
61. Karim, I.; Mubasshir, K.S.; Rahman, M.M.; Bertino, E. SPEC5G: A dataset for 5G cellular network protocol analysis. *arXiv* **2023**, arXiv:2301.09201. [CrossRef]
62. Miao, Y.; Bai, Y.; Chen, L.; Li, D.; Sun, H.; Wang, X.; Luo, Z.; Ren, Y.; Sun, D.; Xu, X.; et al. An empirical study of netops capability of pre-trained large language models. *arXiv* **2023**, arXiv:2309.05557. [CrossRef]
63. Maatouk, A.; Piovesan, N.; Ayed, F.; De Domenico, A.; Debbah, M. Large Language Models for Telecom: Forthcoming Impact on the Industry. *IEEE Commun. Mag.* **2025**, *63*, 62–68. [CrossRef]
64. Xu, M.; Niyato, D.; Kang, J.; Xiong, Z.; Mao, S.; Han, Z.; Kim, D.I.; Letaief, K.B. When Large Language Model Agents Meet 6G Networks: Perception, Grounding, and Alignment. *IEEE Wirel. Commun.* **2024**, *31*, 63–71. [CrossRef]
65. Wang, H.; Abhashkumar, A.; Lin, C.; Zhang, T.; Gu, X.; Ma, N.; Wu, C.; Liu, S.; Zhou, W.; Dong, Y.; et al. {NetAssistant}: Dialogue Based Network Diagnosis in Data Center Networks. In Proceedings of the 21st USENIX Symposium on Networked Systems Design and Implementation (NSDI 24), Santa Clara, CA, USA, 16–18 April 2024; pp. 2011–2024.
66. Chen, Y.; Xie, H.; Ma, M.; Kang, Y.; Gao, X.; Shi, L.; Cao, Y.; Gao, X.; Fan, H.; Wen, M.; et al. Automatic root cause analysis via large language models for cloud incidents. In Proceedings of the Nineteenth European Conference on Computer Systems, Athens, Greece, 22–25 April 2024; pp. 674–688. [CrossRef]
67. Yuan, Y.; Ding, J.; Feng, J.; Jin, D.; Li, Y. Unist: A prompt-empowered universal model for urban spatio-temporal prediction. In Proceedings of the 30th ACM SIGKDD Conference on Knowledge Discovery and Data Mining, Barcelona, Spain, 25–29 August 2024; pp. 4095–4106. [CrossRef]
68. Cao, D.; Wu, J.; Bashir, A.K. Multimodal Large Language Models Driven Privacy-Preserving Wireless Semantic Communication in 6G. In Proceedings of the 2024 IEEE International Conference on Communications Workshops (ICC Workshops), Denver, CO, USA, 9–13 June 2024; pp. 171–176. [CrossRef]
69. Wang, T.; Xie, X.; Wang, W.; Wang, C.; Zhao, Y.; Cui, Y. NetMamba: Efficient Network Traffic Classification via Pre-training Unidirectional Mamba. *arXiv* **2024**, arXiv:2405.11449. [CrossRef]
70. Rong, B.; Rutagemwa, H. Leveraging large language models for intelligent control of 6G integrated TN-NTN with IoT service. *IEEE Netw.* **2024**, *38*, 136–142. [CrossRef]
71. Hao, S.; Liu, T.; Wang, Z.; Hu, Z. Toolkengpt: Augmenting frozen language models with massive tools via tool embeddings. *Adv. Neural Inf. Process. Syst.* **2023**, *36*, 45870–45894.
72. Pang, L.; Yang, C.; Chen, D.; Song, Y.; Guizani, M. A survey on intent-driven networks. *IEEE Access* **2020**, *8*, 22862–22873. [CrossRef]
73. Li, J.; Sun, A.; Han, J.; Li, C. A survey on deep learning for named entity recognition. *IEEE Trans. Knowl. Data Eng.* **2020**, *34*, 50–70. [CrossRef]
74. Garau-Luis, J.J.; Torrens, S.A.; Vila, G.C.; Pachler, N.; Crawley, E.F.; Cameron, B.G. Frequency plan design for multibeam satellite constellations using integer linear programming. *IEEE Trans. Wirel. Commun.* **2023**, *23*, 3312–3327. [CrossRef]

MDPI AG
Grosspeteranlage 5
4052 Basel
Switzerland
Tel.: +41 61 683 77 34

Electronics Editorial Office
E-mail: electronics@mdpi.com
www.mdpi.com/journal/electronics

www.ingramcontent.com/pod-product-compliance
Lightning Source LLC
Chambersburg PA
CBHW040312270926
51850CB00007BC/1112

9783725855254